Comparative Ecology of Microorganisms and Macroorganisms

Brock/Springer Series in Contemporary Bioscience

Series Editor: Thomas D. Brock
University of Wisconsin–Madison

John H. Andrews

Comparative Ecology of Microorganisms and Macroorganisms

With 59 Figures and 14 Tables

Springer-Verlag

New York   Berlin   Heidelberg   London
Paris   Tokyo   Hong Kong   Barcelona

John H. Andrews
Department of Plant Pathology
University of Wisconsin
Madison, Wisconsin 53706, USA

Cover illustration: Kandis Elliot

Library of Congress Cataloging-in-Publication Data
Andrews, John H.
Comparative ecology of microorganisms and macroorganisms / by John H. Andrews
    p.   cm — (Brock/Springer series in contemporary bioscience)
   Includes bibliographical references and index.
ISBN 0-387-97439-3 Springer-Verlag New York Berlin Heidelberg
ISBN 3-540-97439-3 Springer-Verlag Berlin Heidelberg New York
   1. Microbial ecology. 2. Ecology. I. Title. II. Series.
QR100.A53   1991
576'.15—dc20   90-19686

Printed on acid-free paper.

Production and editorial supervision: Science Tech Publishers.
Printed and bound by Edwards Brothers, Ann Arbor, Michigan.
Printed in the United States of America.

9 8 7 6 5 4 3 2 1

ISBN 0-387-97439-3 Springer-Verlag New York Berlin Heidelberg
ISBN 3-540-97439-3 Springer-Verlag Berlin Heidelberg New York

In memory of my mother and father
and for T.A.S.

Writing a book is an adventure: To begin with it is a toy, then amusement, then it becomes a mistress, then it becomes a master, and then it becomes a tyrant, and the last phase is that just as you are about to become reconciled to your servitude, you kill the monster and strew him about to the public.

—WINSTON CHURCHILL

Preface

The most important feature of the modern synthetic theory of evolution is its foundation upon a great variety of biological disciplines.

—G.L. STEBBINS, 1968, p. 17

This book is written with the goal of presenting ecologically significant analogies between the biology of microorganisms and macroorganisms. I consider such parallels to be important for two reasons. First, they serve to emphasize that however diverse life may be, there are common themes at the ecological level (not to mention other levels). Second, research done with either microbes or macroorganisms has implications which transcend a particular field of study. Although both points may appear obvious, the fact remains that attempts to forge a conceptual synthesis are astonishingly meager. While unifying concepts may not necessarily be strictly correct, they enable one to draw analogies across disciplines. New starting points are discovered as a consequence, and new ways of looking at things emerge.

The macroscopic organisms ('macroorganisms') include most representatives of the plant and animal kingdoms. I interpret the term 'microorganism' (microbe) literally to mean the small or microscopic forms of life, and I include in this category the bacteria, the protists (excluding the macroscopic green, brown, and red algae), and the fungi. Certain higher organisms, such as many of the nematodes, fall logically within this realm, but are not discussed at any length. Most of the microbial examples are drawn from the bacteria and fungi because I am most familiar with those groups, but the concepts they illustrate are not restrictive.

Because the intended audience includes individuals who are interested in 'big' organisms and those interested in 'small' organisms, I have tried to provide sufficient background information so that specific examples used to illustrate the principles are intelligible to both groups without being elementary or unduly repetitious to either. Evolutionary biologists in either camp may also find the work of interest because the central premise is that organisms have been shaped by evolution operating through differential reproductive success. Survivors would be expected to show some analogies in 'tactics' developed through natural selection. The only background assumed is a comprehensive course in college biology. Hence, the book is also appropriate for advanced undergraduates.

The need for a synthesis of plant, animal, and microbial ecology is apparent from an historical overview (e.g., McIntosh 1985). Plant ecology was originally almost exclusively descriptive. It remains so to a large degree. A strongly predictive science has emerged, but one based largely on correlation rather than causation (Harper 1984). Within the past few decades it has branched into three main streams: phytosociology, population biology, and the physiological or biophysical study of individual plants (Cody 1986).

In contrast, animal ecologists have tended to study groups of ecologically similar species (i.e., guilds; Cody 1986). Although animal ecologists have probably been the most influential in developing the field of population ecology, many of their mathematical models derive from work based on microorganisms (e.g., Gause 1932), and from medical epidemiology (again involving microorganisms). Zoologists have also been the main force behind ecological theory, particularly as it applies to communities. Significantly, in both plant and animal ecology, the processes underlying patterns are usually inferred: The evidence marshalled is typically correlative rather than causative, and frequently the difference between the two is overlooked, if recognized.

Many microbial ecologists study systems. Ecology as practiced by microbiologists, however, has been mainly autecology, is typically reductionist in the extreme (increasingly often at the level of the gene), and lacks the strong theoretical basis of macroecology. Microbiology is a diverse field and microbial ecologists may be protozoologists, bacteriologists, mycologists, parasitologists, plant or animal pathologists, epidemiologists, phycologists, or molecular biologists. Approaches and semantics differ considerably among these subdisciplines. Nevertheless, microbial ecology is in general like plant ecology in its strong descriptive element. For example, often the goal is to identify and quantify the members of a particular community. Despite increasing emphasis on field research in the last few decades, microbes are usually brought into and remain within the laboratory for study under highly controlled conditions to determine the genetic, biochemical, growth rate, or sporulation characteristics, or the pathological attributes of a particular population, and how these are influenced by various factors. Thus, the major operational difference between microorganisms and macroorganisms is that the former must generally be cultured to understand their properties. It is this requirement that brings the microbial ecologist so quickly from the field to the laboratory.

To oversimplify, one could say that macroecology consists of phenomena in search of mechanistic explanation, whereas microbial ecology is experimentation in search of theory.

The foregoing picture is reinforced by observations on the parochialism evident in science. Examples include the streamlined makeup of most professional societies, the narrow disciplinary affiliation of participants at a typical conference, or the content of many journals and textbooks. Microbes as illustrations of ecological phenomena are omitted from the mainstream of ecology texts and are mentioned in cursory fashion only insofar as they relate to nutrient cycling or to macroorganisms. Discussion is left for microbiologists to develop in books on 'microbial ecology', which in itself suggests that there must be something unique about being microscopic. This is not meant as criticism, but rather as an observation on the state of affairs and an illustration of the gulf between disciplines. Finally, if additional evidence were necessary, one need only reflect on the current furor about the release of genetically engineered organisms. To a large degree this contentious issue has polarized along disciplinary lines: Ecologists express concern that microbiologists have little real knowledge of 'ecology'; microbiologists respond that ecologists really do not understand microbial systems. While acknowledging obvious differences between the two groups of organisms, one must ask whether their biology is so distinct that such isolationism is warranted. This book addresses that question.

A synthesis of microbial ecology and macroecology could be attempted at the level of the individual, the population, or the community. There would be gains and losses with any choice. Even the use of all three hierarchical levels would not ensure absence of misleading statements. Relative to the individual, comparisons based on populations or communities offer considerably increased scope and allow for inclusion of additional properties emerging at each level of complexity (e.g., density of organisms at the population level or diversity indices at the community level). Parallels based, for example, on supposed influential (i.e., organizing) forces in representative communities would be precarious, however, because unambiguous data implicating the relative roles of particular forces such as competition are usually lacking. The data that do exist have been variously interpreted, often with intense debate. Moreover, the same type of community may be organized quite differently in various parts of the world. For example, barnacle communities in Scotland (Connell 1961) are structured differently from those in the intertidal zone of the California coastline (Roughgarden et al. 1988). Entirely different pictures could emerge depending on the communities of microbes and macroorganisms arbitrarily selected for comparison.

Thus, I have decided to focus comparisons at the level of the individual. This implies neither that the individual is the only level on which selection acts, nor that shortcomings of the sort noted above do not exist. However, I consider the individual (defined in Chapter 1) through its entire life cycle to be operationally the fundamental unit of ecology in the sense that it is the *primary* one on which natural selection acts. Strictly speaking, natural selection acts at many levels, and a proper analysis of this issue requires that a distinction be made between what is transmitted and what transmits (Gliddon and Gouyon 1989). Efforts to relate the former (genetic information, broadly construed to include genes and epigenetic information) to a particular organizational level in a biological hierarchy extending from a nucleotide sequence to an ecosystem have engendered lively debate (e.g., Chapter 6 in Bell 1982; Brandon 1984; Tuomi and Vuorisalo 1989a,b; Gliddon and Gouyon 1989; Dawkins 1989). Without extending this controversy here, I note that at least my choice of the individual as the relevant ecological unit in this context is consistent with the conventional *ecological* viewpoint expressed by naturalists from Darwin onward (e.g., Dobzhansky 1956; Williams 1966; Mayr 1970, 1982; Begon et al. 1986; Ricklefs 1990). Comparisons of traits at the level of the individual can then be carried either downward to the corresponding genes—which have passed the screen of natural selection—or upward to assess the role of selection of traits at the group level.

This book proceeds first, by developing a common format (in Chapter 1) for comparing organisms; second, by contrasting (in subsequent chapters) the biology of macro- and microorganisms within that context; third (and also in the subsequent chapters), by drawing *ecologically useful* analogies, that is, parallels of interpretive value in comparisons of traits or 'strategies' between representative organisms. The first two phases can be accomplished relatively objectively; the third is necessarily subjective. There are situations where macroorganisms provide the wrong model for microbes or vice versa. I have tried to avoid forcing the parallels and have stated where close parallels simply do not exist.

All citations appear in a bibliography at the end of the work. Where books are cited in the text, I also specify, when appropriate, the specific chapters or pages concerned, in order to assist the reader in locating the relevant passages. Finally, at the conclusion of each chapter, a few suggestions are made for additional reading on each particular topic.

This is a book mainly of ideas. I find that the study of analogs is instructive and productive, not to mention interesting and frequently exciting. There is considerable speculation and a strong emphasis on generalizations, which I hope will prove stimulating and useful to the reader. Inevitably, there will be exceptions to any of the general statements. I am not particularly concerned by these oddities and think that it would be a mistake for the book to be read with the intent of finding the exception in each case in an effort to demolish the principle. However, this is not the same as advocating that it be read uncritically! MacArthur and Wilson (1967, p.5) have said that even a crude theory, which may account for about 85% of the variation in a phenomenon, is laudable if it points to relationships otherwise hidden, thus stimulating new forms of research. Williams (1966, p.273) points out that every one of Dalton's six postulates about the nature of atoms eventually turned out to be wrong, but because of the questioning and experimentation that resulted from them, they stand as a beacon in the advance of chemistry. Finally and most explicitly, Bonner (1965, p.15) states,

> It is, after all, quite accepted that in a quantitative experiment, a statistical significance is sufficient to show a correlation. The fact that there are a few points that are off the curve, even though the majority are on it, does not impel one to disregard the whole experiment. Yet when we make generalizations about trends among animals and plants, such as changes in size, it is almost automatic to point out the exceptions and throw out the baby with the bath [*sic*]. This is not a question of fuzzy logic or sloppy thought; it is merely a question whether [*sic*] the rule or the deviations from the rule are of significance in the particular discussion.

I started this book during a sabbatical leave in 1986-1987 with John L. Harper, to whom I am deeply indebted for his generosity and kindness. His enthusiasm, innumerable ideas, and critical insight have contributed much to what is presented here. I thank R. Whitbread and the faculty of the School of Plant Biology at the University College of North Wales in Bangor, Wales, U.K. for putting up with me for a year; in particular I acknowledge with appreciation the hospitality of N.R. and C. Sackville Hamilton, the help of Michelle Jones in many ways, and the wise counsel of Adrian Bell. Finally, I thank the many colleagues elsewhere who have helped in various ways: G. Bowen, J. Burdon, D. Ingram, and J. Parke read a draft prospectus for the book. Aspects I have discussed with P. Ahlquist, J. Handelsman, and D. Shaw. I am grateful to the following individuals who have commented on one or more of the chapters: C. Allen, A. Bell, G. Carroll, K. Clay, A. Dobson, I. Eastwood, A. Ellingboe, R. Evert, J. Farrow, T. Forge, J. Gaunt, J. Harper, M. Havey, R. Jeanne, D. Jennings, L. Kinkel, W. Pfender, G. Roberts, N. Sackville Hamilton, B. Schmid, and K. Willis. Errors of fact or interpretation remain mine.

I thank my editor, T.D. Brock, for constructive criticism and for encouraging me to write this book. I am indebted to Kandis Elliot for the outstanding artwork, Ruth Siegel for copyediting, Steve Heinemann for assistance with the computer graphics, and Steve Vicen for some of the photographs. Finally, I thank my family for their forbearance.

J.H.A.
Bangor, Wales
and
Madison, Wisconsin

Contents

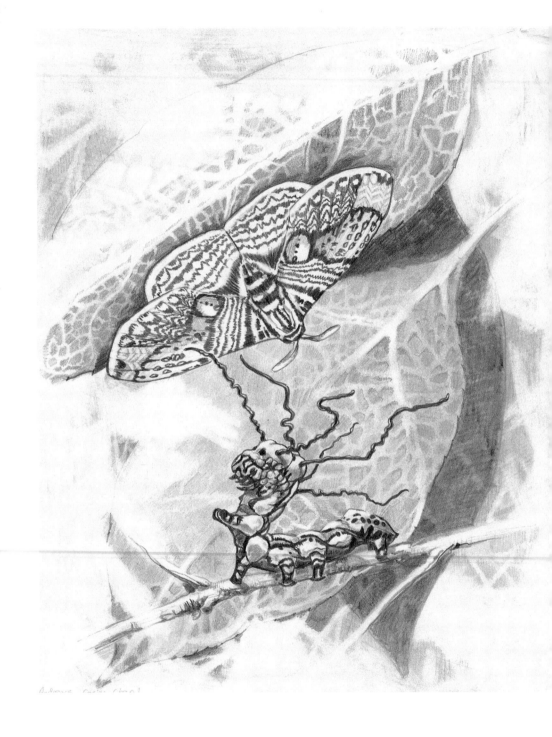

1

Introduction: Prospects for a Conceptual Synthesis

Nothing in biology has meaning except in the light of evolution.

—T. Dobzhansky, 1973, p. 125

1.1 Differences and Similarities

To even a casual observer, the great diversity among organisms in size, form, locomotion, and color is evident. If the microscopic world is also considered, the variety becomes overwhelming: With respect to size (mass) alone, for example, the range extends approximately 21 orders of magnitude, from the smallest bacteria at 10^{-13} grams to blue whales, which exceed 10^8 grams (see Chapter 4).

A common, informal demarcation of living things is at about the level of resolution of the human eye into microorganism or microbe (from micro = small and bios = life) and macroorganism. Macroscopically visible green, red, and brown algae, as well as the plants, and most of the animals constitute the latter category. Microorganisms include most of the protists, or Protoctista (Margulis et al. 1990; e.g., the flagellates, amoebae and relatives, sporozoans, ciliates, many of the green algae), the bacteria (including the cyanobacteria), the fungi, and certain microscopic invertebrates, such as many of the nematodes. In practice, most biologists probably do not consider small animals such as the nematodes and rotifers to be microorganisms, although strictly speaking they fall within the microscopic realm and 'see' the world to some extent as do the fungi and bacteria. The distinction between 'microorganism' and 'macroorganism' is obviously an arbitrary and occasionally hazy one. Some portion of the life cycle of every creature is microscopic; conversely, many microbes produce macroscopic structures and stages. Whatever the

1

terms may mean to different people, a major subdiscipline of biology and entire university departments are devoted to the study of '*microbiology*'. Operationally, the major general difference between microorganism and macroorganism is that, in general, members of the former group have to be cultured in order for their properties to be studied. Although the culturing requirement for identification purposes will diminish with the availability of molecular methods (e.g., Ward et al. 1990), study of microorganisms under controlled, laboratory conditions will remain a distinctive attribute of microbial ecology.

Of course, microorganisms as a group differ from macroorganisms other than in size alone. Although the cells of all macroorganisms are eukaryotic, the fundamental division of organisms as prokaryotes or eukaryotes is irrelevant in our context because among the microbes only the bacteria are prokaryotic. The key characteristics that distinguish microbes quantitatively or qualitatively from macroorganisms appear below and are discussed at length in subsequent chapters:

- Capacity for dormancy and frequently an extended quiescent state. Most plants and some animals such as the insects share this trait at least to a degree.

- High metabolic and potentially high population growth rates. For unicellular microorganisms, because there are many cells and because each is capable of forming more cells, beneficial mutations can be rapidly established in a population by natural selection. This is true also for the filamentous fungi. Fast response to changing environments is thus possible.

- Active growth (population increase) typically within a relatively narrow range of environmental conditions (which are usually distributed discontinuously, but very widely and often globally).

- High metabolic versatility, broadly speaking the capacity for rapid physiological adjustment, frequently involving entire biochemical pathways.

- Direct exposure of individual cells to the environment, rather than enclosure within a multicellular, homeostatic soma. Hence, organizational simplicity (fewer cell types and interactions; division of labor nonexistent or poorly developed).

- Usually an extremely high number of 'individuals' (as used in this case to mean physiologically independent, functional units; see Section 1.3) per species and, as a group, per unit of actively occupied or colonized area.

- Predominant or exclusively haploid condition in the vegetative part of the life cycle.

- Smaller role for true sexuality as a genetic recombination mechanism

relative to the one it plays in macroorganisms; gene transfer and assortment by various 'unconventional' means such as parasexuality in
the fungi, or in fragmentary fashion, such as for transformation, transduction, and even conjugation in the bacteria.

Given these appreciable differences, one might well ask what micro- and
macroorganisms have in common. The most general *intrinsic* property is that
every living thing is an island of order in a background of entropy or disarray.
The individual bacterial cell of a single cell type and the blue whale of about
120 cell types (Bonner 1988, p.122) are constructed in orderly fashion. Cell
structure differs in detail among life forms, but the cells of essentially all
organisms consist of a lipid bilayer membrane (the archaebacteria being the
sole known exception) surrounding a cytoplasm. All cells take up chemicals
from the environment, transform them, and release waste products. All can
conserve and transfer energy, direct information flow, and differentiate to
some degree. ATP is essentially *the* currency of biologically usable energy in
all organisms. The metabolic machinery, including the enzymes, pathways,
program for cell division, and sequence of the reactions, is remarkably similar
(although the subcellular sites and controlling mechanisms differ). For example, the assembly of a rod-shaped virus particle in an infected cell proceeds
through much the same stepwise process as does a microtubule in that cell.
    The genetic code and its associated parts (various RNAs and translation
machinery) are fundamentally the same, even down to the four nucleotide
building blocks and the specific triplets—which code for the same amino acids
in a human and a bacterium. (The code is not quite universal, because most
but not all the code words are the same.) For all life forms the genetic code
is distributed such that, with minor exceptions, every cell of the organism
has a complete copy of the DNA blueprint. All can communicate by chemical
signals and all respond to some extent to environmental signals by switching
genes on and off. Although the switching mechanisms vary, and are much
more complex in eukaryotes than in prokaryotes, all organisms are able to
receive and to react to stimuli. Moreover, the general features of gene regulation are similar in bacteria and higher organisms (Ptashne 1988).
    The fundamentally identical cellular biochemistry among organisms led
Bonner (1965, pp.129-130) to state that this characteristic is nonselectable in
the sense that it is essential to life, and as such has not been altered by selection
since at least Precambrian times. In a similar vein, Monod (quoted by Koch
1976, p.47) has said that "what is true of E. coli is also true of the elephant,
only more so". These commonalities brought interdisciplinary research teams
together at the start of the 'DNA era' in the late 1950s. The question whether
a comparable parallel exists at the level of the individual remains open and
is the basis for this book.
    The most important *extrinsic* property common to all extant organisms
is the filter of natural selection. As survivors, all are 'successful' (see below),

even if only transiently so. While microbial evolution differs in dynamics and detail from that of macroorganisms (Chapter 2), all life forms share an evolutionary history extending over some 3 billion years. It could be said that the 1.4 million or so known species represent 1.4 million responses to natural selection. A more informative statement would address what, if any, *commonality* was evident among the survivors. As Jacob (1982, pp.14-15) has said, natural selection is more than just a sieve. Over time it "integrates mutations and *orders them into adaptively coherent patterns* [and] gives direction to changes [and] orients chance". . . . (emphasis added). The central question that sets the unifying theme for this book is thus the extent to which the ecology of microorganisms and macroorganisms has been similarly molded by differential reproductive success operating among individuals.

Since the premise for this work is that natural selection is the organizing principle of biology, it is worthwhile to diverge briefly to consider some important aspects of Darwin's theory and their implications for what follows. This pertains especially to the terms 'fitness', 'success', 'optimum', and 'maximize', which are common in the biological literature and some of which are used throughout this book. According to Darwin (1859, p.81) natural selection was, "This preservation of favourable variations and rejection of injurious variations". In contemporary terminology we could say that natural selection is evolutionary change in the heritable characteristics of a population from generation to generation resulting from the differential reproductive success of genotypes. Natural selection is thus only one of the means by which evolution can occur, and it hinges critically on the proposition that different individuals leave different numbers of descendants, as determined both by the characteristics of the individual and those of the environment with which it interacts.

We must remember first, that the organisms we see about us today are the outcome of natural selection acting on *past* organisms in *past* environments. Thus it has been argued that a better term than *a*daptation is '*ab*aptation' (i.e., organisms seem to be adapted only insofar as current environments resemble past environments; Harper 1982). Equivalently, the effects of current natural selection will become manifest in the nature of organisms in the *future* in *future* environments. In other words, natural selection acts to improve fitness of descendants for the environment of their parents, not for their own environment. Since the environment inevitably changes (Chapter 7), the truly 'optimum state' can never be reached. In essence it is 'tracked' through time by the organism, or as van Valen (1973) and Lewontin (1978) have phrased it, environments are tracked by organisms.

Second, at best natural selection can only operate on the best of what can be produced by mutation and recombination of existing genotypes. The choice is restricted to what is available. Mayr (1982, p.490) likens natural selection to a statistical concept: possessing a superior genotype does not ensure survival, it only offers a higher probability of survival. For reproductive success it is merely sufficient to be better (or momentarily lucky!), not perfect or 'optimal'. Every genotype is necessarily a compromise between opposing selection forces (Dobzhansky 1956; Mayr 1982). Thus, *Darwin's theory does not predict an optimum or perfection* (although the terms are often used loosely), merely that some individuals will leave more descendants than others (e.g., Sober 1984; Begon et al. 1986, pp.6-7). For these reasons it is technically incorrect to say that fitness is 'maximized' (see comments in Chapter 4 on phylogenetic, allometric, and ontogenetic constraints).

Finally, for the sake of brevity and simplicity, and in keeping with traditional ecological terminology, expressions such as 'choosing', 'strategy', 'tactic', and so forth are used throughout the text. They carry no teleological implications for the organisms concerned.

1.2   A Framework for Comparison

Competing demands, trade-offs, and resource allocation   All organisms can be viewed simply as input/output systems (Figure 1.1; Pianka 1976, 1988: Chapter 6). Each acquires some kind of resource (energy and materials) as input. Progeny are the output. The principle common to all is that under natural selection, each creature will tend to allocate limited inputs 'optimally' among the competing demands of growth, maintenance, and reproduction (Gadgil and Bossert 1970; Abrahamson and Bossert 1982). As developed below and in later chapters, this entails trade-offs because there are finite resources to meet these competing needs. Increased allocation to one demand (e.g., growth to a large size) is at the expense of some other life function (e.g., early reproduction).

As is the case for all models, the input/output concept is a simplification. It is worthwhile to consider briefly what is *not* represented explicitly. For example, metabolic products as output may be invisible (e.g., oxygen in the case of plants; various secretions in the case of animals; extracellular poly-saccharides and organic acids for microbes), but are not unimportant, especially in the microbial world! Responses of the individual to predation or

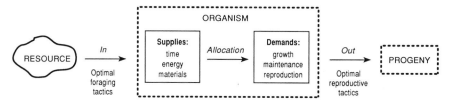

Figure 1.1   The organism as an input/output system. Redrawn from Andrews (1984); after Pianka (1976).

competition, while not recognized explicitly, would be reflected by altered biomass-allocation patterns and in the output term as altered number of progeny. While foraging tactics involve the obvious maneuvers such as predators hunting in packs rather than alone, they include also such devices as protecting a resource by various secondary compounds including antibiotics or mycotoxins (e.g., Janzen 1977), or directly poisoning competitors. In the same vein, attracting mates or pollinators, or defending a breeding territory from competitors or predators, are a part of optimal reproductive tactics. In other words, fitness can be increased by high acquisition and efficient allocation of your own resources, but also by interfering with those same activities of a competitor.

Fitness or 'success' is measured by the progeny or output over time. By definition, the fittest individuals leave the most *descendants*. (The current abbreviated wording of the theory is tautological in that the fittest must be the survivors, which are then declared to be the most fit *because* they survived. However, this apparently circular argument is not applicable to the form in which the concept of natural selection was originally expressed by Darwin. For discussion of this issue see Dawkins [1982, pp.181-182] and Mayr [1982, pp.518-519].) In population genetics terms, the number of descendants amounts to the proportion of the individual's genes left in the population gene pool (Pianka 1976). Given the diversity of life forms, what constitutes an 'individual' is not clear and is discussed below in Section 1.3 and Chapter 5. In addition to being transient, success is thus assessed in a *relative* way, against other members of the population. It is also important to realize that high numbers of *progeny* do not ensure high numbers of *descendants*. The issue is the fitness or competency of the offspring, that is, the likelihood that they will go on to reproduce and perpetuate the cycle.

Acquisition (input)   Three general points will be made here concerning resources (see also Figure 1.1). First, a basis for grouping all life forms is how they meet requirements for organic carbon compounds and energy for growth, maintenance, and reproduction (Chapter 3). With respect to carbon, organisms are either autotrophic ("self-feeders") or heterotrophic ("fed from others"). With respect to energy source, organisms are either phototrophic (source is light) or chemotrophic (source is chemical substances). If inorganic chemicals are oxidized for energy the organisms are said to be lithotrophic; organic chemicals are oxidized by organotrophs. These characteristics set very broad limits on what organisms can do and consequently where they are found, or at least where they can grow actively to competitive advantage. One example pertinent to each requirement will suffice to make the point. The need for sunlight restricts most aquatic plants to relatively shallow depths (photic zone) of oceans and most lakes. The distribution of those bacteria that obtain their energy by oxidizing specific inorganic compounds such as sulfur or iron (lithotrophs) may mirror that of the particular deposit. Members of the genus

Sulfolobus, for example, live mainly in hot springs and in other geothermal habitats rich in sulfur. In short, specialization can be a constraint as well as an advantage.

Assignment of all organisms to broad resource categories also establishes the well-known trophic structure of communities, which is informative in terms of energy flow and nutrient cycling. Since a high percentage of energy is dissipated by organisms in maintenance (and as heat and motion), not only is it probable that the number of links is set ultimately by energy losses at each, but the density of individuals and total biomass typically decrease at each successive trophic level (for caveats and details, see Chapter 3).

The second generality concerns *how* resources are presented to the organism. All organisms must contend with variation in the distribution and abundance of resources. Environments can be characterized with respect to resource availability (and other attributes) from the *organism's viewpoint* as relatively heterogeneous (discontinuous, patchy) or homogeneous (uniform). Patches are dynamic in that they vary in size, time, space, and because they can be superimposed in various permutations with respect to such characteristics as type of resource, time scale of availability, and micro-environment.

The idea of environmental *grain* (Chapter 7; see also Levins 1968) relates the size of a patch to the size of the individual and to the space within which the organism is active. *Coarse-grained* environments are sufficiently large that the individual either comes by fate to spend its entire life in a patch (for instance, when deposited there as a seed or spore) or chooses among them (for example, a millipede living only within a rotten log). Where the patches are so small that they appear uniform to the individual, or when the individual encounters many states of the heterogeneity during its lifetime, the environment is *fine-grained*. The larger, the more mobile, or the longer-lived the organism, the more likely it is to 'see' its surroundings as fine-grained. Plants in a field would appear fine-grained to us, but coarse-grained to an insect larva and immensely so, even at the level of the individual leaf, to a bacterium. A butterfly flits across several hundred square meters of a field in fine-grained fashion, while a slug, sliding slowly on its mucus trail across only a few meters of that same field, experiences its surroundings as coarse-grained. Over its life span of several hours or days, a bacterial or yeast cell on the skin of an animal or in a plant exudate confronts a coarse-grained environment, whereas typical changes on this order of time would appear relatively fine-grained to the microbe's host. In modular organisms (see below), the genetic individual commonly samples many environments concurrently. Sporadic and cyclic changes in resource availability (and other environmental characteristics) have immense influence on life history features such as dispersal, migration, and dormancy (Chapters 6 and 7).

Third is the issue of *efficiency* of resource utilization. This has been expressed in optimal foraging theory (Figure 1.1), developed by animal ecologists (Chapter 6 in Pianka 1988). When the concept is interpreted broadly and in conjunction with optimal digestion theory (Chapter 3), however, it is easy to see implications for all organisms. The basis for the theory is that there are both benefits (matter and energy) and costs (exposure to predators or parasites; time and energy diverted from other activities such as reproduction) associated with foraging. It seems reasonable to suppose that natural selection will act to maximize the difference between benefits and costs, so organisms presumably have evolved some sort of efficient strategy in terms of allocating time among alternative activities.

To a species of bird, for example, optimal foraging strategy might concern net gain or loss involved in the catching of insects of various sizes at various distances from a perch. How small a prey item is too small, or how far away is too far, to make a trip worthwhile? Under what circumstances would it be better to adopt a 'sit-and-wait' tactic as opposed to actively searching for prey? Plants 'forage' from a more-or-less fixed location in the sense that they display leaf canopies for photosynthesis and root systems to collect water and nutrients. How branch, leaf, and root architecture has evolved to meet these needs and in the face of competitive and predation pressure is of much interest (Chapter 5; see also Horn 1971; A.D. Bell 1984; Givnish 1986b). Analogously, sedentary filter-feeding invertebrates (e.g., bivalves, brachiopods, ectoprocts, and phoronids), and both attached and motile microbes 'forage', although in this sense foraging may entail a sit-and-wait strategy (quiescent spores of a root-infecting fungus in soil; sessile, stalked bacteria of the genus *Caulobacter*).

The form of living things as it pertains to resource utilization can be generalized further. Plants and other sessile, modular organisms such as corals, bryozoans, fungi, and clonal ascidians (as opposed to unitary organisms; Chapter 5) capture space and other resources by a branching habit of growth. In so doing they create "resource depletion zones" (Harper et al. 1986). These zones might represent, for instance, a shaded area resulting from interception of light by a plant or a volume of soil or water from which nutrients had been partially removed. Many biological systems exhibit dichotomous branching. Why? A 'challenge' to all modular organisms regardless of size is development of a branching pattern that most effectively captures resources. As detailed in Chapter 5, a continuum can be visualized between two extreme growth forms, phalanx (closely packed branches; resource site densely occupied) and guerrilla (infrequent branching; rapid extension; much unoccupied intervening space) (Lovett-Doust 1981). Thus, there is a correspondence between optimal foraging theory and optimal branching strategy. The issue is not straightforward because selection for resource capture cannot proceed independently of other selection pressures.

Reproduction (output)   Having acquired resources, the organism must then allocate them among the competing needs of growth, maintenance, and reproduction. To delay reproduction will be advantageous only if more descendants are ultimately contributed to future generations. For example, formation of large fruiting structures by the basidiomycete fungi is undoubtedly expensive in diverted materials, time, and energy, but offers the multiple advantages of sexual recombination and widespread aerial dispersal of millions of propagules.

Use of time and energy by all organisms varies seasonally (Chapter 6) as is apparent to anyone who has collected mushrooms in the autumn or followed the varying activities of birds over a few months. Reproductive activities, to the extent that they are associated concurrently with *dispersal of progeny*, are typically well synchronized to periods of the year when the (physical) environmental conditions are favorable for survival of offspring (resources available; climate hospitable). Birds become highly territorial and nest in the spring; parasites are remarkably well coordinated to the behavior or phenological development of their hosts. For instance, the timing of ascospore release from the apple scab pathogen, *Venturia inaequalis*, corresponds both to optimal environmental conditions for dispersal and infection, and to maximum susceptibility of emerging leaf and floral tissues. The fact that sexuality in algae and fungi is frequently triggered by environmental adversity is not inconsistent with the above generality because the resulting zygote is enclosed in a thick-walled, resistant structure that remains dormant for more or less long periods of time.

Maintenance and repair activities are continuous at the cellular level (Kirkwood 1981), but are especially apparent at the level of the individual as stages of rest. In one form maintenance is recognizable in the circadian sleep cycle of most higher animals, in another as seasonal inactivity (winter hibernation; plant dormancy) often associated with unfavorable environmental conditions (Chapter 7), and frequently in conjunction with dispersal (plant seeds; fungal spores). Although obviously essential for survival, maintenance represents lost time, energy, and materials in that limited resources are diverted from reproduction.

Optimal reproductive tactics (Figure 1.1) seemingly have evolved by natural selection acting on organisms which as a result have improved long term reproductive success. As was the case for foraging, the issue here is again one of relative allocation: resources and time assigned to reproduction are diverted from other activities. Also, what is allocated to reproduction may be invested in various ways. For instance, a partial suite of either/or reproductive 'choices' includes whether to: 1) engage in only one round of reproduction (semelparity) or several (iteroparity) during a lifetime; in the latter case a further question is how many episodes to have; 2) choose one mate or several or none (clonal or asexual reproduction); 3) reproduce early or late in the life cycle (or season); 4) produce few large, highly 'competent' progeny or many small, relatively 'incompetent' progeny; related to this is the length of gestation; 5) reproduce locally or risk migration to new breeding grounds; 6) maintain separate sexes (sexual dimorphism; dioecy; heterothallism) or unite sexual function within one individual (hermaphrodism; monoecy; homo-

"Possessing both sexes, the Hypoplectrus gutavarius mates with another Hypoplectrus gutavarius, each switching roles after each encounter."

Drawing by Booth; © 1989. The New Yorker Magazine, Inc.

thallism); related to this is the capacity to vary the sex ratio and the even larger question of why there should be two, rather than some other number of sexes; 7) disperse progeny widely or concentrate offspring near the parents.

Longevity is tied closely to reproduction. In evolutionary terms, senescence should occur wherever the reproductive value of the individual declines with time. Under relatively stable environmental conditions and where the reproductive value of the individual increases with age, as it typically does for modular (unlike unitary) organisms, senescence should be delayed or not occur (Harper et al. 1986). Under these circumstances, the genetic individual (Section 1.3) lives for an indefinite if not an infinite time, and there is the potential for exponential increase in offspring (Chapter 6).

The interrelationships of the chapters in this book are shown in Table 1.1. The ecological issues summarized in Figure 1.1 are developed in subsequent chapters as follows: Since variation provides the raw material for evolution, the stage is set in Chapter 2 with an overview of the genetics and mechanisms of genetic variation of different kinds of organisms. While ob-

Table 1.1   Topics and interrelationships of chapters[1]

Basis for comparing organisms	Implications or attributes; relevant chapter	Some ecological consequences
Genetic variation; recombination mechanisms ⟶	Variable mechanisms/time of occurrence vs fixed pattern and occurrence; Chapter 2 ⟶	Fluidity of genetic individual; distribution of mutations
Nutritional mode; biochemical diversity/ flexibility ⟶	Autotrophy vs heterotrophy; generalist vs specialist (substrate diversity); Chapter 3	Distribution pattern; energy flow; trophic structure; allocation patterns
Size ⟶	Structural & biophysical limits; complexity; Chapter 4	Growth rate; biomass; homeostasis; colonizing potential; r/K selection; mutation rate
Growth dynamics and form ⟶	Modular vs unitary; population growth potential; Chapter 5	Mobility; senescence; death; reproductive value of individual; resource exploitation
Life cycle pattern; senescence ⟶	Ploidy level; phenotypic expression; longevity; Chapter 6 ⟶	Dormancy; dispersal; seasonal activity
Environmental interactions ⟶	Temporal & spatial grain; habitable sites; Chapter 7	Ontogeny; life cycle programming; resource acquisition and allocation; gene flow; dispersal mechanisms

[1]Arrows indicate the major linkages among themes.

viously there is interaction between genomic expression and the form and function of an individual, genes must operate within broad, predetermined constraints imposed by lineage. A bacterium can never be a pine tree, regardless of how, when, or to what extent its genes recombine or are switched on or off. Worded differently, this means that the life history options available to any organism are subject to phylogenetic (i.e., taxon-related) and allometric (size and shape) constraints. I then take up in Chapter 3 the supply side of Figure 1.1, namely the kinds of resources different organisms use, and how they acquire and allocate them in the face of competing demands. This leads to a consideration of how size (Chapter 4) and growth form (Chapter 5) influence foraging and reproductive tactics. Chapter 6 concerns the life cycle, in essence how the developmental play is acted out under the direction of the genes. The role of the environment in shaping the plot (life history features) is discussed in Chapter 7. Some general conclusions are summarized and perspectives are advanced in Chapter 8.

1.3   What is an Individual?

The 'individual' figures prominently in this book. However, unless defined, the term can be ambiguous because it may be used by authors in at least three contexts (which are not mutually exclusive) (Figure 1.2). First, an individual is often taken operationally in a *numerical or quantitative* sense to mean a representative of a particular species, something that can be counted (as in one deer; one maple tree; or in microbiology, one propagule or one colony-forming unit, CFU). Individuals by this definition are supposedly discrete and functionally independent units (Cooke and Rayner 1984, p.56). For colonial organisms such as bacteria, fungi, various algae, bryozoans, and coelenterates, it is not at all clear what level of cellular aggregation fulfils these criteria (e.g., Larwood and Rosen 1979). For example, the reason the term CFU is used in bacteriology and to some extent in mycology is that in practice it is generally unknown whether growth is produced from a unicell or a cluster of cells, or whether all living cells seen in a sample will multiply in a given medium.

Second, an individual may be used in the *genetic* sense to mean the genetic unit, abbreviated "genet" (Kays and Harper 1974), resulting from growth of a zygote (Figure 1.2; this idea is developed in Chapter 5; see also Harper 1977, p.26). This usage is consistent with the numerical definition when applied to unitary organisms such as most mobile animals (Chapter 5), that is, those in which the zygote develops into a determinate body repeated only when a new life cycle starts. A deer is both a numerical individual and a genet. The concept is meaningful in evolutionary terms because it focuses attention on the genotype through time. However, for organisms such as microbes, corals, and those plants with clonal (asexual) growth, the number of discrete, countable 'individuals' is not the same as the number of genets. The term given to each of these countable units is "ramet" (Chapter 5; see also Harper 1977, p.24). Each is a member of the genet but is actually or potentially capable of independent existence as a separate or physiological individual (Figure 1.2). An entire hillside of bracken fern may constitute a single genetic individual (one genet or clone), being nothing more than the multiple phenotypic representation (many ramets) of a single highly successful genotype (Harper 1977, p.27). Of course in practical terms, if several genets intermingle, as is often the case, they may well *appear* identical, and usually there is no easy way to differentiate among them.

Third, in an *ecological* sense, the individual is taken implicitly to be the organism through its entire life cycle (Figure 1.2). Within the life cycle context, an organism is a whole, and generally easily visualized in its various genetic,

Organism	Context			
	Numerical	Genetic (genet)	Physiological (ramet)	Ecological (life cycle)
 elephant	one	one	NA[a]	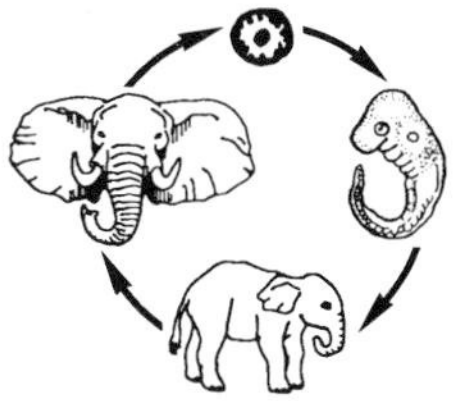
 strawberry	many	one	many	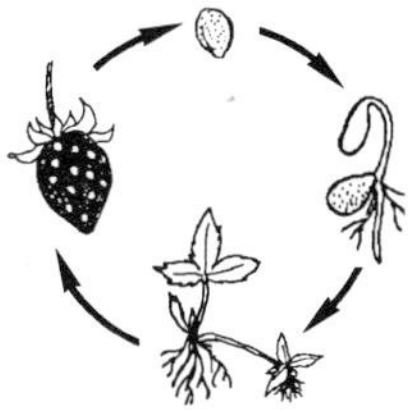
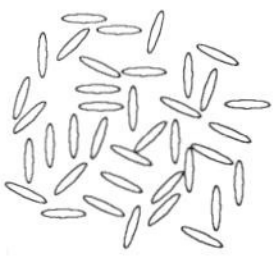 bacterial colony	many	one (if clonal) or more[b]	many[b]	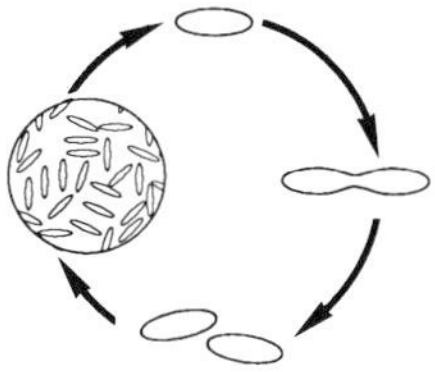
 fungus	many	one (if clonal) or more[b]	many[b]	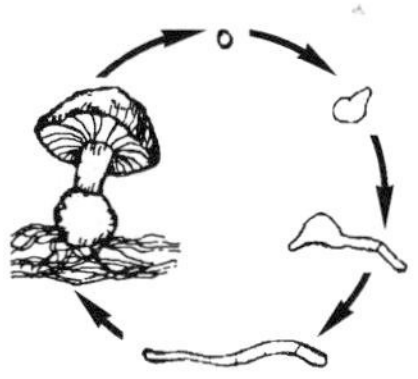

[a] NA = Not Applicable because ramet designation refers to a unit of clonal growth
[b] See text for qualifications

Figure 1.2 Relationships among different concepts of an 'individual' for various kinds of organisms.

physiological, morphological, or developmental states, from birth to death. For unitary organisms, there is also a direct correspondence between life cycle and genet, from zygote to zygote. Genetically different individuals (i.e., new genets) are generated at one or more specific points in the life cycle. Germ cells in the parental genet undergo meiosis to produce gametes, sexual reproduction occurs, and the new genets emerge as developing zygotes. The parent typically continues to live for some time; consequently, generations overlap. This situation is also true (although more complicated to visualize) for those modular organisms where the genet undergoes fragmentation, such as the free-floating aquatic plants (*Lemna* spp. [duckweeds], *Azolla* spp. and *Salvinia* spp. [water ferns]) and the corals. Old modules are sloughed and are free to move about passively. Kariba weed (*Salvinia molesta*), which propagates entirely by clonal means, has come to occupy vast areas and may weigh millions of tons (Barrett 1989). Thus, it considerably exceeds the mass of the blue whale as the largest organism on earth!

However, the genetic individual is not continuous through the life cycle of those plants that have a free-living haploid stage. A good example is the life cycle of the fern, which consists of morphologically and ecologically distinct genets, namely a sporophyte (2n) and a gametophyte (n) phase. The genetic individual is also discontinuous through the life cycle of most microorganisms (Figure 1.2) because of their irregular timing of genetic exchange and their unorthodox systems of genetic recombination (Chapter 2), such as transduction in the bacteria and parasexuality in the fungi (see below and also the discussion of complex life cycles in Chapter 6).

There is no simple, unambiguous concept of an individual that can be used as a benchmark in the comparative ecology of micro- and macroorganisms. Comparisons can be made conceptually to a greater or lesser extent with any of the three definitions, but in a strict sense not functionally (experimentally). The numerical usage is rejected because it means nothing in evolutionary terms and operationally in microbiology is quite contentious. Direct enumeration of microbes by microscopy in situ is subject to several qualifications and sources of error; even more error attends counts of 'individuals' tallied as colony-forming units on some sort of growth medium (usually necessary to enable identification of the microbe). These issues have been reviewed extensively (e.g., Brock 1971; Dickinson 1971). Thus, to arbitrarily choose, for example, bacterial cells, fungal colonies, and slime mold amoebae, and attempt to compare them as 'individuals' with trees, insects, and birds would be futile.

The genet concept, though attractive, cannot be used for microbes without substantial modification because an individual can change genetically quite separately from the conventional process of reproduction with meiosis. Thus, in many microbial situations it is impossible to know when the genetic individual starts to exist. As seen above, for unitary organisms this is at the time of genetic fusion, in other words, when genetic recombination occurs at zygote formation and the contribution from each parental genome is diluted by exactly one half. Moreover, the point in the life cycle (reproductive phase; sexual maturity) when karyogamy can occur is fairly standard and easily recognized. This is not the case for bacteria and fungi which display indeterminate, highly plastic development associated with novel genetic events. On one hand, a bacterial cell may divide repeatedly, yielding descendants that are in effect reproductively isolated as a genetically homogeneous, clonal microcolony. On the other hand, transformation, transduction, or conjugation (mating) may occur at various times and to various degrees. During conjugation, one or more plasmids may be transferred with or without chromosomal DNA from donor to recipient. How many genes have to mutate or be acquired in a haploid organism for the original genet to cease to exist? Any answer would be arbitrary (see below).

An analogous phenomenon occurs in the fungi that have a dikaryotic (n + n) phase or exhibit parasexuality (Chapters 2 and 6). For instance, *Puccinia graminis tritici*, the parasite causing stem rust of wheat, has five spore stages with associated mycelial growth involving barberry as well as wheat as hosts. The life cycle includes haploid, dikaryotic, and diploid phases. Shortly after meiosis, the fungus becomes dikaryotic, but karyogamy is delayed and true diploidy is a transient stage. The dikaryotic phase, technically haploid but functionally diploid, is of variable duration. Furthermore, in absence of the barberry as an alternate host but in the presence of a favorable environment, the fungus may cycle indefinitely in the dikaryotic condition. What, then, is the genet for the wheat rust fungus? Whatever the designation, it corresponds neither to a single numerical entity nor to the entire life cycle.

The genet concept could be brought into line with what intuitively seems to physically represent the organism by allowing for a certain amount of genetic 'flux' after recombination in the zygote, or at the start of the life cycle in the case of haploid and asexual organisms. This would acknowledge that, while the genetic basis for the individual is established by genetic recombination in some form at the outset, the genotype could be molded by additional genetic events which might periodically occur during the life cycle. Clearly this happens for microbes, as has been driven home forcefully by the rapidly increasing information on mobile genetic elements. For macroorganisms, the comparable event would be insertion and subsequent expression of foreign genes in the *germ* as opposed to the somatic cells. This can indeed occur (Benveniste 1985), with potentially far-reaching implications. However, the full evolutionary consequences remain to be seen. Thus, one might envision a 'fluid mosaic model' of the genet in the same sense that a fluid mosaic model exists to depict the structure of membranes. This idea of the genet may be appealing to the imagination, but is of not much pragmatic value.

The individual depicted as an organism passing in organized fashion through stages in its life cycle provides a universal basis for ecological comparisons. Unless noted otherwise, this is the frame of reference that is used throughout this text. The scheme is applicable to all of the various phases of diverse organisms, and accommodates changes in ploidy, morphology (phenotype) including size, and variation in the time and extent of genetic change. Bonner (1965) has interpreted the life cycle as proceeding sequentially from a state of minimum size (generally at the point of fertilization) to maximum size (generally the point of reproductive maturity). The transition involves growth over a relatively long period to adulthood, followed by return to minimal size over a relatively short period, accomplished by separation of buds or gametes from the adult. For single-celled organisms, the cell cycle is the life cycle of the physiological but not of the genetic individual.

1.4   Summary

Although every organism is unique in detail, all share fundamental properties. The cellular chemistry of life forms is basically the same. All represent order as opposed to disorder. All display cellular organization, growth, metabolism, reproduction, differentiation to some degree, the ability to communicate by chemical signals, and a hereditary mechanism based on transfer of information encoded in DNA. Of more significance ecologically is the fact that all have been shaped by evolution operating through differential reproductive success. All are survivors and would be expected to show some analogies in 'tactics' developed in response to natural selection.

The term 'individual' has various connotations: numerical, genetic, physiological, and ecological. Ecologically, the best way to view an individual is as the organism through its entire life cycle. Based on this definition, a comparative ecology of micro- and macroorganisms is developed within the general framework of the individual as an input/output system. Each individual acquires resources as input, marshals them for growth, maintenance, or reproduction, and produces progeny and metabolic products as output. The latter, in the form of chemicals such as antibiotics and organic acids, are an ecologically important output for microorganisms. Analogous products for macroorganisms (excretions and secretions) are relatively less important and are only explicitly considered in the most detailed of models. In practice, resources and time are limiting and the alternative demands on them are conflicting. Thus, overall, selection should favor organisms 'choosing' optimal allocation, thereby increasing fitness as measured by the number of *descendants*. The major factors bearing on this model are the subject of the following chapters and are presented in a way chosen to highlight the analogies and distinctions between the ecology of micro- and macroorganisms. These subsequent topics include genetic variation and the means by which different organisms transfer genetic information (Chapter 2); the nutritional mode; size; growth dynamics and growth form of the individual (Chapters 3, 4, and 5, respectively), the life cycle (Chapter 6), and interaction between the individual and the environment (Chapter 7). Chapter 8 consolidates the book and returns us to the theme developed here.

1.5 Suggested Additional Reading*

Bonner, J.T. 1965. Size and cycle: An essay on the structure of biology. Princeton Univ. Press, Princeton, N.J. The biology of organisms from the life cycle perspective. An outstanding synthesis; stimulating; full of ideas.

Dawkins, R. 1989. The selfish gene, 2nd ed. Oxford University Press, Oxford, U.K. A well-written, speculative, stimulating thesis that the fundamental unit of natural selection is the gene.

Larwood, G. and B.R. Rosen (eds.). 1979. Biology and systematics of colonial organisms. Academic Press, New York. Collected papers on representative organisms posing a challenge to the definition of 'individual'.

Pianka, E.R. 1976. Natural selection of optimal reproductive tactics. Amer. Zool. 16: 775-784. Optimal foraging theory, optimal reproductive tactics, and the model of the organism as an input/output system.

---

*A complete bibliography is given at the end of this book.

2

Genetic Variation

Diversity is a way of coping with the possible.

—F. Jacob, 1982, p. 66

2.1 Introduction

Because variation (among entities at any level) provides the raw material on which evolution acts, this chapter sets the stage for later comparisons in the book by reviewing briefly the generation and maintenance of genetic variation in different kinds of organisms. While some sections of the text are controversial, for the most part the chapter is more factual and less speculative than those which follow.

The following metaphor allows us to categorize the major kinds of genetic variability. If the text of this paragraph is viewed as a molecule of DNA from which copies are made, then there are basically three sorts of changes that can occur in the transmission of information. First, individual letters or, second, entire words and phrases could be deleted, added, or shuffled about. A third possibility is that the text might remain intact but unreadable if portions were typeset in white rather than black. These characteristics correspond crudely to point or other structural gene mutations, recombinations, and epigenetic controls, respectively. The outline shown in Figure 2.1 gives the approximate relationships among the principal categories and provides a framework for discussion. The important role that the cellular and external environments play in reciprocal interactions with the genome is developed in Chapter 7.

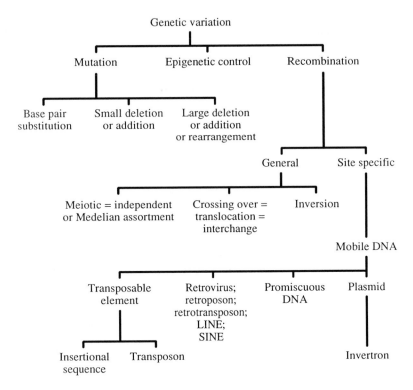

Figure 2.1   A simplified overview of the major sources of genetic variation.

2.2   Mechanisms

Mutation   Mutations are the fundamental source of variation and, by definition (Chapter 2 in Hartl 1988), consist of changes in the genetic message that are heritable and, implicitly, detectable. It should be noted in passing that molecular biologists speak of inherent mutation rates (replacement rates of one DNA base by another), whether or not there is a change in the wild-type phenotype. This fundamental level of mutation may go undetected if the protein is not functionally altered, if two or more mutations compensate for each other giving a pseudo wild-type, or if another gene product takes over the function of the altered protein. Thus, there are three mutation rates: 1) the inherent rate of uncorrected change in DNA sequences; 2) the rate at which the mutants survive; 3) the rate at which survivors are detected in a population by virtue of their phenotypic differences.

Mutations are mechanically inevitable—copying and proofreading mechanisms are not and cannot be perfect. For this reason, Williams (1966, p.12) has argued that mutations are not adaptive, that is they should not be considered to be a means for ensuring evolutionary plasticity. That they still occur is largely in spite of natural selection rather than because of it (Williams 1966, p.139). However, this does not mean that there cannot be selection for a particular mutation *rate*. Overall, because most of these alterations are deleterious, there has been strong selection pressure (reduced reproductive fitness) over millennia to reduce rates in organisms.

Nevertheless, there is a cost in maintaining an error-avoidance and correction system, and also some value in having background genetic 'noise' (Drake 1974). The latter is illustrated by the essential function of mutation (somatic hypermutation) in antibody genes of the B lymphocyte cells, which contributes to antibody diversity in vertebrates (French et al. 1989). Analogous genetic rearrangements provide the operon structure needed for expression of *nif* genes in a nitrogen-fixing cyanobacterium, *Anabaena* (Golden et al. 1985). The terminal step in heterocyst differentiation involves excision of DNA in response to an environmentally triggered, site-specific DNA recombinase (Haselkorn et al. 1987). Much the same rearrangement process is thus used in a prokaryote and an advanced eukaryote. The case for adaptiveness of individual mutations (as opposed to rates) is furthered by preliminary evidence that mutation in *E. coli* is nonrandom: When challenged by a particular environment, bacterial cells may be able to 'choose' which mutations occur, thereby enhancing their fitness (Cairns et al. 1988; Hall 1988; but see Davis 1989; Lederberg 1989). The genes involved fit into Campbell's (1982) "contingently dynamic" category of gene behavior (Table 2.1). This heretical idea follows the view first expressed by Steele (1981) and colleagues on directional gene mutations in the germline and somatic cells of higher animals (Section 2.4). At this point, the work as it pertains to both groups of organisms remains unconfirmed.

The mutant phenotype can arise from mutations differing in extent of alteration of phenotype. Mutations can be grouped into three arbitrary categories spanning the range from 1) changes at the molecular level (point mutations) and 2) small deletions and inversions to 3) gross chromosome abnormalities visible with the light microscope (Chapter 12 in Fincham 1983; see also Raff and Kaufman 1983, pp.64-65). It should be emphasized that point and other small changes in the *noncoding* regions (introns) are very significant in evolutionary terms—perhaps as much or even more so than those in the coding segments (exons) themselves. Thus, *it is not necessary to destroy a gene in order to change the phenotype*. A related phenomenon concerns suppressor mutations which counteract changes in the same or in a related gene. This is a matter of regulating gene expression and is taken up later in this chapter under **Epigenetic controls**.

Classification schemes are more-or-less subjective because the degree of change occurs as a continuum. With the discovery of transposable elements in both prokaryotes and eukaryotes, the distinction at the molecular level between mutation and recombination is often one of semantics. For instance,

Table 2.1   Genes classified by their degree of dynamic activity ("autonomy")[1]

Gene class[2]	Description	Examples
Classical	Mutate only rarely and randomly; inactive in the autonomous sense	Classical Mendelian traits
Profane		
Sporadic alteration	Have target sequences; structure deliberately modified by organism using gene-processing enzymes	Insertion sequences (bacteria); mating type (yeast—cassette switching)
Programmed alteration	Gene processing enzymes active only at certain times in life cycle	rRNA multigene family (frog— amplification in oocyte)
Self-dynamic	Have target sites and code for their own specific gene-processing enzymes	Retroviruses; controlling elements (corn); certain transposons (bacteria)
Contingently dynamic	Sense environment and change in response to it	Certain transposons (bacteria)
Automodulating	External condition causes gene to alter future sensitivity to that condition	Antibody genes (splicing and mutation— vertebrates)
Experiential	Modification induced in soma transmitted to germ line	Drug-induced diabetes; immune tolerance (mice) ?
Anticipatory/ cognitive	Change in anticipation of their usefulness	?

[1]From Campbell (1982).
[2]Arranged in ascending order of dynamic complexity.

where relatively large segments (hundreds or thousands of nucleotides) of DNA are rearranged, for example, by inversion or exchange between chromosomes, the event is often called mutation by molecular biologists (Chapter 12 in Watson et al. 1987; see also Bainbridge 1987). Yet it falls within the domain of recombination (in the sense of mixing of nucleotides, see **Recombination**, later) as defined by eukaryote geneticists. Thus, some scientists prefer to define mutation as *any* change in genotype, in which case recombination would simply be a mechanism by which it occurs. I recognize here both recombination and mutation as categories of genetic change. Maintaining the former as a separate entity establishes as distinctive that orderly process of genetic events surrounding gametogenesis in higher diploid organisms, wherein genes are reassorted (recombined) by crossing over so that all chromosomes in each gamete contain both maternal and paternal genes.

Before the details are presented, three important general points about mutations need to be made. First, they occur in somatic as well as in germ line cells, the implications of which will be discussed in Section 2.4. Suffice it to say here that, with few exceptions (such as among the somatic cell lines that produce antibodies), high rates must be avoided in *both* lineages, a fact that is often overlooked in consequence of the focus by evolutionary biologists on the germ line. Second, the impact of a mutation will depend on the environment, broadly construed (Chapter 7). A mutation deleterious in one set of circumstances may be beneficial in another. Third, the impact of a mutation will depend on the stage of development of an organism. Mutations occurring in early ontogeny, that is at the blastula stage, are much more likely to be lethal because of their far-reaching impact on subsequent differentiation. In contrast, those which happen later may affect relatively superficial properties such as eye color (Bonner 1965, pp.123-128; 1988, p.168). As Bonner notes, it follows from this that any mutation at *any* stage in the life of a unicellular organism is more likely to be lethal than would a similar mutation during most of the life of a macroorganism.

The first mutational class consists of base pair (intragenic or 'point') substitutions in the nucleotide sequence. Because of the degeneracy feature of the genetic code ('wobble' in the third codon position), often there is no phenotypic effect, that is, no change in the amino acid sequence ('synonymous' or 'silent' mutation). Mutations may also be silent if a related amino acid is inserted or if the affected region of the gene product is unimportant. If the change occurs in a coding region, usually a phenotypic alteration is evident (altered amino acid sequence or 'nonsynonymous' substitution), hence the term *missense* mutation. This is manifested either as an altered but functional ('leaky') protein product or as a defective protein. The latter category is very common, as in sickle cell anemia, Tay-Sachs disease, and phenylketonuria. One form of retinitis pigmentosa is apparently caused by a C to A transversion in codon 23, corresponding to a proline to histidine substitution (Dryja et al. 1990). Alternatively, if translation is terminated (*nonsense* mutation), a shortened and almost certainly defective protein results.

Second, small deletions or insertions of one or a few base pairs can occur. When this happens within regions coding for polypeptides the mutations are called *frameshift* (Chapter 12 in Fincham 1983) if the reading frame of the codons is shifted. A frameshift mutation may result in a truncated, possibly unstable protein, or it may generate a completely nonfunctional product, depending on its position in the structural gene. In general, the foregoing changes except deletions are revertible to wild-type. Reversion of deletions depends on the nature of the deletion and the encoded gene product.

The third group comprises the larger scale changes such as bigger deletions, inversions, duplications, and transpositions that intergrade with recombination, discussed later. Where a deletion is large enough to remove or disrupt a gene coding for a critical enzyme, the result is likely to be death of the cell or organism. Plants, however, are distinctive in being able to tolerate imperfectly matched chromosomes that result from loss or rearrangement (Walbot and Cullis 1983). Gene duplications seem not only to be well tolerated among organisms in general, but may contribute significantly to the evolution of complexity (Stebbins 1968; see also Chapter 4). In contrast, genes can be inactivated by insertion of mobile DNA elements into the chromosome. These may be fragments (generally 1,000-2,000 nucleotides) that do not code for any characters beyond those needed for their own transposition (simple transposon, or insertion sequence; Figure 2.2a). Alternatively, they may be longer elements (complex transposon; Figure 2.2b) which code for products such as antibiotic resistance in bacteria. In a hierarchy of gene behavior, such genes are classified as "profane" (Table 2.1 and Campbell 1982) because organisms can modify their behavior sporadically or in programmed fashion with special gene-processing enzymes. If the genes code for their own enzymes (as in the case of transposons), they are called "self-dynamic".

Mutation rates can be estimated by recording spontaneous phenotypic

(a) Insertion sequence

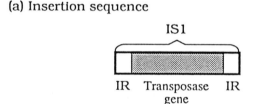

(b) Transposon

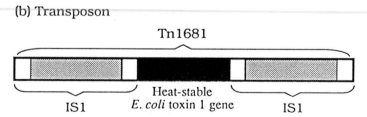

Figure 2.2   The two main classes of transposable elements in bacteria. (a) An insertion sequence (IS1), also called a simple transposon, codes only for its own transposition. The transposase gene is typically enclosed by inverted repeat (IR) sequences. (b) A complex transposon contains genetic material in addition to that needed for transposition. Here the transposon Tn1681 from *E. coli* carries a gene coding for a heat-stable toxin. Redrawn from Watson et al. (1987, p.203), Molecular Biology of the Gene, 4th ed. vol. 1. © 1987. Benjamin Cummings Co.

changes over time in a population of organisms in nature (animals or plants) or in the laboratory (tissue culture or bacteria). Note that this method detects only survivors and, among the survivors, only those which are phenotypically distinct from the wild-type. Because the frequencies of most mutations are very low in natural populations and can only be detected in large sample sizes, a modification of this approach is to examine specific proteins electrophoretically (Lewontin and Hubby 1966), or DNA sequences (e.g., by studying restriction fragment length polymorphisms). This involves comparing the amino acid or gene sequences in a particular highly conserved protein (such as hemoglobin), or base sequences in a gene, in several species (Wilson et al. 1977; see "molecular evolutionary clock", below). For protein comparisons, each discrepancy is assumed to represent a stable change in a codon corresponding to the altered amino acid. In this analysis, the original (ancestral) state, being extinct, is of course unavailable for comparison. So the versions of the sequence that appear in two (or more) living relatives of the phyla of interest are compared and expressed as units of accumulated amino acid or nucleotide changes. The organisms are then arranged in a branching diagram drawn to minimize the number of changes needed to describe all the permutations from the ancestral sequence. The number of stable genetic changes can then be related to chronological estimates of evolutionary time, obtained from fossil records, since the species diverged from the common ancestor. Rate calculations by the phenotypic and molecular clock methods are in general agreement *when expressed on a per unit time basis*. This amounts to $10^{-8} - 10^{-9}$ per locus per day (Table 2.2).

Mutation rate calculations convey several pieces of interesting information. The inherent rate for *different* genes varies substantially *within* a single

Table 2.2 Point (spontaneous) mutation rates per locus per generation for various organisms[1]

| Organism | Mutation rate/locus/gen.[2] | | Generation time (days) | Modal mutation rate per day |
	Range	Midpoint		
Bacterium	$10^{-10} - 10^{-6}$	10^{-8}	10 hr ($10^{-0.4}$)	$10^{-7.6}$
Fly	$10^{-7} - 10^{-5}$	10^{-6}	3 wk ($10^{1.3}$)	$10^{-7.3}$
Mouse	$10^{-6} - 10^{-5}$	$10^{-5.5}$	2 mo ($10^{1.8}$)	$10^{-7.3}$
Corn	$10^{-6} - 10^{-4}$	10^{-5}	1 yr ($10^{2.6}$)	$10^{-7.6}$
Human	$10^{-6} - 10^{-3}$	$10^{-4.5}$	20 yr ($10^{3.9}$)	$10^{-8.4}$

[1]After Ricklefs (1979, p.335).
[2]Rate per generation modified after Lerner (1968, p.198); see also Drake (1974). Per generation means per cell generation, i.e., per base pair replication event for bacteria, and per generation of newborn offspring for the other organisms.

species (Wilson et al. 1977), whether expressed on a per generation or per unit time basis, because of 'hot spots' and 'cold spots' of mutational activity along a chromosome. (In this context, per generation means per generation of newborn offspring for macroorganisms, or per cell generation = per DNA replication event for single-celled microorganisms; Table 2.2.) The substitution rate is much faster and more uniform for synonymous than for nonsynonymous nucleotide changes (Hartl 1988, pp.155-162). So, different proteins evolve at different rates.

The overall inherent mutation rate also varies appreciably *among* species when expressed on a per *generation* basis. In this latter case the pattern is consistent and rates generally decline with declining generation time. Thus, the inherent rate for bacteria is 1,000- to 10,000-fold lower per locus per generation than it is for humans (Table 2.2). This is not particularly surprising because unfavorable mutations occurring in haploid organisms will be exposed immediately to selection pressure whereas they will not be in diploids unless both alleles are affected or the mutant gene is dominant. The proportion of mutations that are neutral will also be much smaller in microorganisms rather than macroorganisms. Neutral mutations are those which have little effect as far as natural selection is concerned or, stated more precisely (Crow 1985, p.6), "a mutant allele is operationally neutral if its selective advantage or disadvantage is small relative to the reciprocal of the effective *population number*" (emphasis added). Thus, by definition, if neutrality is to be maintained, where populations are large—as is typically the case for micro- as opposed to macroorganisms—the selection coefficient for such mutations must be vanishingly small.

Longer generation times usually, though not necessarily (depending on the cause of the mutations and error correction mechanisms), mean more mutations can accumulate per individual per generation (Klekowski and Godfrey 1989). There is also a higher probability that a change can occur at any given locus, because of the longer exposure to environmental mutagens. Some mutagens such as X-rays and alkylating agents can cause changes in nonreplicating DNA; others (base analogs) cause mutation only if growth, that is, DNA replication, occurs (Koch 1981). The human female carries her eggs all her life. On a generation time basis this is 20 years or about 350-fold longer than the 3-week generation time of a fly. So, while rates are approximately constant per codon per unit time, they are proportional to genome size and generation time.

The phenomenon that the average rates of amino acid or nucleotide substitution are approximately constant among taxa (Table 2.2 and Kimura 1987) has been called the "molecular evolutionary clock" (Zuckerkandl and Pauling 1965). This presumably reflects the fact that the same functional constraints exist for a given gene or gene product in different organisms. How and why the rates have stabilized where they have is unknown, and the observation has caused some excitement. An inference is often made that the clock is centrally standardized, ticking away uniformly for all species, much like a radioactive decay process, which is not strictly the case (Jukes 1987). Substitutions are nonlinear in that irregularities occur which may or may not reconcile over time. Also, for species of both microorganisms and macroorganisms, the clock can tick at different rates among genes, even when only silent substitution rates are considered (Sharp et al. 1989). Nevertheless, among other interesting observations are the following by Ochman and Wilson (1987): 1) the silent mutation rate in *Salmonella typhimurium* and *Escherichia coli* is comparable to that in the nuclear genes of invertebrates, mammals, and flowering plants; 2) the average substitution rate for 16S rRNA of eubacteria is similar to that of 18S rRNA in vertebrates and flowering plants; 3) the rate for 5S rRNA is about the same for eubacteria and eukaryotes. Although more data are needed for other species, the present indications are that the rates of DNA divergence for a gene encoding a given function are approximately the same, regardless of the organism. The major implication is that key macromolecules change at roughly constant rates in different phylogenetic lineages. The variation in a protein such as cytochrome c between zebras and yeasts gives us one measure of how different the organisms are and when they diverged on the evolutionary tree.

Finally, it should be noted that the rate of phylogenetic change is evidently not controlled by the mutation rate. Rather, the evolutionary history of the major groups of organisms appears to depend on ecological opportunities afforded the simple or complex genetic variants (Wright 1978, pp.491-511). Although there are several exceptions, evolution, overall, has been toward increased size and complexity (Chapter 4). Evolutionary rates, unlike mutation rates, vary greatly even along single phyletic lines. For some existing genera (e.g., the lungfishes and the opossum among animals; horsetails [*Equisetum*] and club mosses [*Lycopodium*] among plants) evolution has been virtually at a standstill for hundreds of millions of years (Wright, p.493). Where genetic variation has provided for a major and entirely new way of life, swift adaptive radiation follows (Wright, p.498). Examples include the development of feathers, wings, and temperature regulation in the case of the birds; efficient limbs, hair, temperature regulation, and mammary glands in the mammals. Other ecological opportunities are presented when new forms colonize an area where either the niches are unoccupied or are better filled by the mutant than the wild-type (Wright, p.509).

Recombination   A broad yet succinct definition of recombination is that it is a process leading to the rearrangement of nucleotides (Chapter 5 in Alberts et al. 1989). Recombination has been divided into two categories, general, and site-specific.

General recombination includes: 1) independent or Mendelian assortment of entire, nonhomologous chromosomes during metaphase I and anaphase I of meiosis and 2) reciprocal recombination of chromosome segments (crossing over) that occurs between homologs during meiosis. An analogous phenomenon also occurs in bacterial transformation and conjugation, where exogenous chromosome fragments are integrated into the genome of the recipient cell by homologous recombination (see **Sex in prokaryotes** in Section 2.3). For eukaryotes, assortment is quantitatively more significant than crossing over in all organisms with a haploid chromosome number exceeding two (Crow 1988). A third kind of general recombination involves repair of haphazard breakages (see below) which may affect a single chromosome (e.g., via an inversion) or several chromosomes (arising from segmental interchanges).

Meiotic recombination will be discussed later in Section 2.3 with respect to sex. Nonmeiotic (nonhomologous) recombination occurs because chromosomes in all organisms can break spontaneously and are inclined to rejoin. Repaired chromosomes can still undergo normal replication. Joining is nonspecific so segments can be interchanged: For instance, if two chromosomes break in one place the four pieces can reassociate in any pairwise combination (Chapter 6 in Fincham 1983). If the exchanges are grossly unequal, an interchange may closely resemble a unilateral transposition (discussed below under site-specific recombination). In any case, reformed chromosomes will be stably transmitted through mitosis if each has a centromere. If the cells involved end up in the germ line and undergo meiosis, the meiotic products usually have deficiencies, depending on the type of interchange.

*Intra*chromosomal structural changes can be illustrated by imagining the three repair options for two breaks occurring within the same chromosome (Chapter 6 in Fincham 1983). First, the resulting pieces may rejoin in the same sequence in which case no phenotypic changes occur. Second, the two end segments may anneal, deleting the middle piece. This may not produce a phenotypic effect in higher eukaryotes, which have a lot of noncoding DNA and consequently in which a functional gene may not be interrupted by the breakage. Third, the middle piece may be reincorporated in an inverted orientation. While an inversion might result in no phenotypic alteration, there is at least one case where it does. In *Drosophila* and some other species, 'position-effect variegation' occurs when gene activation fails in some cells because a gene normally located in the euchromatin region is moved close to heterochromatin as a result of an inversion (Fincham 1983, pp.508-509; Reuter et al. 1990). Also, crossing-over in an inversion results in reduced fertility.

Site-specific recombination is mediated by recombination enzymes, and base pairing is not necessarily involved (Chapter 5 in Alberts et al. 1989). The process is characterized by exchanges of short nucleotides on one or both of the two DNA sequences and seems frequently to involve what Dawkins (1982, p.159) calls a "motley riff-raff of DNA and RNA fragments [that] go by various names depending on size and properties: plasmids, episomes, insertion sequences, plasmons, viroids, transposons, replicons, viruses". A general if less interesting term for this "motley riff-raff" of DNA segments is mobile DNA (Berg and Howe 1989). The diverse elements involved are united by their ability to move to new locations, invert, and become amplified or deleted—all without the requirement for appreciable sequence homology needed for general recombination discussed above. In aggregate, these various forms of dynamic, wandering genetic elements (Table 2.1) within the eukaryotic cell appear to have a direct evolutionary link with prokaryotic sex (Section 2.3).

Insertions and excisions may occur in organized fashion, as with phages and the bacterial genome, or with retroviruses and the vertebrate genome, or in the programmed genic rearrangements required for the generation of immunological specificity (Chapter 23 in Watson et al. 1987). Recombination may also occur haphazardly, which is the case for mobile genetic (transposable) elements in the prokaryotic and eukaryotic genomes (Camerow et al. 1979; Carlos and Miller 1980; Berg and Howe 1989). The same mechanism pertains to the integration of plasmids (small, ancillary, self-replicating chromosomes) and "promiscuous" (organellar) DNA into the nuclear chromosomes. To date, plasmids (see Section 2.3) are known to occur ubiquitously in bacteria; they are also found in many fungi, and in some higher eukaryotes, often in association with mitochondria (Tudzynski 1982; Wickner et al. 1986). The term "invertrons" has been coined (Sakaguchi 1990) to describe those "linear DNA plasmids which have identical sequences in inverted orientation at their termini". For simplicity, I have categorized these as a subgroup of plasmids (Figure 2.1), although the invertron group includes also adenoviruses, as well as representatives from other types of genetic elements, including various transposable elements from eukaryotes, transposons from bacteria, and bacteriophage.

Promiscuous DNA has been detected in numerous organisms, including plants, filamentous fungi, yeasts, and invertebrates (Sager 1985). The term originated with Ellis (1982) who used it to refer to DNA which appeared to move from chloroplasts to mitochondria. Subsequently, evidence has accrued for a broader process, including the insertion of mitochondrial and chloroplast DNA sequences into nuclear DNA (reviewed by Palmer 1985). Transpositions of promiscuous DNA have not been shown to produce functional transcripts (Sager 1985), but if these are documented in the future there are considerable evolutionary implications because of the different modes of inheritance of a nuclear as opposed to an organellar gene. Presumably such transpositions can also interrupt nuclear gene function, depending on where they insert.

Transposable elements (transposons), or "self-dynamic genes" (Campbell 1982), are so named because they can replicate and insert copies of themselves into chromosomes by jumping to new sites (Table 2.1; Figure 2.3; for a review of their role in different organisms see Berg and Howe 1989). McClintock (1956) first discovered transposable elements in maize in the late 1940s and 1950s, and called them "controlling elements" because, although distinct from

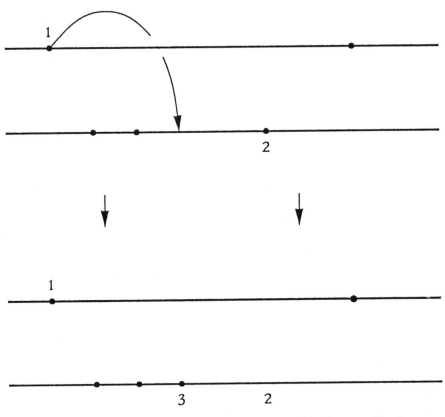

Figure 2.3 General model for the behavior of transposable elements. The lines are two, not necessarily homologous, host chromosomes. The circles are sites of insertion of transposable elements. At 1, the element replicates and inserts a new copy at 3. The element at 2 is lost by excision. From Charlesworth (1985).

genes, they could modify gene expression. Transposable elements have since been well documented in many other taxa (Shapiro 1983; Kingsman et al. 1988; Lambert et al. 1988), including: bacteria (phage Mu; insertion sequences; transposons conferring antibiotic/metal resistance or surface antigen variation); yeasts (Ty and mating type elements of *Saccharomyces*); and animals (*Drosophila* transposable elements and hybrid dysgenesis determinants; vertebrate and invertebrate retroviruses). The consequences of insertion are far-reaching. Transposition always disrupts gene function at the insertion site. Previously dormant genes flanking the site may be activated. Chromosome inversions, deletions, duplications, and fusions can occur (e.g., Simon et al. 1980). In diploid organisms, a pattern of gene expression characteristic for the particular interaction between controlling element and gene typically results (Fedoroff 1983). At least in maize, and presumably in other higher life forms, transposition occurs at predictable times and frequencies in the ontogeny of the organism. In maize, a controlling element can have a similar effect on genes governing different biochemical pathways and at different places in the genome (McClintock 1956; Fedoroff 1983, 1989). Moreover, a single element can control more than one gene concurrently.

The evolutionary implications of transposable elements are exciting. As demonstrated elegantly with the maize system, they provide a novel mechanism for controlling gene expression. Further, a new, highly variable means is generated to provide for restructuring nucleic acid sequences. Beneficial mutations such as drug resistance in bacteria can result, and an additional mechanism is provided to help organisms (especially sessile organisms) respond adaptively to the vagaries of their environment (Chapter 7; see also Adams and Oeller 1986). Frequently, however, the changes are deleterious; for this reason and because the elements replicate themselves independently of host chromosomes, they have been called "intra-genomic parasites" or "selfish DNA" (Charlesworth 1985; also see Section 2.3).

To what extent should mobile DNA be favored by natural selection? At the level of the gene, selection is presumably for these mobile elements, especially in eukaryotes with excess DNA, or in bacteria where they add unique features as plasmids. However, once essential gene functions are disrupted, selection at the level of the gene will be offset by counterselection at the level of the physiological individual. So the tendency should be towards some balance in opposing forces. Note, however, that deleterious genes can still spread in a population by over-replication or if they alter reproductive mechanisms to favor themselves (Campbell 1981; Chapter 6 in Bell 1982). Certain transposable elements (retroviruses, see below) provide an independent mechanism for moving genetic material between distant, reproductively isolated genomes.

Perhaps the most intriguing subcategory of site-specific recombination involves the retroviruses. These may be a category of retrotransposon*, or alternatively, they may represent the original form from which vertically transmitted retrotransposons are derived. Regardless, this retroviral group is unique in having an RNA genome that replicates by reverse transcription through a DNA intermediate, which can then integrate as provirus into host chromosomal DNA. Retroviruses are considered with transposons because they have similar structures and functional properties. They do not transpose in the same way that bacterial transposons do, but are analogous in that they can be viewed as intermediates in the transposition of viral genes from proviral integration sites in the host chromosomes (Varmus 1983; Varmus and

---

*Transposons move by transfer of information from DNA to DNA; retrotransposons transpose by reverse transcription of RNA to DNA which is converted to a double helix and inserted at a new location.

Brown 1989). Retroviruses and viral-like elements have been described from diverse genomes, including those of mammals, the slime mold *Dictyostelium*, yeast, fish, reptiles, birds, and plants (McDonald et al. 1988). The most information is on mammalian retroviruses and because of the interesting evolutionary implications, this is summarized briefly below (see also Doolittle et al. 1989).

There are two retroviral categories (Benveniste 1985; Varmus 1988). *Infectious* or *exogenous* retroviruses occur as a few copies of proviral DNA per cell, only in the genome of infected cells; they are infectious and often pathogenic; and they are transmitted horizontally, i.e., among individuals of a species. *Endogenous* retroviruses occur as multigene families in the host DNA of all cells (both somatic and germ) of all animals of the species of origin (hence they are transmitted vertically, i.e., to all descendants) (Benveniste 1985). Endogenous retroviruses are usually not infectious to cells of the species of origin, but are often so to those of other species. In fact it is this property of being able to replicate in heterologous cells that sets them apart from conventional cellular genes.

Both types of retroviruses can cause host cell genes to mutate, or can carry host genes with them. This is significant because of the spreading of somatic variation through the soma, and the possibility of introducing variation directly to the germ line (Section 2.4). From an evolutionary standpoint, the endogenous group is particularly intriguing because the members have been transmitted between distantly related vertebrate species, and once established may subsequently be incorporated into the germ line and transmitted vertically. Benveniste (1985, p.362) reviews evidence that retrovirus transfers have included those "from ancestors of primates to ancestors of carnivores, from rodents to carnivores, from rodents to primates, from rodents to artiodactyls, from primates to primates, and from primates to birds". A specific example is the baboon type C viruses which are transmitted vertically in primates, and which were transferred millions of years ago to ancestors of the domestic cat, where they were incorporated into germ cells and inherited thereafter in conventional Mendelian fashion (Figure 2.4) (Benveniste and Todaro 1974, 1975). Benveniste (1985) proposes that retroviruses may promote genetic interaction above the species level, much as do plasmids in bacteria (discussed later).

Are retroviruses a major force in evolution? It is too early to say. Retroviral elements do seem to play a major role by influencing gene regulation (McDonald 1990). The more exciting and novel message, however, is that *gene transfer occurs horizontally between distantly related species and that the genes involved have been shown in some instances to be incorporated into the germ line.* The only analog that comes close to this is the transfer and incorporation of bacterial DNA into plant chromosomes. Crown gall and hairy root diseases of plants involve transfer of a plasmid (Ti plasmid) from *Agrobacterium tumefaciens* to the plant where a fragment (T-DNA) is covalently

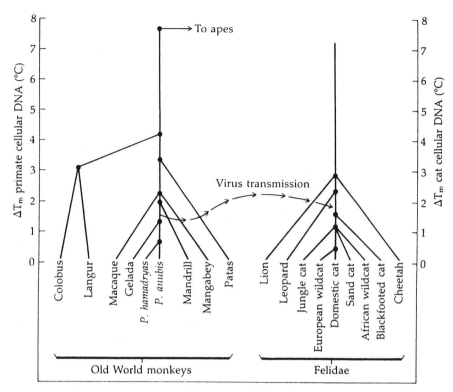

Figure 2.4 Molecular phylogenetic tree for Old World monkeys and the Felidae showing horizontal transfer several million years ago of an endogenous retrovirus (RD-114) from ancestors of the baboons to ancestors of the domestic cat. Genetic relatedness within each tree is inferred from hybridization of labelled, unique sequence DNA from reference species (e.g., baboon = *Papio* spp., or domestic cat, respectively) to each species being compared with it in the tree. Thermal stability, T_m, is an index of base-pair mismatching; mismatched DNA:DNA hybrids melt at a lower temperature than do matched hybrids. The greater the difference in T_m between the heterologous and homologous hybrids, the lower the genetic relatedness between the compared species. Also, sequences partially homologous to the endogenous baboon type C retrovirus (RD-114) are evident in *all* Old World monkeys but in only *some* closely related Felidae. The time of transfer is postulated by comparing donor and recipient species for presence of sequences similar to the retrovirus and, if present, the degree of homology of such sequences to the virus. Thus, transfer from the primates evidently occurred after divergence of the mandrill. Acquisition by the Felidae occurred after separation of the lion, cheetah, and leopard lineages; the six species diverging after the leopard (i.e., below it in the figure) all contain the baboon virus. From Hartl (1988, p.207; based on Benveniste 1985).

integrated into the host nuclear genome. In essence, the agrobacteria use genetic engineering methods to force the infected plant to synthesize nutrients (opines) which the bacteria utilize (Zambryski 1989). (Although the process of *Agrobacterium* oncogenesis is often considered with processes involving transposable elements, strictly the T-DNA is not a transposon because it does not jump about the chromosome).

Finally, there is increasing evidence in plants and animals for reverse transcription being associated with systems other than the classical retroviruses and retrotransposons noted above. This includes many kinds of repeated DNA elements and nontranscribed, 'processed pseudogenes' whose duplication evidently involves reverse transcription (Rogers 1985, 1986; Boer et al. 1987). These dispersed copies of RNAs, called retroposons (Rogers 1985), may be functional or nonfunctional, but do undoubtedly provide material for further evolution. Closely related repeated sequences in the chromosomal DNA of eukaryotes include the so-called LINEs (long interspersed nucleotide sequences; Hutchison et al. 1989) and SINEs (short interspersed nucleotide sequences; Deininger 1989).

Epigenetic controls   Epigenesis is the translation of genotype and environment into phenotype (Cheverud 1984). Epigenetic controls are particularly striking in the unfurling of the developmental program. During ontogeny, unspecialized cells of the *same* genetic constitution undergo a process of determination (i.e., their developmental fate becomes predictable) followed by commitment (their developmental fate becomes irreversible) and ultimately differentiation into distinct tissue types. For instance, one of the pair of X chromosomes in female mammals becomes inactivated during embryogenesis, and liver cells somehow come to act differently from nerve cells or the rod and cone cells of the eye. These examples all illustrate regulation of gene expression, and thus constitute the third source of genetic variation.

In bacteria, one method of control involves proteins which inhibit or activate transcription of specific genes (see discussion of the *lac* operon in Chapter 3). In eukaryotes, various mechanisms have been proposed, including the packaging of portions of the genome into heterochromatin (transcriptionally inactive chromatin); the operation of cooperatively bound groups of gene regulatory proteins which may control gene expression by opening or closing large DNA domains (Reuter et al. 1990); and the methylation of DNA (Chapter 9 in Alberts et al. 1989). For instance, in mammalian cells, patterns of methylation have been correlated with gene activity (inactivation = sites largely methylated; activation = sites not methylated) (Razin and Riggs 1980; Holliday 1988). Inactive genes can be reactivated by the demethylating agent azacytidine (Jones 1985). The pattern of methylation is heritable through the maintenance of DNA methylase. Consequently, changes in DNA methylation patterns could be used to switch on and off genes in a particular cell lineage.

Holliday (1988) has proposed that loss of methylation may contribute to aging of the soma (see **Senescence** in Chapter 6); in the germ line, it may play a role in signalling the onset of meiotic recombination.

There is a class of genes (proto-oncogenes) that appears to have a normal housekeeping function within the cell but which, if altered by mutation, can lead to malignant transformation. In the human Ha-*ras* gene, a single point mutation within the fourth intron can cause a tenfold increase in gene expression and transforming activity (Cohen and Levinson 1988). This is only one example of many that show there are several ways to change gene expression without changing the gene itself (see also McDonald 1990). Cells of more complex organisms in particular have a large repertoire of mechanisms to generate genetic variability! Undoubtedly all ways of altering expression are important in evolution and may explain why humans have so many genes in common with other organisms. Perhaps humans differ from chimpanzees largely because of differences in gene expression, rather than in gene structure. That there are many gene copies in the more complex life forms provides the opportunity for alterations in the genome by changing introns, exons, or both. New gene products can be exposed to natural selection while the organism is buffered through continued function of the unaltered product encoded at another site(s). Genes that have been rendered silent can be brought back, and other genes turned on or off, all with the phenotypic expression of a point mutation in a coding region, but accomplished simply by changes in gene expression.

2.3   Sex and Meiotic Recombination

To establish the basis for a definition of sex, it should be noted first that reproduction is considered here, broadly speaking, to be the production of offspring. (For a more restrictive definition of reproduction, involving the distinction between "reproduction" per se versus "growth", see Harper [1977, pp.26-27].) Any form of reproduction in which offspring inherit the parental genotype precisely is *asexual* (Williams 1975, p.111). Definitions of sex vary considerably (Michod and Levin 1988). I consider sex to be the bringing together in a single cell of genes from two (or rarely more) genetically different individuals (Maynard Smith 1978). In most but not all organisms sex is tied to reproduction. Prokaryotes and certain parasexual events in the fungi, described later, are the major exceptions. The most important consequence of sex is thus that the zygote, or its functional equivalent in prokaryotes, acquires new genes (mutations in the germ cells of one or both parents) and new gene combinations. Notice that while sex is often loosely construed as equivalent to recombination or mixis, it is actually only one (though admittedly an important one) of several ways in which nucleotides can be rearranged. Also, except for meiotic reassortment, all of the recombination mechanisms discussed previously (Section 2.2) can operate to generate variation in the somatic line.

The bringing together of different sources of genes has been accomplished by bacteria in several ways, discussed below. In eukaryotes the process has been formalized to varying degrees as discrete life cycle stages involving protoplast fusion followed by nuclear fusion (plasmogamy and karyogamy). This leads to the onset of the diploid phase, which is followed more or less rapidly by meiosis, which restores the haploid condition. Phylogenetically, there is a tremendous range in the number of cell divisions separating fertilization and meiosis: bacteria and most fungi are exclusively haploid; the n and 2n phases are about equally prominent in the sea lettuce (*Ulva lactuca*); while in most animals and plants the haploid condition is brief and represented only by gametes. Representative life cycles and the nuclear states (n; 2n; n + n) of the key stages (diploid phase; products of meiosis; haploid phase; gametes; dikaryotic phase, zygote) are summarized in Table 2.3. Clearly, evolution has favored the diploid state. Presumably this is because it masks deleterious somatic mutations and alleles, and increases the variation in genetic resources available to the individual. As Stanier (1953, p.2) summarizes the issue quite vividly, "haploidy renders microorganisms highly exposed to the chill winds of selection". It seems more than coincidental that the haploid condition is found almost exclusively among either single-celled organisms—where defective cells would not adversely affect others—or among lower plants—where defective cells can be relatively easily discarded and replaced (Ricklefs 1979, p.346).

While the processes and pathways involved in meiotic recombination are likely to be unique in detail for specific organisms, genetic analyses of mating both in microorganisms and macroorganisms indicate that the result is the same (Meselson and Radding 1975). The main features (Chapter 5 in Alberts et al. 1989) can be summarized as four points: 1) synapsis of homologous chromosomes is followed by breakage and rejoining of the ends to form intact but genetically recombined double helices; 2) the sites of exchange can be anywhere in the homologous sequences; 3) the sites of exchange occur as a base-paired, staggered joint, which may involve thousands of nucleotides; 4) there is no loss, gain, or alteration of nucleotides in the process.

In terms of antiquity, sex can be traced to about the origin of life itself. In fact it has been argued that sexual reproduction predates asexual reproduction because the latter requires an existing mechanism to preserve genetic individuality (Williams 1966, p.133; 1975, pp.111-123). In other words, for the asexual process to be conducted accurately there would have had to be the equipment and biochemical machinery to ensure that offspring inherited exactly the parental genotype, not a variable amount of it. In contrast, primitive sex was probably initiated in rather sloppy fashion when "information-bearing molecules" escaped from early living systems, managed to avoid

Table 2.3 The life cycles of some representative micro- and macroorganisms[1]

Organism	2n; diploid phase	n; products of meiosis	n; haploid phase	n; gametes	n+n; dikaryon	2n; zygote
Escherichia coli (bacterium)	—	—	Entire life cycle	Direct gene exchange	—	—
Neurospora crassa (filamentous fungus)	Ascus (within fruit body—perithecium)	Ascospores, four-spore pairs	Filamentous mycelium bearing asexual spores (conidia)	Ascogonium (in protoperithecium); conidium (opposite mating type)	Ascogenous hyphae within perithecium	Ascus initial
Schizophyllum commune (bracket fungus)	Basidium (borne on gills of fruit body)	Basidiospores, borne externally on basidium in tetrads	Monokaryotic mycelium	Mycelial cells of compatible mating types	Mycelium—the major growth phase—forming fruit bodies (brackets)	Young basidium
Ulva (multicellular green marine alga; two-layered sheet of cells)	Sheet-like thallus	Zoospores formed in tetrads	Sheet-like thallus (similar to diploid)	Motile cells—opposite mating types, morphologically identical	(Karyogamy immediately follows gamete fusion)	Germinates to give haploid thallus
Any moss	Capsule borne on haploid plant	Spores, initially in tetrads, liberated from capsule	Leafy plant bearing archegonia and antheridia at fertile shoot apices	Egg (in archegonium); sperm (liberated from antheridium)	(Karyogamy immediately follows gamete fusion)	Fertilized egg develops into diploid capsule
Zea mays (angiosperm herbaceous plant)	Maize (corn) plant	Megaspores in ovules; microspores (pollen) in anthers of flowers	Embryo sac (eight nuclei including egg nucleus); pollen tube (three nuclei including two gamete nuclei)	Egg (in embryo sac); pollen tube nucleus	(Karyogamy immediately follows gamete fusion)	Fertilized egg develops into dormant embryo in maize kernel
Drosophila melanogaster (fruit fly)	Larval stages leading through pupal stage to adult fly	Eggs; spermatozoa	—	Eggs; spermatozoa	(Karyogamy immediately follows gamete fusion)	Fertilized egg develops into larva

[1]Modified slightly from Fincham (1983, pp.52-53).

degradation in the surrounding environment or in a recipient cell, and were preserved because they enhanced fitness of their new host (Dougherty 1955; Parker et al. 1972; Williams 1975, pp.111-120). This is the analog of what we know today as transformation in bacteria. When asexual reproduction evolved, sex for some organisms then became associated with periods of environmental stress or change (Chapter 7), alternating with rounds of asexual reproduction.

While the origins of sex seem relatively straightforward, the reasons for its almost universal phylogenetic distribution and maintenance are not. This controversial issue has been explored in numerous books (e.g., Williams 1975; Maynard Smith 1978; Bell 1982; Michod and Levin 1988) and a lengthy discussion is clearly beyond the scope of this one. A brief overview of the major theories and how they relate to the ecology of micro- and macroorganisms follows.

The common point of departure for all enquiries into why sex is maintained is the observation that the loss of half of one's genes in meiosis gives sexual individuals about a 50% disadvantage relative to their asexual counterparts whose genome would be passed on in full. Observe that, genetically speaking, your children are only half you, your grandchildren one quarter, your great-grandchildren one eighth, and so forth. If you were a clonal organism your descendants would have all your genes and be identical copies of yourself. This genetic dilution is variously called the "cost of meiosis", the "cost of sex", or the "cost of producing males". Crow (1988) lists other disadvantages of sexual reproduction. Nevertheless, many organisms reproduce exclusively sexually. In most others, sexual and asexual modes coexist stably, evidently fostering local adaptation as in the clonal and seed mechanisms of many plants (Silander 1985). The major ideas on why sex does not disappear can be formulated as four hypotheses: 1) the Variation Hypothesis; 2) the Parasitic DNA (Selfish Gene) Hypothesis; 3) the Morphogenesis Hypothesis; 4) the Repair Hypothesis (Table 2.4).

The Variation Hypothesis formed for many years the dogma that the contribution of sex is to directly promote recombination, that is, to generate new gene combinations. Of course, if favorable new gene combinations can be assembled, this implies that existing good combinations can also be destroyed: Genetic exchange can increase mutational load as easily as decrease it! However, at the population level, natural selection can then act on the resulting genotypes, eliminating inferior individuals and thereby decreasing mutational load overall. The three major versions of the Variation Hypothesis are that sexual recombination produces the occasional genotype of exceptionally high fitness; that it establishes a safety margin of genotypic variability against adverse environments and hence extinction (or, stated slightly differently, that it allows for efficient utilization of a variable environment); and that the rates of gene substitution and hence phylogenetic change are in-

Table 2.4   The major concepts about the maintenance of sex

Hypothesis[1]	Description	Key references
Variation	Sex promotes genetic variation	Williams 1975; Maynard Smith 1978; Crow 1988
Parasitic DNA	Sex forced on organisms for benefit of extraneous DNA	Orgel and Crick 1980; Doolittle and Sapienza 1980; Hickey and Rose 1988
Tissue differentiation	Sex maintained because it cannot be uncoupled from morphogenesis	Margulis et al. 1985
Repair	Sex maintained because of meiotic repair of DNA; where outcrossing occurs, harmful mutations masked	Bernstein et al. 1985; 1988

[1]Note that only in the Variation Hypothesis is sexual recombination the object of natural selection; in the other three it follows as a consequence of selection for other features.

creased. For a discussion of these, see Williams (1975: Chapter 12) and Bell (1989: Chapters 8-11).

According to the Parasitic DNA (Selfish Gene) Hypothesis, sex is forced on organisms for the benefit of DNA sequences whose spread through host populations is facilitated by gamete fusion (Orgel and Crick 1980; Hickey 1982; Chapter 9 in Dawkins 1982; Hickey and Rose 1988). While sex may have been parasitic in origin (see transformation and transduction in bacteria, below), parasitism per se seems to provide insufficient grounds for its maintenance. By this theory, the original sequences promoting 'syngamy' have long since become part of the genome, in which they are selectively neutral or deleterious. Hickey and Rose (1988) suggest that sex continues because it facilitates transposition which in turn, they argue, may foster chromosomal recombination (it could also possibly create new gene control mechanisms or novel proteins). On balance, however, transposons seem generally to do more harm than good, at least to their eukaryotic hosts, in which case it would seem odd that organisms maintain sex for the reason of transposition. For bacterial populations, Condit et al. (1988) propose that under certain conditions (among these are replicative as opposed to conservative transposition; infectious transfer of transposons hitchhiking on self-transmissible plasmids), parasitic DNA will be maintained even if the elements are deleterious to their hosts.

The Morphogenesis Hypothesis proposes that meiotic sex occurs because it is inextricably linked to orderly morphogenetic reproduction of the individual (Margulis et al. 1985). Meiosis is seen as providing a quality control function by ensuring that the genes are functional prior to morphogenesis.

Any mixis aspect of sex is thus viewed as being distinctly secondary (an "epiphenomenon") to the role of meiosis as the controller of differentiation. This hypothesis does not address the reason for sex in prokaryotes. Also it is a difficult idea to reconcile with meiotic recombination in the simple, relatively undifferentiated eukaryotes, many of which are single-celled organisms (e.g., yeasts, diatoms, flagellates) or, conversely, the apparent absence of meiotic sex in many fungi. Finally, it is the most abstract, and the least testable hypothesis of the group.

The Repair Hypothesis states that sex arose and is maintained because of the repair of genetic damage by recombination (Bernstein et al. 1985, 1988). Recombination appears to be initiated by breaks in DNA. Enzymes involved in crossing over seem to be responsible for repair of damage to DNA, a process that evolved in bacteria.

Maintenance is further promoted wherever the two DNA molecules come from different individuals (outcrossing). The key to this idea is the distinction between damage to DNA and mutation, and the implications of both events. Damage is a physical alteration of the DNA molecule (e.g., crosslinks, breaks) which is usually not replicated or inherited. If only one strand is affected, it can usually be fixed by excision repair, in which the undamaged strand serves as a template for replacement of the excised bases (endogenous repair). If both strands are affected, the template can only be provided by the homologous molecule through an exchange process which occurs during pairing at meiosis (this does not require an open genetic system). Mutations, on the other hand, cannot in general be repaired (because the correct message cannot be recognized from the incorrect) and are inherited (see Section 2.2). At the level of the individual, they can be masked, however, by complementation when a correct copy of the affected sequence is present in the same cell. Hence, the Repair Hypothesis includes both the fixing of nucleotides (endogenous mechanisms) and complementation (exogenous mechanisms which in the extreme case involve open genetic systems). In this sense, the exogenous form is essentially a restatement of the Variation Hypothesis, and is viewed as the central factor in evolution of outcrossing and maintenance of diploidy (Bernstein et al. 1985; Chapter 11 in Bell 1989).

Holliday (1988) has developed a modified version of the Repair Hypothesis based on DNA methylation. Several lines of evidence primarily from bacteria, *Drosophila*, and yeast systems, support the role of repair. For example, recombination-deficient *Escherichia coli* mutants (*recA⁻*) have low viability because they cannot produce a protein essential for repair of damaged DNA (Chapter 12 in Fincham 1983). The repair concept also provides an explanation for senescence of the soma (little capacity for repair) versus continuity of the germ line (meiotic repair, hence rejuvenation) (see **Senescence** in Chapter 6; also Bernstein et al. 1985).

In overview, none of the four major hypotheses can be dismissed, but none appears to provide the sole answer. Indeed, they are not mutually exclusive and part of the impasse appears to be that a universal explanation which can be held forth as a paradigm is being sought. For example, while there is certainly a strong case to be made for repair in sex, much is made of the fact that there is no meiotic recombination in some taxa (e.g., the males of many species of Diptera). Assuming these observations are correct, then the possible explanations are that damage in the germ lines of such organisms may not occur (which seems inconceivable), that it is unusually rare (unlikely), that it is not repaired (affected cells consequently die), or that it is fixed by some other, unidentified mechanism (quite possible). Whatever the explanation, it does not exclude repair as a basis for sex in other, perhaps most, lineages. The predominant view today among biologists is probably that sex originated first as a repair mechanism (in both prokaryotes and eukaryotes) and that recombination evolved from it. Margulis and Sagan (1986, p.60) phrase it this way, "genetic recombination began as a part of an enormous health delivery system to ancient DNA molecules . . .".

Sex in prokaryotes: the bacteria   Not only does sex occur in bacteria, but if one means by the term that genes from different sources are recombined in a single entity, then "bacteria are particularly sexy organisms", to use the words of Levin (1988). While the three processes involved—transformation, transduction, and conjugation—are distinct from each other and from eukaryotic sex, all produce the same end result: recombination of DNA from genetically different individuals. Unlike recombination in macroorganisms, which generally involves two complete genomes, recombination in bacteria is asymmetrical, involving a relatively large and a small donation, respectively, from the two partners. It also seems to occur much less often than in eukaryotes (Bodmer 1970). Each mechanism is quite complicated in detail and beyond the scope of this discussion. An overview is attempted to provide a basis for comparisons between prokaryotic sex and that which occurs in higher organisms; specifics are available in general microbiology texts and advanced treatises (e.g., Campbell 1981; Levin 1988).

In transformation, a bacterial cell takes up naked DNA from the surrounding medium (originating usually from a lysed or decomposing cell) which is then integrated into and replicates with the recipient's genome (Figure 2.5). The mechanism hinges on several conditions, among them development of a transient ability (competent state) by the recipient to be transformed. Transformation occurs naturally and has been documented in many genera, including *Streptococcus, Staphylococcus, Haemophilus*, and *Pseudomonas* (Trevors et al. 1987; Levin 1988). It would appear to be potentially most significant in habitats where DNA can be protected from digestion (e.g., by adsorption to a matrix such as clay particles) and where cells occur densely.

Transduction occurs when bacterial DNA is packaged within the protein capsid of a bacteriophage particle and injected into a recipient cell during the viral infection process (Trevors et al. 1987; Levin 1988). In generalized trans-

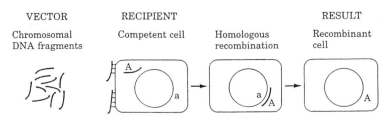

Figure 2.5 Gene transfer in bacteria by acquisition of free DNA (transformation). Incoming chromosomal fragments (dark lines) bind to bacterial cell (rectangle); one enters the bacterium and is incorporated into the genome (light circle) by homologous recombination. From Levin (1988).

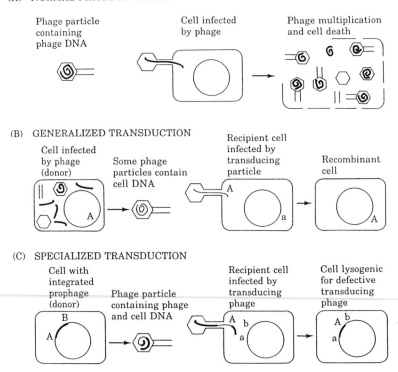

Figure 2.6 Three possible results of infection of bacteria by bacteriophage. Heavy lines = phage genome; light lines = host genome. (A) In normal infection by lytic phages, replication of the phage leads to packaging of phage DNA in phage particles. (B) In generalized transduction, a few of the progeny phage contain random portions of bacterial DNA instead of phage DNA. These progeny phage then transfer the host DNA into new cells where it replaces the recipient's genes. (C) In specialized transduction, part of the phage genome is replaced by adjacent host genes when the phage excises from the bacterial chromosome; these are inserted by site-specific mechanisms when the virus enters the genome of its new host. From Levin (1988).

duction, chromosomal genes from the donor bacterium are transferred randomly when a small proportion of progeny phage carry some bacterial DNA together with phage DNA. In specialized transduction, temperate phage move specific bacterial genes when they excise imprecisely from their prophage state in the bacterial chromosome at the onset of the lytic cycle (Figure 2.6). Transduction has been shown to occur in various soil and aquatic bacteria under nonsterile experimental conditions (Trevors et al. 1987; Levy and Miller 1989).

Conjugation involves cell to cell contact and results in the transmission of plasmid (extrachromosomal) DNA alone or both plasmid and chromosomal DNA from donor to recipient (Figure 2.7). Plasmids are one form of replicating pieces of DNA (replicons); in the older literature those plasmids which could integrate into the chromosome were called episomes. Some plasmids (F particles) mediate their own transfer and bacterial cell populations can be classified as F$^+$ ('male') or F$^-$ ('female'). One of the proteins specified by the F particle is for the sex pilus, a temporary projection which joins the F$^+$ and F$^-$

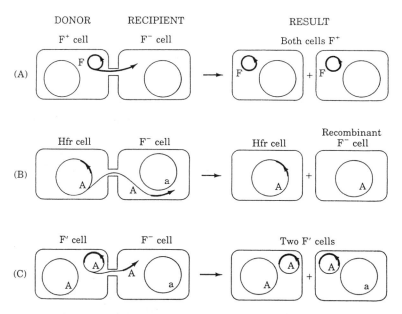

Figure 2.7   Three mechanisms for gene transfer in bacteria as mediated by F plasmids (conjugation). Heavy lines = plasmid genome; light circle = host chromosome. (A) Only the F plasmid is copied and transferred. (B) The F plasmid is part of the host chromosome (in Hfr cells, see text). During conjugation a copy of some of the donor chromosome is transferred and replaces part of the recipient's chromosome. (C) Some of the host chromosome is carried along with the plasmid. In this case the plasmid may not integrate into the host chromosome and the recipient becomes diploid for the newly acquired chromosomal genes. From Levin (1988).

A Case Study: Transferable Drug Resistance in Bacteria

A central message of this chapter is that although all living things have essentially similar means of generating and transmitting genetic variation, the potential evolutionary rates of microorganisms are much higher than those of macroorganisms. This is a function of their short generation times, hence large population sizes (Chapter 4), and efficient means of genetic exchange.

A classic example of natural selection in action is the evolution of antibiotic resistance in bacteria. Such resistance is either encoded by chromosomal genes or plasmid-borne (the elements responsible are one class of conjugative plasmids, the resistance or R factors; see Section 2.2). Plasmids are especially significant ecologically because resistance to several different antibiotics can be combined in a single sequence which can also serve a role as an efficient vector, as well as mediating genetic rearrangement within and between bacterial cells (Koch 1981). The plasmids may exist in multiple copies per cell and so the antibiotic-resistance genes can be amplified when necessary and deamplified when not needed.

Resistance to many if not most antibiotics is known, and frequently the genes are inserted within transposons. In the presence of the antibiotic(s), the drug-resistant phenotype has a selective advantage. Conjugative plasmids typically replicate during transfer; hence, acquisition by the recipient is not at the expense of loss from the donor. The genes responsible spread so quickly (often exponentially) within and between populations that the phenomenon has been called *infectious* drug resistance. For instance, following the introduc-

cells and through which plasmid DNA moves. In rare F⁺ cells (Hfr = high frequency of recombination types), where the F particle is integrated into the bacterial chromosome, chromosomal genes are also transferred. The F particle can also mobilize a class of nonconjugal plasmids when both occur in the same donor cell. However, to put these events in perspective, it should be realized that Hfr formation, even under laboratory conditions, is a comparatively rare event. Such Hfrs, once formed, are relatively unstable because the F factor excises at a high frequency. Therefore, the contribution of this type of genetic exchange to variation of bacterial populations in nature is unclear.

Plasmids are ubiquitous among bacteria. They can be transferred by trans-tion of antibiotic therapy with streptomycin, chloramphenicol, and tetracycline from 1950-1965 in Japan, the proportion of drug-resistant *Shigella* (the bacterium which causes bacillary dysentery) increased from about 1% to 80% of the isolates (Figure 2.8 and Mitsuhashi 1971).

The phenomenon of transfer of drug resistance is not only important within the context of basic microbial ecology, but obviously has profound implications in practical terms of how antibiotics can be over-prescribed in medicine. Furthermore, because drug-resistant bacteria of animal origin can cause serious diseases in humans (Holmberg et al. 1984), the use of antibiotics as animal feed supplements should be restricted. For additional reading on the ecology of transferable drug resistance, see Anderson (1968); Koch (1981); Campbell (1981); Clewell and Gawron-Burke (1986); Levy and Miller (1989); and Condit and Levin (1990).

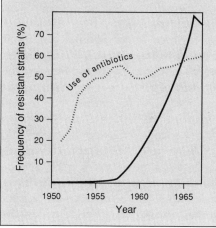

Figure 2.8   Increase in the percentage of antibiotic-resistant *Shigella* strains (solid line) in Japan following generally increased production of the antibiotics streptomycin, chloramphenicol, and tetracycline in the years indicated. The average production (use) for the three antibiotics is plotted (dotted line). Based on Mitsuhashi (1971, p.8).

duction and transformation as well as by conjugation. Bacteria lacking them generally multiply normally; hence, plasmid DNA does not encode essential functions. Plasmids are best regarded as expendable accessory sources of traits. Their DNA contributes to the genetic plasticity of the carrier and confers many characteristics of adaptive value in diverse environments. The best known of these is antibiotic or heavy metal resistance (R factor plasmids; see the case study); others include the ability to induce tumors in plants (*Agrobacterium tumefaciens*), nitrogen-fixing capability (*Rhizobium* spp.), increased virulence (*Yersinia enterocolitica*), and antibiotic synthesis (*Streptomyces* spp.). Transfer of plasmids has been observed between bacterial strains, species, and, in some instances, even between unrelated genera (Beringer and Hirsch

1984; Freter 1984; Levy and Miller 1989). In the case of the crown gall disease, the mobilization function of a bacterial plasmid promotes its transfer to *plant* hosts: Not only can this plasmid move among bacteria, but plants evidently have access to the gene pool of at least some bacteria (Buchanan-Wollaston et al. 1987). While plasmids may be the key means by which bacterial genes are transferred in nature (Reanney et al. 1983; Trevors et al. 1987), it is not yet clear that conjugation is the mechanism involved, although this is generally inferred to be the case.

Ecological attributes of bacteria that can be construed directly from their genetic properties are:

1. Bacteria are considered to be haploid because generally there is only one copy of the genome present (although there may be multiple copies of the chromosome). Any genetic change is thus immediately expressed and exposed to natural selection.

2. The bacterial genome is compact (on the order of about 2000 genes), efficient, and designed to fit into a small cell (see Chapter 4). This contrasts with the eukaryote genome, characterized by multiple gene copies, and extensive noncoding regions. Under optimal conditions bacteria divide on the order of minutes, as fast as their DNA replicates. This is facilitated by the relatively small size of their genomes, the close spacing of genes, the organization of genes into functional groups which are transcribed together (operons), and the clustering of related operons (see Chapter 3). Selection pressure is thus continuous to remove noncoding regions which enables faster replication, division, and displacement of slower growing cells from the population. In contrast to large mammals which have generation times of years, many species of bacteria have hundreds or thousands of generations per year. Thus, at the population level, deleterious mutants are removed quickly; likewise, favorable mutants or recombinants spread quickly. Because the potential for evolutionary change in organisms depends on the variability of the population and its turnover rate, the above points together indicate that the evolutionary potential should be relatively high for bacteria.

3. Although genetic exchange occurs between bacterial cells, the extent to which this actually occurs in nature (e.g., Brenner and Falkow 1971) may be quite restricted and is a debatable issue. Intraspecific matings may be relatively rare (Levin 1981). Many more studies are needed under realistic conditions in situ to confirm the extent of genetic mixing. One view of horizontal exchange (Loomis 1988, p. 215) is that 'species' barriers were not recognized and that individual genes were promiscuous over the Precambrian period of evolution. Promiscuity subsequently gave way to general but by no means absolute fidelity, es-

pecially as regards horizontal movement of episomes (plasmids) (Shapiro 1985; Slater 1985) as bacteria evolved the ability to recognize and degrade foreign DNA (Boyer 1971). Perhaps it is not surprising that some recent studies by molecular hybridization techniques have shown, contrary to earlier belief, that a prokaryotic species does indeed possess "genetic cohesiveness" (Hartl 1985) similar to that of eukaryotes.

The foregoing requires at least passing consideration of what a 'species' is. Although the species concept is more or less an abstraction, to most macroecologists a species would consist of a set of individuals which, though variable, are united by strong phenotypic resemblance, compatibility of genomes, and reproductive isolation. This idealized notion does not hold for bacteria. Cowan (1962, p.451) concludes quite bluntly "the microbial species does not exist; it is impossible to define except in terms of a nomenclatural type, and it is one of the greatest myths of microbiology"! One problem is that sexual reproduction occurs relatively infrequently compared with asexual reproduction and, when it does, absolute barriers do not exist. Plasmids are not necessarily host (species) specific, and transfer may be mediated by transposons (Brisson-Noel et al. 1988). Duncan et al. (1989) observed bidirectional genetic exchange between *Bacillus subtilis* and *B. licheniformis*, both in broth and soil. The transferred genes were expressed in the recipient cells, although the stability of continued expression of these foreign genes was variable. Thus, bacterial species might better be thought of as "distinctive arrays of varying but coadapted gene complexes which are periodically reshuffled" (Duncan et al. 1989, p.1586; for general remarks see Koch 1986). So, at one end of the continuum we have gene sets ("species") which in effect are fairly isolated because the entry of foreign DNA would be disruptive and selected against; at the other is the view that all bacteria are basically cousins assimilated into a common melting pot by virtue of the many genetic changes among them. This extreme stand is that bacteria in aggregate comprise a single planetary entity or superorganism existing as "a unique, complex type of clone composed of highly differentiated (specialized) cells" (Sonea and Panisset 1983, p.22). Probably bacterial "species" are neither as genetically delimited as are eukaryote species nor as promiscuous as the superorganism complex would have us believe. Regardless of the extent to which it may occur in nature, bacterial sex is distinct from sex in eukaryotes by being nonmeiotic, irregular in occurrence, fragmentary in extent of genetic exchange (only partial reassortment), and separate from the event of reproduction.

4. For bacteria in nature, one can say that generally mutation must be the major source of variation. This is unlike the case for higher or-

ganisms where shuffling of mutations by meiotic recombination plays the major role.

5. Bacteria display all the capability of eukaryotes in generating genetic variability. The major difference seems to be in the relatively unordered fashion in which this is transmitted, compared with a more organized manner, culminating in the meiotic process, of higher organisms.

Sex in some simple eukaryotes: the fungi   For the fungi, sex followed by meiosis is, in very general terms, similar to that in the higher eukaryotes. Three distinctions, however, need to be emphasized: 1) Variation among taxa in the stages of sexual reproduction—plasmogamy, karyogamy, and meiosis—establishes four basic life cycles in addition to what is generally referred to as an exclusively 'asexual' category (discussed below). 2) Fungi have 'alternative' genetic systems, most notably heterokaryosis and parasexuality (described below). Both sexual and vegetative fusions enable other DNA molecules such as virions, plasmids, and mitochondrial DNA to be exchanged. Furthermore, septa transversing the hyphae are rarely complete in areas other than where reproductive structures are borne. Apart from the obvious physiological implications pertaining to cytoplasmic streaming, solute movement, and so forth, this means that multiple, potentially genetically distinct nuclei exist within a common cytoplasm. 3) Sex is associated with resting spore formation and is frequently triggered by adverse environmental conditions (see Chapter 7). For these reasons and because the fungi display the genetic characteristics of a transitional group between the prokaryotes and the more complex organisms, the following overview is informative.

The basic life cycle categories of the fungi (Figure 2.9) are as follows (Raper 1954; Chapter 1 in Burnett 1975): 1) 'Asexual'—an artificial assemblage of species (Fungi Imperfecti or Deuteromycetes) united by their apparent absence of conventional sexuality. In these fungi the parasexual cycle (below) can and frequently does occur, although it is probably less important than originally believed (Caten 1987). An example is *Candida albicans,* a yeastlike pathogen of humans. 2) Haploid (haploid monokaryotic)—the life cycle is predominantly haploid because meiosis follows immediately upon nuclear fusion. This cycle is typical of most of the lower fungi (Zygomycetes, e.g., the bread mold *Rhizopus stolonifer*) and certain Ascomycetes. 3) Haploid (haploid/dikaryotic)—the cycle is similar to (2) except that "paired, potentially conjugant kinds of nuclei persist in close physical association in the same hyphal segment (hence dikaryon) and divide synchronously" (Burnett 1975, p.10). Classic examples are the rust fungi such as *Puccinia graminis*, discussed in Chapter 6. The dikaryotic phase may be transient (many Ascomycetes) or exist for most of the life cycle (likely several centuries for the fairy-ring fungi). 4) Haploid/diploid—the cycle alternates regularly or irregularly between the two nuclear conditions as in many yeasts, for example, the common baker's

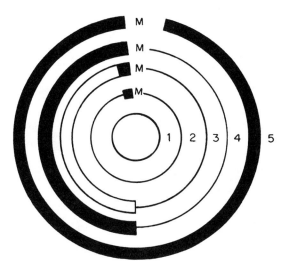

Figure 2.9   The five types of life cycles among the fungi: 1) asexual; 2) haploid; 3) haploid/dikaryotic; 4) haploid/diploid; 5) diploid. Each circle is a life cycle and should be read clockwise starting at 12 o'clock. M = meiosis. See text for example of each type. Thin line = haploid phse; double line = dikaryotic phase; thick line = diploid. After Burnett (1965, p.6; based on Raper 1954).

yeast, *Saccharomyces cerevisiae* (Saccharomycetaceae of the Ascomycetes). 5) Diploid—analogous to most higher organisms where the haploid phase is inconspicuous and relegated to the gametes; most Oomycetes (now considered by many biologists to be protists rather than fungi [Raven and Johnson 1986, Chapter 22]). The different kinds of nuclear states theoretically possible are shown in Figure 2.10.

Most fungi belong either to category (2) or (3), that is, they reproduce both sexually and asexually and are haploid for most of their life cycle. The haploid, asexual phase of the cycle is generally repeated numerous times annually, typically by spores but potentially also by other asexual methods such as by budding or fragmentation of the soma. The sexual phase normally occurs only once a year, and may more or less overlap the asexual state. Although a cycle of haploid and diploid states is thus conventional in the fungi, the alternation of generations is not distinct, unlike the case in many higher organisms. Some fungi that engage in sex are hermaphroditic in that a single thallus can function as both 'male' and 'female'; others are dioecious, in which case a given thallus produces only male or female organs. However, not all hermaphrodites are self-fertile (homothallic); most fungi are heterothallic, requiring a partner compatible in mating type. In the absence of a partner, sexual spores may be formed parthenogenically.

The nuclear situation in the fungi is further complicated because different

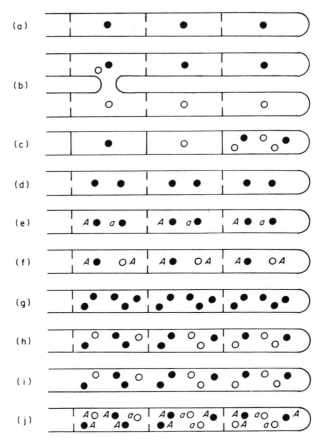

Figure 2.10   The different nuclear states that can exist in fungi. The fungus in (a) is a monokaryon; (d), (e), (f) are dikaryons; (b), (c) and (g)-(j) are heterokaryons. The letters [*A*, *a*] designate mating-type factors. From J.H. Burnett (1975, p.87), Mycogenetics. © 1975. Reprinted by permission of John Wiley and Sons, Ltd.

haploid nuclei, as well as different haploid and diploid nuclei, may exist in the same soma. *Heterokaryosis*, wherein two or more genetically different nuclei occur within an essentially common cytoplasm, is widespread among the Ascomycetes, Basidiomycetes, and Fungi Imperfecti (Davis 1966). The condition is established following fusion between *vegetative* cells, most commonly hyphae. A single fungal colony may thus arise in nature either from the germination of one spore or from several spores whose germ tubes have fused (anastomosed). In general, within a colony or strain, fusions between hyphae occur easily (and do not result in heterokaryosis). Genetically distinct heterokaryons usually reject each other. They can provide nuclei to homokaryons (as in 'di-mon' matings, below), but do not receive them. Several homokaryons can mate, but the resulting heterokaryons (each containing only two of the nuclear types) then grow away separately from the zone of nuclear interchange (Rayner et al. 1984).

Usually the component nuclei of a heterokaryon are not closely coordinated—they are distributed more or less randomly within a hypha and vary in relative number. Where they differ in mating type, however, the nuclei are closely and stably associated in pairs in each cell, and often divide synchronously. This particular condition is termed *dikaryosis*, and is the prominent somatic condition in the Basidiomycetes. Strict nuclear balance in the dikaryon precludes flexibility in somatic variation characteristic of heterokaryosis (below). Barring anastomoses (Raper and Flexer 1970), variation in dikaryotic fungi occurs not during the protracted vegetative phase, but at the point of nuclear fusion, meiosis, and recombination (Jinks 1952). Thus, the dikaryon is a balanced yet flexible system of two associated genomes, which as a unit can accomplish functionally all that its component strains can do separately. It is, as Raper and Flexer (1970, p.419) say, ". . . something of a biological oddity and an evolutionary cul-de-sac, although a highly successful one . . . ".

The significance of heterokaryosis is that it endows the mycelium with considerable genetic and hence physiological plasticity. Variation no longer depends on a conventional sexual cycle (Jinks 1952). In essence this imparts a mechanism of outbreeding for homothallic fungi or for species evidently lacking a 'true' sexual mode. Expression of the genome is analogous to that of the conventional heterozygous state in diploids, yet the nuclei, being separate, have some independence. They can segregate during growth in response to changing nutritional or other conditions, thereby changing the nuclear ratio. Heterokaryons of *Penicillium cyclopium* have outperformed both of their component homokaryons (Jinks 1952). Even if there is minimal nuclear migration after hyphal anastomosis, the entire mycelium may behave as a heterokaryon. For instance, in *Verticillium dahliae*, fusion between auxotrophic strains led to functional complementation in a transient heterokaryon in which the entire resulting colony behaved as if prototrophic, evidently sustained by scattered heterokaryotic cells (Pualla and Mayfield 1974). In population genetics terms, J.L. Harper (Univ. Coll. North Wales; personal communication, 1987) has likened the phenomenon of heterokaryosis to mate competition theory. Among polygamous and promiscuous animals there is often intense competition among males for females, which has led to the evolution of various sexual selection strategies. In the fungi it would seem advantageous for an organism to preempt rivals by sequestering the nuclei of a compatible mate through the mechanism of rapid and extensive heterokaryon formation.

So, what we see in heterokaryosis is the ability of an organism to adjust the proportion of different sets of genes in response to environmental variation (e.g., available substrates). This is distinct from the formal mitotic-meiotic system of macroorganisms where the genotype (apart from somatic mutation) is continuous throughout the soma (Jinks 1952). The heterokaryotic fungus adapts genetically and physiologically literally as it grows.

Based on experimental fusions in the laboratory and the occasional observation from nature, heterokaryosis was once presumed to be common in vivo, the implication being that genetically distinct fungal colonies became linked together by anastomoses into a ramifying, functionally cooperative, genetically mosaic unit. In these situations, as for bacteria, it is very difficult to define a genetic individual (see Section 2.4 and Chapter 5). Although the extent and significance of heterokaryosis are still largely speculative, the view that heterokaryosis is pervasive or even beneficial is being questioned increasingly (Chapter 2 in Cooke and Rayner 1984; Rayner et al. 1984; Carlile 1987; Caten 1987) because of recent evidence that fusion between individuals differing genetically at more than a few loci is prevented by somatic incompatibility. Incompatibility appears to be very common. A good example is *Neurospora* (Ascomycetes), one of the most exhaustively studied eukaryotic microorganisms. Research on strains from nature (Perkins et al. 1976; Perkins and Turner 1988) shows that in the heterothallic species there are multiple loci for vegetative incompatibility that prevent heterokaryons. Haploid clones thus persist as genetically distinct individuals. For other fungi, even where fusions do occur, they may not be followed by nuclear migration, or heterokaryotization may only be transient, unilateral, or to varying degrees incomplete (Chapter 5 in Burnett 1975). Thus, fusibility among fungi is typically restricted to close relatives because the governing loci consist of many alleles which must match if fusion is to succeed.

Vegetative incompatability in the fungi may have arisen to prevent competition or replacement ("parasitism") of a genetic cell lineage by alien genes (Caten 1972). This has been proposed for colonial hydroids and other lower metazoans (reviewed by Buss 1987, pp.149-158). Allorecognition mechanisms (those discriminating among individuals of the same species) serve to limit vegetative fusion in organisms with a sessile habit which makes them vulnerable to genetic invasion. These recognition mechanisms are analogous to restriction enzyme systems in bacteria, noted above, which prevent invasion by foreign DNA. Regardless, the important point with respect to heterokaryosis is that the evolutionary implications of migration of new genes through an existing genet, followed by change in phenotype and outgrowth of a new genet, may not be addressed by conventional modular theory (Chapter 5). The situation does not arise with unitary organisms because the germ cells are segregated from the soma, and within modular life forms this kind of gene migration would be rare, if not unique.

Parasexuality, as the term implies, involves genetic recombination outside the usual sexual mechanisms, and is best known among the filamentous fungi. There are four phases, each of which occurs relatively rarely (Pontecorvo 1956; Roper 1966): 1) two genomes come to be associated in the same cell, that is, either two homokaryotic nuclei associate or a heterokaryotic condition is established, as described above; 2) diploidization (nuclear fusion) occurs; 3) mitotic crossing over occurs leading to intrachromosomal recombination; 4) haploidization follows as chromosomes are lost from the diploid nuclei in successive mitoses—in this process there is interchromosomal recombination by the independent assortment of nonhomologous chromosomes. The four phases are unrelated. Parasexuality, in being a consequence of these uncoordinated, fortuitous events, is a process distinct from sexual recombination (although meiotic-like recombination can occur; Ellingboe and Raper 1962). Presumably it carries the most genetic significance for the Fungi Imperfecti, although parasexuality cannot replace true sex as a means for recombining genes and in any *one generation* contributes insignificantly to variation (Caten 1987). Considerably more information is needed on the extent and significance of the phenomenon in nature. Despite its apparent rarity, the process provides yet one more means for genetic recombination in certain organisms, and it illustrates how mitosis can play a role in genetic variability.

In overview, it is evident that the fungi are a transitional group which span the gamut in means of transmitting genetic variability. They exhibit some of the orderliness (meiotic mechanisms) of the macroorganisms, together with haphazard variation mechanisms akin to those of the bacteria.

2.4 Somatic Variation and the Concept of the Genet

The concept of the genet (Section 1.3 in Chapter 1; also see Section 5.2 in Chapter 5) builds on classic Weismann doctrine, the main components of which (Weismann 1885; summarized by Buss 1987, p.13) are that: 1) the zygote produces somatic cells mitotically and germ cells meiotically; 2) genetic variation developing during ontogeny cannot be inherited; and hence 3) heritable variation can occur only in the zygote or during meiosis in the formation of gametes. To what extent is the genet a useful common denominator in phylogenetic comparisons? More specifically, one might ask what kind and extent of genetic change occurs in the soma; whether somatic variation is heritable in different kinds of organisms; and, most importantly, whether it is significant in evolutionary terms.

Somatic variation occurs in all organisms. As noted earlier, short of meiotic recombination the same kinds of genetic alterations that affect germ cells can cause somatic variation. These include mutation, various gene and chromosomal rearrangements, (somatic) crossing over, and several types of regulatory change (Silander 1985). Indeed, somatic mutation rates (10^{-3} to 10^{-5}/locus/generation) can be as high or higher than those in germ cells, and are presumed to be a significant source of phenotypic and genotypic variation in plant populations (Silander 1985; Klekowski 1988).

There is no doubt that somatic variation can affect the life of the organism. For instance, the main changes that lead to cancer are now believed to involve several somatic mutations. The cells in most forms of cancer have aberrant chromosomes and every cell in a particular tumor type has the same kind of change (Watson et al. 1987, p.1007). Accumulating somatic mutations also form the basis of one of the theories of aging (see Chapter 6). Other sorts of somatic change may be beneficial, either in allowing the organism to survive transient environmental adversity or by concomitant adjustment to specific genotypic alterations: Bateson (1963) uses the hypothetical example of a pre-giraffe which happened to carry the mutant gene, "long neck". Expression of this new feature would obviously entail associated changes, that is, in the circulatory and nervous systems. These adjustments would have to be accommodated initially by somatic change (e.g., perhaps by gene amplification).

In organisms that retain cellular totipotency throughout their bodies and hence are able to reproduce asexually (clonally), somatic variants can be transmitted to progeny. In the fungi, for instance, mutants arising in any tissue can be transferred sexually or asexually. Because asexual reproductive rates of fungi are so high, favorable mutants (e.g., those containing virulence alleles; Burdon et al. 1983), are rapidly increased through natural selection (Caten 1987). Thus, by virtue of clonal growth, the fungi can evolve significantly by mutation alone in the absence of recombination. Among plants, the phenomenon of somatic variation has been exploited for centuries by horticulturists (Orton 1984; Silander 1985). Various characteristics of plants related to their modular growth form (Chapter 5) allow mutations to accumulate. Long-lived plant species have higher mutation rates per generation than do short-lived species, evidently because of the greater number of cell divisions before gamete formation (Klekowski and Godfrey 1989). Somatic mutations were the origin of most varieties of fruit trees, potatoes, sugar cane, bananas, not to mention countless vegetatively propagated ornamental and floricultural plants. Considerable genetic variation within clones of certain asexually propogated plants (e.g., lemon, cherry, grape, chrysanthemum) has been reported (Shamel and Pomeroy 1936) and in plants that clone naturally, for example, *Lolium perenne* (Breese et al. 1965).

More interesting is the question of whether somatic variation can be inherited by descendants produced sexually, and this depends on the onto-genetic program of the organism. For dipterans, as illustrated by *Drosophila*, it would be highly improbable. The totipotent lineage in *Drosophila* is restricted to only the first 13 nuclear divisions per generation (Buss 1987, pp.13-25)—a fleeting opportunity for the origin of a somatic variant. Similarly, in humans, germ cells established in the 56-day-old embryo remain sequestered for one to about three decades (Buss, p.100). Based on current evidence, both dipterans and humans come as close as any organism does to being a homogeneous genetic entity. That the period of accessibility to the germ line is short for vertebrates has been confirmed elegantly in recent studies where foreign genes are introduced during early embryogenesis (Robertson et al. 1986; Jaenisch 1988). For instance, early embryonic cells can be infected with retroviruses in vitro and reintroduced to the embryo at the blastocyst stage of ontogeny. The infected cells contain integrated provirus which contributes to both the somatic and germ cell lineages, as confirmed biochemically and by the chimeric phenotype of the transgenic animal and its progeny. Infection of preimplantation stage mouse embryos results in transmission to the germ line, whereas infection at the postimplantation stage (between days 8 and 14 of gestation) results in transmission to the somatic but generally not to the germ line (Soriano and Jaenisch 1986).

At the other extreme, the zygote in *Hydra* divides to produce an interstitial and a somatic cell lineage. The former remains totipotent and mitotically active throughout the life of the organism. By the time gametes are differentiated it is highly likely that somatic variation will have arisen in the forerunners of those cells (Buss 1987, pp.16-19). Likewise, corals are totipotent and in Buss's words (p.107), "a 20,000-year-old reef coral had [*sic*] passed uncounted millions of fruit fly generations". Thus, individuals in these taxa do not closely approximate genetically discrete units.

Plants have rigid walls and in part because of this undergo a different ontogeny from that of the relatively simple metazoans discussed above. They share the somatic embryogenesis (absence of a distinct germ line) mode of development (Buss 1983), that is, somatic meristems can give rise to gametes. Because of the clonal aspect of development, somatic mutations in precursors of a floral lineage can potentially be passed on through gametes. This does happen as has been shown in the work with transposable elements of maize (Fedoroff 1983, 1989). For example, if a genetic change occurs during the first embryotic cell division, a plant with genotypically and phenotypically distinct halves is created. Each half will go on to produce different gametes. If a similar change is delayed until ears form, two different sectors with correspondingly distinct kernels will develop. Indeed, the order of genetic events can be surmised from the timing in appearance of the sectors. The important point here, however, is that *somatic changes can be reflected in the gametes*. Hence, somatic changes can not only alter the fitness of the carrier in which they arise, but they can, at least in some instances, be passed on to offspring produced sexually, thereby extending the conventional forms of genetic variation discussed previously (Section 2.3).

Whether or not somatic variation is transmitted sexually in any given instance, somatic mutation in plants is considered to have potentially greater survivability than a comparable mutation in the gamete pool of most animals. This is because it can be maintained by naturally occurring means of asexual reproduction (Whitham and Slobodchikoff 1981). Consider plants as an example. They are modular organisms (Chapter 5) which contain many meristems and iterated parts originating from those meristems (branches, leaves). Because of somatic mutation, plants can develop as mosaics where one component, say a branch, is genetically quite different from another. Among the interesting implications (Whitham and Slobodchikoff 1981) is that beneficial somatic mutations (e.g., resistance to parasites or insect grazers) could spread easily whereas at least some kinds of deleterious mutations would be inconsequential because the affected part could be shed (hence, it is argued, no increase in mutational load would occur). These and related considerations apply more broadly to all modular, as opposed to unitary, organisms (Chapter 5). The main problem is that the story is largely speculative. In particular, there is virtually no solid evidence on the magnitude of gametic as opposed to the asexual transfer of somatic variation.

Based on two models which compared the implications of somatic or gametic mutations for a population of trees, Slatkin (1985) concluded that somatic mutation rates must be appreciably higher than gametic rates to be of evolutionary significance. Even where this is true, the prediction was that the former will act mainly to increase genetic diversity among rather than within individual trees. However, the possibility is left open that in long-lived, clonal organisms, variation within a lineage (or even, it appears, within a ramet) could be comparable to that between genets. To date, the models that do exist contain numerous simplifying assumptions and there are often simply insufficient data from either observational or experimental sources to draw any strong conclusions.

Fueling interest as well as much controversy on the evolutionary role of somatic variation are reports of somatic selection in cells of the animal immune system (Gorczynski and Steele 1980; 1981) that parallel those described above for plants. The background (Steele 1981) starts with the central paradox of immunology, namely (p.15), "How do we reconcile the inheritance of the germline of a characteristic which by an established set of criteria only arises in the soma?" Steele (1981, p.55-56) has summarized the putative role of a retroviral mechanism in the the somatic selection hypothesis (Figure 2.11) succinctly as follows:

> The first step is the chance generation of somatic mutations and their clonal selection under favourable environmental stimuli [i.e., hyperimmunization; see below]. The second step includes transfer of this clonally packaged information to an endogenous RNA viral vector [retrovirus; see Section 2.2] which allows integration of the somatically selected mutant gene into the germline DNA. The last step involves classical Darwinian selection operating on members of the progeny generation.

The experimental protocol involves initially unimmunized parental animals (rabbits or mice within the same pedigree) which are mated to produce litters. The parents are then immunized (technically, hyperimmunized, so that the somatically variable idiotype(s) [those conveying individual antigenic specificity] in the parents can become clonally dominant), mated again, and more offspring produced. The progeny of the two matings are immunized and their sera compared for presence or absence of parental idiotypes. If somatic selection is involved, an appreciable proportion (Steele [1981] sug-

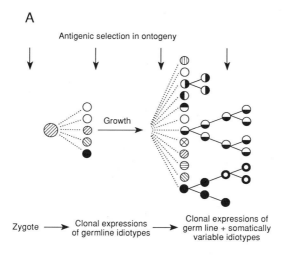

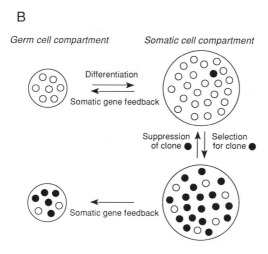

Figure 2.11 The clonal selection and the somatic mutation theories of the immune system. (A) The circles represent different lymphocytes and their clonal lineages, which are unique because each possesses specific antibody receptor molecules. The germline idiotypes (individual antigen specificities) are predictable within a given animal pedigree. The somatically variable idiotypes comprise a much larger set elicited in response to the same antigen yet are unpredictable from animal to animal. During ontogeny, germline idiotypes are expressed phenotypically, as are somatic variations, in response to the antigen. Mutational frequency would normally be low, but if stimulated by hyperimmunization, the mutant may be clonally selected, becoming dominant. From Steel (1981, p.33). (B) Steele's controversial hypothesis is that there is genetic feedback from the somatic cells to the germline. The gene responsible for antigenetic specificity in the clonally selected somatic line is conveyed via a retrovirus to the germline DNA, where it is integrated and eventually acquired by offspring. From Steele (1981, p.51).

gests about 10%) of sera from the second lot of progeny but not the first, should have parental markers. (For details, including experimental controls, see Chapter 4 in Steele 1981). There is some experimental evidence to support the hypothesis (Gorczynski and Steele 1980, 1981).

Steele (1981) develops the concept largely in the Lamarckian vein (i.e., acquired characteristics can be inherited) because progeny animals, without being directly exposed to antigen themselves, benefit from the immunized challenge of their parents. Dawkins (1982, pp.164-173) argues correctly however that the theory is fundamentally Darwinian because it entails selection at the cellular level for the mutated genes they contain. This means that our concept of the germ line needs to be expanded considerably, since any somatic cell which may provide genes via a proviral route to a conventionally defined germ cell would qualify (Dawkins, p.169). Either the potential germ lineage must be expanded conceptually, or the "Steele" doctrine is the antithesis of the Weismann (1885) doctrine:

The Weismann Doctrine: one-way travel of information

$$\text{germ cells} \;\underleftarrow{\quad/\!/\quad}\overrightarrow{\quad}\; \text{somatic cells}$$

The Steele Doctrine: gene transfer from somatic to germ cells

$$\text{germ cells} \;\underleftarrow{\qquad}\overrightarrow{\qquad}\; \text{somatic cells}$$

Not surprisingly, because of its Lamarckian tones, Steele's most intriguing idea has been attacked, and indeed he foresaw many of the criticisms (Steele 1981, pp.108-116). Two among them are whether the hypothesis can be generalized to tissues other than the immune system, and why the mechanism does not result in "genetic chaos" due to disruption of conventional Mendelian inheritance patterns. The frequency of expression of a somatically selected mutant in the germ line—which would be a product of the somatic mutation rate and the rate of acquisition of a single locus by the retrovirus—would also be very small. Clearly the work needs substantiation; an early independent effort at confirmation failed (Brent et al. 1981). The present status in Steele's own words (personal communication, 1989) is as follows:

> The somatic selection theory outlines a plausible process for higher animals—we await a definitive experiment showing that a somatic gene can enter the somatic germ-line via an RNA intermediate through the agency of some viral vector (such as an endogenous retrovirus). Every step of the postulated process has, by itself, been demonstrated in isolation. . . . It is a very difficult experiment to design or perform at this stage. . . .

As things stand, the theory can hardly be said to have entered the mainstream of current evolutionary thinking, even in the restricted immunological context. Nevertheless, the central message—that somatic gene mutations can be genetically inherited—is intrinsically plausible, although unlikely. Moreover, as recent research is showing, the retroviruses, by their ability to carry cellular genes and to integrate stably as proviral DNA into the chromosomes of somatic and germ cells, provide the necessary link. There are analogous events in other organisms, including yeasts and plants, where the environment—apparently acting through mobile gene mechanisms—induces heritable changes (Chapter 7; Cullis 1983, 1987; Walbot and Cullis 1985).

Finally, in the controversy about the role of somatic mutation it seems to have been overlooked that the *ability* to generate somatic variation *is* heritable. This is illustrated by the relatively high mutation rates in the B lymphocyte somatic lineage which is instrumental in contributing to antibody diversity. That this ability per se is heritable seems more significant in an evolutionary context than any specific somatic change which occasionally arises.

The preceding relatively lengthy foray into somatic variation is important because it establishes that genetic variation occurs at many levels, including those of the organelle and the cell, as well as those of the so-called 'physiological' and 'genetic' individual. This challenges the dogma that the developing product of the zygote (by which is implied a single entity) is *the* unit of variation.

Thus, the concept of a genet is an ideal that is more or less approximated in various phyla. The unitary organisms, most clearly illustrated by the higher animals, come closer than do modular life forms (lower animals, plants, fungi) in behaving as genetic individuals. Exponentially increasing information in the realm of molecular biology is showing that mobile genetic elements can move among chromosomes of a cell, between cells, and between the somatic and germ lines. One consequence of this fluidity is increased somatic variation and a direct route from soma to gametes. Even in the case of unitary macroorganisms, our emerging concept of the genetic individual must be a much more flexible one than has been the case heretofore.

Microorganisms have always posed problems for the genet idea: Technically, even a single mutation will signal the end of a genet in any haploid organism. A single fungal thallus may exist as a genetic mosaic. By way of heterokaryosis and parasexuality, mutational and recombinational events can be expressed, tranferred clonally, and exposed to natural selection independently of fertilization. Similarly, in bacteria, any form of recombination marks the end of the original and the onset of the new genet in the recipient cell. Conceptually this is clear but operationally it means nothing because, unlike the case with most other organisms, formation of the bacterial recombinant is not tied to a particular divisional event, a morphological structure, or a characteristic life cycle stage such as dormancy or dispersal. The recombinant cells are not even evident in a mixed population unless identifiable phenotypic traits such as auxotrophic markers are involved. Thus, while genets can be visualized abstractly for bacteria, the concept can only by applied loosely. Consequently it loses much of its appeal.

2.5   Summary

The principal general categories of genetic variation are mutation, recombination, and epigenetic controls. Mutation, by giving rise to different alleles of a gene in a population, is the fundamental source of genetic change in all organisms. Generally, the inherent rates of nucleotide substitution for a given DNA sequence among taxa are approximately constant *per unit time*, 10^{-7} to 10^{-8} per locus per day (the "molecular evolutionary clock"). While mutation rates are approximately time-dependent and constant, the rates of phylogenetic change are far from constant, and are apparently governed in large part by the ecological opportunity afforded the genetic variants.

Recombination, the mixis or rearrangement of nucleotides, is accomplished by general (meiotic assortment; crossing over; segmental interchanges) and site-specific (retroviruses; plasmids; transposons; promiscuous DNA) mechanisms. The potential of site-specific mechanisms to effect genome restructuring and to affect genetic expression in diverse taxa is only beginning to be fully appreciated. Sexual recombination, the bringing together in a cell of genes from two genetically different sources, is thus merely one of the means by which recombination can occur. Sex is usually but not always associated with reproduction, that is, the generation of offspring. For example, sex in bacteria is never an obligatory aspect of reproduction. Sex is essentially ubiquitous among organisms. Virtually all clonal life forms (those which give rise to genetically identical offspring) can also reproduce sexually, although the sexual stage may occur infrequently. Apparent absence of sex in certain living things (some of the fungi and undoubtedly some of the protists) may simply be a consequence of insufficient observation or an overly restrictive concept of sex, which precludes events (transduction, transformation, plasmid-mediated conjugation, anastomosis and resulting heterokaryosis, parasexuality) which are in effect sexual, but do not involve the fusion of gametes. Retroviral gene transport in mammals is analogous to transduction in bacteria. Meiotic recombination in eukaryotes has an obvious parallel in conjugal transfer in prokaryotes.

Sexual reproduction probably predated the asexual process. It evidently arose as a mechanism to repair DNA, and the recombination function evolved from this. In diploid organisms, sexual reproduction entails a dilution of 50% of the genome per generation. Given this as well as other disadvantages relative to an asexual alternative, why sex has been consistently maintained is unclear. It is probably retained for several reasons which may well be different or of differing importance in the various phyla. If this is the case, efforts to find a singular role for sex are unlikely to succeed.

The best evidence for epigenetic controls (those influencing the developmental program) is that cells of the same genetic consitution differentiate into different tissue types. Transcription of bacterial genes may be controlled by repressor or activator proteins, among other mechanisms. In eukaryotes, various mechanisms have been implicated, including the sequestering of portions of the genome into transcriptionally inactive regions (heterochromatin); existence of cooperative groups of regulatory proteins; and methylation of DNA. Suppressor mutations act to suppress changes in the same or another gene thereby altering gene expression. Because of numerous controls on gene expression, it is not necessary to destroy a gene to get a change in phenotype.

Overall, bacteria appear to have essentially all of the capability of macroorganisms for generating genetic variability. The major difference seems to be that this is transmitted in relatively unordered fashion, as opposed to the organized manner of higher organisms which culminates in meiosis and gametogenesis. The fungi are an intermediate group, spanning the gamut in variability-generating mechanisms and orderliness of transmission.

The concept of a genetic individual or genet originated in Weismann's doctrine which viewed the zygote as the seat of all heritable variation and the germline as being insulated from the soma. It has long been recognized that somatic variation occurs and that this can markedly influence the carrier and potentially its clonal descendants. Whether somatic variation can be transmitted to offspring produced sexually depends on the ontogenetic program of the organism. Where the totipotent lineage is strictly limited, as in most higher animals, the chance that somatic variation can occur and be passed on to the developing germ cells is remote. Where cells remain totipotent and mitotically active, as in plants, fungi, and the less complex metazoans, transmissible somatic variation is probable. It has been demonstrated strikingly, for example, in the case of mobile genetic elements of maize. Thus the genet is a much less discrete unit in some taxa than in others. Nevertheless, the evolutionary signficance of somatic variation remains poorly documented and controversial. Even more contentious is the extent to which there may be direct genetic feedback from the soma to the germline by a mobile gene (e.g., retrovirus) mechanism. What is clear and perhaps most important in evolution is that the *ability* to transmit somatic variability *is* heritable. Obviously, the genet must be a much more fluid entity than was conceived originally, in part because of our increasing awareness of the role of mobile elements in genetic rearrangement and expression.

The main conclusions from this chapter, which sets the stage for the rest of the book, are that: 1) All organisms possess several and in principle similar means of generating and transmitting genetic variation on which evolutionary processes, in particular natural selection, can act. The main differences are the relatively unordered (microorganisms) versus ordered (macroorganisms) manner in which the variation is transmitted. 2) Because of their short generation times and large population sizes, microorganisms have higher evolutionary rates as species than do macroorganisms. 3) Because of the variation in ontogenetic programs among taxa—which means that in some cases somatic variation can be transmitted to the germline—and the ubiquity of mobile DNA, the concept of the genetic individual (genet) cannot be applied usefully for some organisms (e.g., the bacteria) and needs to be used guardedly or reassessed for others (plants, fungi, simple metazoans).

2.6 Suggested Additional Reading

Buss, L.W. 1987. The evolution of individuality. Princeton Univ. Press, Princeton, N.J. The history of life as a transition between different units of selection.

Dawkins, R. 1982. The extended phenotype. The gene as the unit of selection. Oxford Univ. Press, Oxford, U.K. A continuation of Dawkins' stimulating examination, begun in *The Selfish Gene*, of the levels in the hierarchy of life at which natural selection acts. The focus is on "selfish genes" and not "selfish organisms".

Loomis, W.F. 1988. Four billion years: An essay on the evolution of genes and organisms. Sinauer, Sunderland, MA. The evolution of life, from prebiological chemistry through higher organisms, from a personal, speculative point of view.

Michod, R.E. and B.R. Levin (eds.). 1988. The evolution of sex: An examination of current ideas. Sinauer, Sunderland, MA. The most stimulating recent collection of broadly based essays on the evolution of sex.

3

Nutritional Mode

It's a very odd thing
As odd as can be—
That whatever Miss T eats
Turns into Miss T.

—WALTER DE LA MARE, 1913

3.1 Introduction

Not much can be said about the biology of an organism until one knows how it provides for itself. Thus, nutritional mode, along with size (Chapter 4) and architecture (Chapter 5) provide the basis for ecological comparisons because they impose broad constraints as well as opportunities. In this chapter I examine how living forms can be grouped depending on which resources they use and how they go about harvesting them. Broadly speaking, the issue is that portion of Figure 1.1 (Chapter 1) concerning resource acquisition.

3.2 What is a Resource?

The term resource is used throughout this book as meaning that quantity which can be reduced in amount by the activity of an organism (Begon et al. 1986, p.75). Typical of definitions of resources, this one involves 'food'. I use 'food' synonymously with nutrients and construe it broadly here to be an energy and/or electron donor and/or source of carbon and other bioelements. This is an unfortunately wordy and perhaps intimidating definition, but is necessary in order to cover the various permutations presented by the living world! Space also can be considered a resource in the sense that it defines an area or volume (a 'habitable site', Chapter 7) in which food is 'captured', but more so because space itself is a resource in some circumstances. For

65

instance, suitable nesting sites or hiding places are resources (Chapter 3 in Begon et al. 1986). Similarly, room on a rotten log for decay fungi to form aerial fruiting structures, or available crevices on natural substrata such as rocks, teeth, skin, and plant surfaces are resources (quite apart from any nutritive role they may play) for colonizing microbes.

Microbiologists use the term substrate to refer to the nutrient component of a resource. Substrates used in laboratory culture are often well defined chemically and their physiological function is usually known. Accordingly, one speaks of a particular organic or inorganic chemical as a carbon source or an energy source or a nitrogen source, and so forth. Depending on the particular microbe, resources might be O_2 or NO_3^- (electron acceptors), NH_4^+ (nitrogen source), H_2 (inorganic electron donor), or glucose or acetate (which are both organic carbon and energy = electron donor sources). This convention provides a convenient format for classifying all organisms based on their sources of carbon and energy, as developed below.

Space as a resource component will be discussed in Chapter 7, where the influence of environmental factors on organisms is considered. The focus in the present chapter is on aspects of nutrition. In this regard the three major 'questions' that every organism must 'ask' are "What do I eat?", "How do I get it (and keep it) before someone else does?" and "How do I metabolize and allocate it efficiently so as to effectively propagate my genes?". Rephrased in a more sophisticated way, these questions concern resource *categories*, resource *acquisition*, and optimal *allocation* of resources, respectively. They set the framework for some perspectives on nutritional mode. The key resources that will be used for illustrative purposes are energy and carbon.

3.3   Some Fundamental Resource Categories and Their Implications

Energy in the form of ATP is probably *the* common currency of all living things, as it is required for almost every activity. Energy can be harvested in three ways: as light energy directly from the sun by photosynthesis, or as chemical energy from inorganic compounds or organic compounds. If all organisms are considered, there are still only two basic mechanisms for generating ATP. The first, electron transport phosphorylation (oxidative phosphorylation or respiratory-chain phosphorylation), involves the flow of electrons 'downhill' from inorganic or organic donors with a relatively negative redox potential (higher energy) to those of a relatively positive (lower energy) potential, tied to the synthesis of ATP from ADP and inorganic phosphate. This may occur in cyclic and noncyclic photophosphorylation (photosynthetic organisms only) and in respiratory chains (most organisms). The second, substrate-level phosphorylation, is not associated with the process of electron transport, and occurs when organic substrates containing high-energy-level phosphoryl bonds are degraded. For example, metabolism of some intermediates in glycolysis (Embden-Meyerhof pathway), such as the catabolism of phosphoenolpyruvate to pyruvate, is associated with the transfer of the phosphate group to ADP to generate ATP.

Carbon is also a common denominator for both macro- and microorganisms, because it plays an essential role in the protoplasmic structure of all organisms, and additionally in the energy metabolism of many, as described below. It is acquired either directly from CO_2 or from organic compounds.

A fairly detailed, often inconsistent, terminology is used to describe the energy and carbon dynamics of organisms. Definitions that appear in ecology and biology texts are more general than those used by most microbiologists. For example, the terms autotroph and heterotroph may be used with respect to the energy source, or the carbon source, or both (that is, unspecified). Presumably this is because a large group of organisms, the animals (not to mention most microbes), use organic carbon both as a carbon source and as an electron donor (energy source). Usage varies, even within microbiology, for instance between bacteriologists and protozoologists. The scheme set out briefly here and summarized in Table 3.1 follows largely that within microbiology and is the one adopted because it is universally applicable.

Organisms deriving their *energy* from light, which induces a flow of electrons, are **phototrophs**; those deriving electrons from a chemical energy source are **chemotrophs**. In either instance the electron donor may be an inorganic or an organic substance. **Lithotrophs** use electrons from inorganic sources such as H_2O, H_2S, H_2, Fe^{+2}, S^0, or NH_3. **Organotrophs** use organic substrates such as wood or meat as electron donors.

With respect to *carbon* source, some organisms can use CO_2 as the sole carbon source, in which case they are **autotrophs**. Organisms using mainly organic compounds rather than CO_2 to supply cell carbon are **heterotrophs**. (As alluded to above, authors differ in their use of these terms; occasionally, for example, heterotroph and organotroph are used synonymously, e.g., see Whittenbury and Kelly 1977; Chapter 16 in Brock and Madigan 1988.) The nutritive options arising from the energy/carbon source permutations form the basic classes into which all creatures fall. Technically, any organism can thus be described by a three-part prefix designating energy source/electron donor/carbon source (Table 3.1). Plants would be photolithoautotrophs and bears would be chemoorganoheterotrophs. This terminology is cumbersome, to say the least, so it is usually abbreviated. Because most photolithotrophs also use inorganic carbon, they are usually simply called photoautotrophs. Likewise, most chemoorganotrophs use organic substrates, both as a source of carbon and electrons, and so are simply called chemoheterotrophs.

Two observations on the information in Table 3.1 are particularly noteworthy. First, the apparent wide spectrum in patterns of energy and carbon source is really a result of the metabolic diversity of bacteria. In terms of

Table 3.1 Classification of organisms, with representative examples, based on energy source, electron donor, and carbon source[1]

Energy source	Substrates by electron donor		Examples by carbon source	
	Inorganic	Organic	Carbon dioxide	Organic compounds
Light	Photolithotrophs use H_2O, H_2S, S, H_2	Photoorgano-trophs use succinate, acetate	Photolitho-autotrophs: plants most algae cyanobacteria some purple and green bacteria	Photoorgano-heterotrophs: some bacteria (Rhodospiril-laceae)
Chemicals	Chemolitho-trophs use H_2, H_2S, NH_3, Fe^{2+}, No_2^-	Chemoorgano-trophs use many organic substrates	Chemolitho-autotrophs: hydrogen bacteria colorless sulfur bacteria nitrifying bacteria iron bacteria methanogenic bacteria methylotrophs	Chemoorgano-heterotrophs: animals most bacteria fungi many protists

[1]Based mainly on Gottschalk (1986); Carlile (1980); Schlegel and Bowien (1989). The demarcations are not absolute and terms are defined differently by various authors. For example, the criterion that all organisms that can utilize CO_2 as a sole C source are autotrophs begs the question as to the status of those that grow better upon the addition of an organic substrate.

species number and biomass, the overwhelming preponderance of organisms are photoautotrophs (plants, broadly speaking) or chemoheterotrophs (animals and most microbes). Second, the resource patterns cross size boundaries. What biological properties belong exclusively to all members of a class, and what are the ecological implications? Is there a fundamental reason why the ability to form complex, multicellular macroorganisms is restricted to the photoautotrophs and the chemoheterotrophs? Why are there no eukaryotic chemolithotrophs, either small or large?

To review, briefly, the most obvious apparent result of food relationships is the well-known trophic structure of ecosystems (Figure 3.1). As a generality, trophic structure is useful and appears in every textbook on ecology. This is not to say that the scheme does not have limitations or that it is not a contentious issue. It is difficult if not impossible to define heterotrophic levels in any objective or measurable way (see e.g., Cousins 1980; Burns 1989). Microbes are usually overlooked (see below) and are difficult to categorize within the trophic context. As conventionally depicted, the base of grazer (herbivore) chains is formed by phototrophs or primary producers and successive tiers consist of heterotrophic species, for example, herbivore, first carnivore, . . . ,

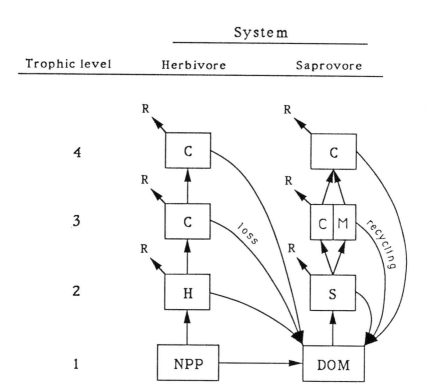

Figure 3.1   A simple model of trophic structure and energy flow in a terrestrial community. Four trophic levels are shown for the herbivore system, but only three (S, C/M, and C) for the saprovore system. Typically, the role of microbes is explicity recognized only as part of the saprovore web, as shown here where saprovores (S) can be either invertebrates or microorganisms. NPP = net primary production; DOM = dead organic matter; C = carnivore; H = herbivore; M = microbivore; R = respiration. Redrawn after Heal and MacLean (1975)

top carnivore (for details see Chapter 11 in Ricklefs 1990; Chapters 26-27 in Krebs 1985).

The chains are short, generally 3-4 links. Of four hypotheses as to why this should be the case (Chapter 6 in Pimm 1982), the most generally accepted pertains to energy loss at each stage coupled with energy requirements of animals at the top of the chain (Hutchinson 1959). Only about 1% of the light energy intercepted by phototrophs is converted to usable energy and of that only about 10% on average is passed through each step in the sequence. Here is another area of dispute, since the 10% level of efficiency is an educated guess and not based on real data: Transformation from level 2 to 3 has not been measured because what constitutes level 2 is often unclear and level 3 and above cannot be clearly defined in any ecosystem.

As one progresses up through a trophic chain the amount of usable energy entrapped at each level (kilocalories/square meter/year), the number of individuals (counts/square meter), and the amount of biomass (grams/ square meter) accordingly usually decrease (energy does so invariably). Ecologists portray these trophic relations as *pyramids* of energy, numbers, and biomass, respectively. Pyramids of numbers and biomass may be inverted if, for example, the producers are much smaller than the consumers (and accordingly have much higher turnover rates; see below).

The mechanistic basis of the energetic hypothesis for the pyramids is interesting, especially in an evolutionary context. The pyramid depiction means that each species at level n + 1 sees only that portion of the resource at level n available to it. The glycolytic reaction used to generate ATP is one of the oldest biochemical processes and is energetically very inefficient (two molecules of ATP are produced per molecule of glucose catabolized). This energetic constraint means that the earliest heterotrophs must have lived only as consumers of the photoautotrophs, rather than of other heterotrophs. In other words, for some time the trophic chain would have consisted of a single link, with a large base of photoautotrophs or chemolithoautotrophs supporting a relatively small number of heterotrophs. It was the evolution of oxidative respiration, energetically an almost 20-fold improvement over glycolysis (36 molecules of ATP per molecule of glucose), that made successive tiers feasible. A possible evolutionary sequence which emphasizes the early role of microorganisms and relates changing food relationships to selection for increase in size is developed in Chapter 4.

For macroorganisms, the pyramids of energy, biomass, and numbers pyramids generally bear overall resemblance. The weight of plants in most terrestrial systems far exceeds that of the herbivores, which in turn is greater than that of the carnivores. This hierarchy derived from biomass is consistent with that based on productivity of the trophic levels. Microorganisms, however, pose several complications. In aquatic systems, phytoplankton are commonly grazed so rapidly that their biomass and numbers are actually lower than the herbivorous zooplankton that the microalgae support (thus, one specific case of an inverted pyramid). Numerically, parasites, pathogens, photosynthetic symbiotic dinoflagellates, and other micro-symbionts, including epifloral or epifaunal microbes, far exceed their hosts. Yet, apart from the microscopic plankton, microbes are rarely considered part of grazer chains. Trophic structure, as conventionally depicted, under-represents the roles played by microorganisms. An exception, which depicts energy flow and standing crop for a eutrophic lake, is shown in Figure 3.2.

The biomass and high activity of gut flora (Hungate 1975; Savage 1977) need to be acknowledged, although it is debatable whether these microbes are best assigned to grazer or saprovore systems. Stanier (1953) gives the picturesque comparison that the numbers of *Streptococcus faecalis* in the intestine of *one* mammal would far outnumber all individuals of that species

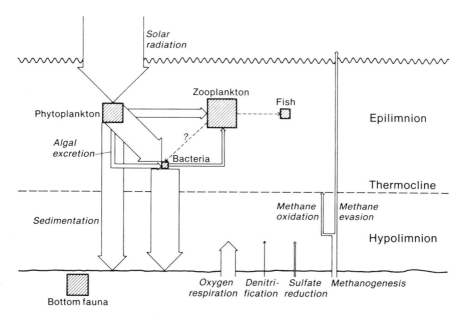

Figure 3.2   Approximate standing crop and energy flow for Lake Mendota, Wisconsin. The size of the boxes and width of the arrows reflect relative differences in magnitude. Note the major role of microbial processes in energy flow from primary producers and in recycling the organic carbon which sediments to the bottom of the lake. From Brock (1985, p.219)

of mammal in the world. Similar if less dramatic situations include mycor-rhizal symbioses, and microbial communities which are supported on the surfaces of plants and animals. Usually microbes are recognized as members of decomposer (saprovore) chains along with vertebrate scavengers (vultures, crows, jackals) and microbivores (earthworms and other invertebrate soil fauna) which are instrumental in the recycling of bodies and feces. These systems always accompany grazer chains (Figure 3.1; Heal and MacLean 1975), and on land they form the major pathway for energy flow (Chapter 13 in Ricklefs 1990). Other microbes (e.g., the lithotropic autotrophs, Table 3.1) are not clearly a part of either the herbivore or the saprovore chain.

Diverse marine animal communities, both in relatively cold water and surrounding thermal vents, appear to be based on the productivity of litho-trophic bacteria which oxidize H_2S or other inorganic substrates such as H_2 and CH_4 (Figure 3.3; Jannasch and Wirsen 1979; Jannasch 1984, 1989). These substrates, electron donors (as well as the electron acceptor, CO_2), originate in the hydrothermal fluid. This exciting discovery means that we now know of *two* systems for primary production. While solar radiation provides energy for photosynthesis, terrestrial thermal energy and pressure provide the energy

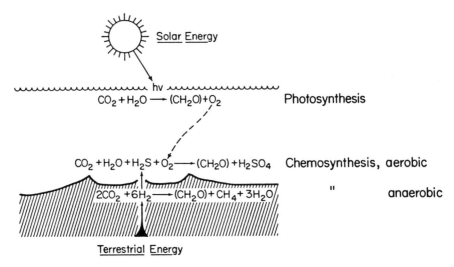

$$CO_2 + H_2O \longrightarrow (CH_2O) + O_2 \qquad \text{Photosynthesis}$$

$$CO_2 + H_2O + H_2S + O_2 \longrightarrow (CH_2O) + H_2SO_4 \qquad \text{Chemosynthesis, aerobic}$$

$$2CO_2 + 6H_2 \longrightarrow (CH_2O) + CH_4 + 3H_2O \qquad \text{''} \qquad \text{anaerobic}$$

Terrestrial Energy

Figure 3.3 Photosynthesis, mediated by light energy, at the sea surface provides the base for herbivore (grazer) and saprovore (decomposer) chains. Bacterial chemosynthesis, mediated by terrestrial energy near deep-sea hydrothermal vents, provides an alternative base for energy flow. From Jannasch (1989).

source on venting water for anaerobic or aerobic chemosynthesis by the lithotropic reduction of CO_2.

The source of energy and carbon also sets broad limits on the distribution of living things. Phototrophic micro- and macroorganisms illustrate this point well, being distributed with respect to light gradients (see below). The distribution of many species of lithotrophic microbes can be said to mirror environmental geochemistry. For example, *Sulfolobus acidocaldarius* lives primarily in sulfur-rich geothermal domains such as hot springs (Brock et al. 1972). The organism is a thermophilic (temperature range 60-85°C), aerobic, obligate acidophile (pH range 1-5). A member of the prokaryotic group Archaebacteria, it can thrive under hot, acid conditions in part because of an unusual cellular membrane which is an exception to the ubiquitous phospholipid bilayer type. In these high-temperature habitats, H_2S oxidizes spontaneously to elemental sulfur (S), which is in turn oxidized to H_2SO_4 by *Sulfolobus*, further acidifying the environment. The bacteria adhere to the crystalline S thereby acquiring lithotropically the few atoms going into solution that they need as an electron source (Figure 3.4; Brock et al. 1972). This, incidentally, is an excellent example of the unique ability of microbes, especially bacteria, to finely partition a resource: One microscopic S crystal, measuring on the order of cubic micrometers, can provide an energy source for thousands of bacterial cells, whereas an entire plant or animal may be only one of many in the diet of a herbivore or carnivore. *Sulfolobus* also

Figure 3.4   The sulfur-oxidizing bacterium *Sulfolobus acidocaldarius* attached to crystals of elemental sulfur. Bacteria appear as bright spots because they have been stained with fluorescent dye acridine orange. Photo courtesy of T.D. Brock.

illustrates how specialization imposes limitations (e.g., a restricted range of resources) as well as opportunities (optimization at using them)!

The distribution of certain heterotrophs can be strictly limited by that of their energy and carbon sources. The Giant Panda eats only bamboo and hence cannot exist in otherwise favorable regions of the world lacking its food source. Obligate parasites are restricted by the availability of their hosts. Heteroecious parasites are those that need two or more hosts to complete their complex life cycles (Chapter 6). The fungus causing black stem rust of wheat (*Puccinia graminis tritici*) requires both wheat (in fact specific, susceptible cultivars of wheat) and common barberry (*Berberis vulgaris*) or other species of wild native barberry or mahonia. In the case of the *Schistosoma* flatworms, there is an obligatory alternation of a sexual generation in humans (occasionally other mammals) and an asexual generation in particular snails.

Terrestrial and aquatic plants and animals (living or dead) provide locally high reservoirs of organic carbon as well as other nutrients, and accordingly act like nutritive islands for colonization by heterotrophic bacteria and fungi. Similarly, the pelagic (open ocean) zone, while impoverished as a whole, contains microenvironments enriched in nutrients. High microbial activity is associated with the relatively large fecal pellets of zooplankton and with 'marine snow', heterogeneous flocculent aggregates containing phytoplankton, detritus, bacteria, and small fecal pellets embedded in mucus (Alldredge and Cohen 1987).

Unlike animals and chemotrophic microbes, obligate phototrophs cannot choose their energy diets, but are of course affected by the intensity, temporal distribution, and spectral composition of light. Although plants vary in growth form, their energy-'gathering' mechanisms for the common commodity, light, are essentially the same. This is in stark contrast to animals which vary strikingly in diet and in anatomical features (e.g., mouth parts and digestive systems) for harvesting their respective food sources. Variation among animals at this level in the process is analogous to the numerous growth forms of plants. However, once gathered and converted to major metabolites such as glucose, the food is handled biochemically and energy is extracted (e.g., via glycolysis and the Krebs cycle) in much the same way by both groups of organisms.

Plants, algae, and phototrophic microbes are limited to areas where light occurs and hence are excluded from such habitats as caves, subterranean layers, or the intestines of animals. In aquatic environments, attached macrophytes and benthic (bottom dwelling) phototrophs are restricted to the photic zone, that region where light intensity is sufficient for net photosynthesis to occur (i.e., photosynthesis above the compensation point, where photosynthesis equals respiration). The depth of this zone varies, but in coastal waters is typically about 100 meters. The energy for phototrophs also arrives essentially in a continuous stream (albeit in quantum units), and does not entail hunting or trapping discrete packages separated in space and time, as the case for animals. Similarly, their energy supply is not subject to the laws governing population biology of prey in the sense that the energy source of chemoheterotrophs is influenced by the population dynamics of the plants and animals on which they live (Harper 1968).

Within the photic zone, the distribution pattern of photosynthetic organisms is related to light intensity and spectral quality which change with depth. Water absorbs proportionately more of the longer (red) than shorter (blue) wavelengths. Algae and cyanobacteria growing aerobically in the upper reaches of the water column further remove the red and some of the blue portions because these are the energy quanta preferentially absorbed by their chlorophylls and accessory pigments. Anaerobic, phototropic purple and green sulfur bacteria commonly develop in a zone or on the muddy bottom (Figure 3.5; Pfennig 1989) if the water is sufficiently clear to allow light penetration to depths which are anoxic, and in the presence of H_2S. Thus, these organisms thrive under relatively specific, ecologically restricted conditions in stagnant water or on mud surfaces where they may form spectacular mats ranging in color from pink to purple-red or green. The phototropic sulfur

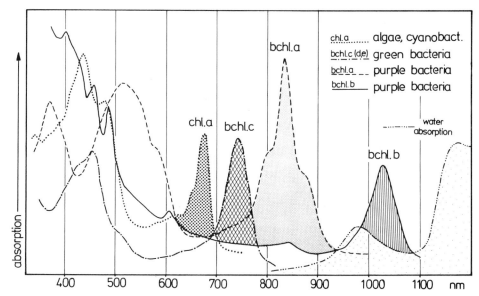

Figure 3.5 Absorption spectra of representative phototrophic microorganisms as measured on intact cells. Note that light absorption by the algae and cyanobacteria as a whole is generally in mid-specturm (stipped area as shown for chl a represented by the green alga *Chlorella*) while that of the green and purple sulfur bacteria is at longer wavelengths (bchl c for *Chlorobium*; bchl a for *Chromatium*, and bchl b for *Thiocapsa*). Thus, light absorbed by the algae and cyanobacteria in shallow water or in floating mats does not appreciably reduce the wavelengths available for the sulfur bacteria From Pfennig (1989).

bacteria absorb primarily in the far red part of the spectrum and so can exploit wavelengths not utilized by the algae (Stanier and Cohen-Bazire 1957; Pfennig 1989).

A similar example of light-limited distribution concerns the conspicuous vertical zonation of intertidal and subtidal seaweeds. Notwithstanding several species that are exceptions, among the benthic marine algae there is a general descending sequence into deeper waters of the green, brown, and red forms, respectively, correlated with light quality (Dawson 1961; Dawes 1981). The ability of red algae (Rhodophyta) to grow at greater depths than can most other algae has been attributed in part to their possession of the accessory photosynthetic pigment r-phycoerythrin which absorbs maximally at the blue-green wavelengths characteristic of deeper waters. By changes in pigment concentration known as 'complementary chromatic adaptation' these plants adjust to prevailing light conditions (Haxo and Blinks 1950). (The same situation is true for the bacteria which, at greater depths reached only by bluegreen radiation, depend relatively more on light absorption by their carotenoid pigments; Pfennig 1989.) Clearly other factors are also involved in algal zonation because some red algae such as *Bangia* and *Porphyra* occur high in the intertidal zone, while some greens such as *Ulva* grow at considerable depth. Vertical stratification with respect to light interception in kelp forests and by understory terrestrial plants is reviewed by Raven (1986a) in the context of the evolution of plant life forms.

In practical terms, knowing the energy and carbon sources of a particular organism is of great value to bacteriologists and mycologists in their attempts to isolate it in pure culture from nature. Isolation generally proceeds by *enrichment*, a process which provides favorable conditions for growth of the desired organism or, conversely, counter-selects against extraneous organisms. For example, use of a simple mineral salts medium with bicarbonate as a carbon source, together with appropriate light conditions, provides the basis for selecting photosynthetic bacteria. If incubation conditions are aerobic, cyanobacteria will be selected; if anaerobic (and in the presence of certain other growth factors and H_2S as an electron donor), purple and green sulfur bacteria will be selected. Likewise, cellulose is commonly used as a carbon and energy source for isolation of cellulolytic fungi.

3.4   Resource Acquisition

Optimal acquisition equals optimal foraging theory plus optimal digestion theory   As originally developed by MacArthur and Pianka (1966), optimal foraging theory dealt with factors that determine for an organism the width of the diet within a habitat. Briefly, this concerned *where* predators should feed to get the greatest 'reward' and *how* they should respond to various types of prey. For instance, is it best to feed indiscriminately, pursuing many food items (the benefits being greater availability of prey, hence less searching, and more time and energy available for other activities) or few (more searching but a 'better' reward, e.g., in terms of nutritional value)? What factors influence the decision to pursue or not to pursue a particular prey? How much time should a honey bee spend collecting nectar from a flower before moving to the next? What is the best size of mussel for a crab to eat? As developed below, the optimality concept casts the organism in the role of economist confronted by endless decisions pertaining to its economic self-interest. Marginal returns overall must exceed marginal costs.

MacArthur and Pianka's paper stimulated much interest among animal ecologists. Numerous subsequent contributions have taken one of two directions: either to further analyze and reinforce the idea mathematically (e.g., Charnov 1976a,b) or to broaden its focus (Schoener 1969; Pianka 1976; Chapter 3 in MacArthur 1972; Pyke et al. 1977; Pyke 1984; Stephens and Krebs 1986). I consider first the terminology as it relates to the original theory and then generalize the concept so that it entails all of acquisition. As noted in Chapter 1, natural selection should also favor plants, microbes, and sessile invertebrates that engage in optimal 'foraging', although their activities are not commonly described in those terms.

At the outset some thought should be given to what is meant by the term 'optimal'. For any particular economic, engineering, technical, or biological problem a number of solutions is possible. Each has a given cost measured by some common denominator such as dollars or energy expenditure. The 'best' or optimal solution is the one which fulfills the requirement with minimum cost. Optimal foraging theory and other optimality concepts in ecology are simply particular applications of general optimality theory and hinge on the longstanding premise that nature pursues economy in all processes (Rosen 1967; Givnish 1986b, pp.3-9). Natural selection should favor organisms that are cost efficient *overall* because they will be competitively superior under given environmental conditions and will leave more descendants. It is important to recognize that cost optimization in one attribute (e.g., growth rate) may be constrained by cost efficiency in another (e.g., reproductive output), so that some general balance is struck. How far each organism can go ultimately is set phylogenetically and by what is available in its particular gene pool, and is complicated by the fact that environments are not constant (Chapter 7). Hence, optimal does not mean the absolute best, but rather the best *compromise* from a *limited range of options* available to the organism in a particular context.

In the framework of foraging theory, optimal is generally taken to describe a strategy that results in *maximal net energy intake per unit time spent foraging* (Pianka 1988, pp.87-88), that is, gross intake less costs of acquisition. Thus, in macroecology the concept is presented in energetic terms. The premise is that energetically efficient foragers will be those selected. (The common use of efficiency in the optimal foraging theory context is ambiguous because strictly speaking efficiency is output/input, while in this ecological context the term is a measure of energy input to the organism less energy output [expended in acquisition], *per unit time*, that is, it really measures a *rate of productivity*.) As a case in point, Wilson has shown in a series of elegant papers (e.g., 1980, 1983) that leaf-cutter ants have evolved morphological features and a complex division of labor system which enable them to attain near optimal efficiency in harvesting and processing leaves for their fungal gardens.

Energy efficiency does not necessarily translate into number of descendants associated with a particular approach to resource acquisition. Subject to the caveats discussed below, however, it appears operationally to be the best currency and forms the basis for a sound working hypothesis. Fitness as the criterion of optimal would be preferable because it would set the issue clearly in evolutionary terms by relating a particular foraging behavior to other activities (interpreted as output, Figures 1.1 and 3.6) of the organism, rather than considering foraging in isolation (as the input term). For macroorganisms, assessments based on fitness rather than energetic efficiency

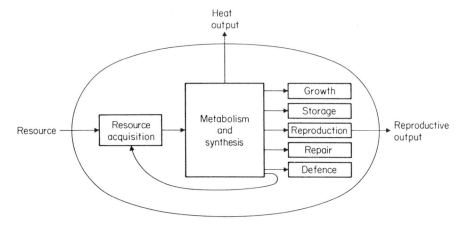

Figure 3.6   Expansion of Figure 1.1 (Chapter 1) to show the relationship among optimal foraging, optimal digestion theory, and optimal reproductive theory. If genes are to be effectively propagated, natural selection should lead to the production of organisms that strike an optimal *compromise* among foraging, digestion, reproduction, and other activities. Efficiency of allocation influences and is in turn influenced by the welfare of the soma. From Calow and Townsend (1981).

would be relatively complicated to conduct due to the long-term nature of experiments, and to interpret because changes noted could not necessarily be ascribed to differences in foraging alone.

There are two important constraints on attempts to generalize optimal foraging theory: First, the common denominator for comparisons is energy, and second, we need to consider what happens to the food after it has been gotten by some foraging mechanism (for details, see Townsend and Calow 1981; Sibly and Calow 1986). Energy is often used apparently for reasons of relatively convenient assessment and because of its universal importance in metabolism, including its role in acquiring and using all other resources. But because efforts just to acquire resources entail costs and are to an extent incomplete, most resources are probably limiting, at least at some point over time. (By definition, a limiting resource is one that can restrain the growth rate of a population and hence is of importance in competition; a nonlimiting or superabundant resource does not constrain growth). So, if an organism is to 'optimize' its foraging behavior, what will it be optimizing for? In reality the direction of selection will be a compromise attempt to optimize foraging for *all* limiting resources. Since all resources are not equally important from the organism's perspective, the compromise will be biased (e.g., Tilman 1982; Silby and Calow 1986) in favor of those resources that: 1) are most extensively used in metabolism and for which substitutes do not exist; 2) are most often limiting; 3) give the greatest return per unit investment in foraging; and 4) can be most easily selected for, given the existing pool of genetic variability in the population. For instance, it may be much easier for a microbe to alter its membrane uptake systems to accommodate some nutrients than others.

Energy is a good currency for expressing efficiency in animal ecology, particularly for homeothermic carnivores, and it is no coincidence that an optimal foraging theory framed in energetic terms was developed by animal ecologists. Carnivores and their prey have about the same biochemical composition (C:N ratios of 8-10:1; Chapter 3 in Begon et al. 1986). It is relatively easy to convert zebra protein to lion protein; the main costs are associated with maintaining a constant body temperature, and with finding and subduing prey, not with digestion. Excretory products are mainly nitrogenous.

In contrast, energy is less useful as a common denominator for assessing other organisms. Herbivores and detritivores, including the heterotrophic microbes, often face shortages of nitrogen and protein but an overabundance of carbohydrate. C:N ratios of plants are 40-20:1. Even if the cell wall fraction is discounted, the ratio still exceeds that of other organisms (see Park 1976). Carbon is respired or passed as fiber in the feces. Heal and MacLean (1975; see their Table 1, p. 91) report the assimilation:consumption efficiencies at 50% for herbivorous vertebrate homeotherms versus 80% for carnivores. In deserts, water is the limiting factor; in oligotrophic lakes, it is nitrogen or phosphorus. The phototrophic and lithotrophic microbes may or may not be predominantly energy limited, depending on the circumstances (e.g., light intensity, solubility of inorganic substrates, respectively). For plants, about 30 resources may be limiting overall (Tilman 1982, p.8), among them potentially carbon and light. For instance, the photosynthetic mechanism of terrestrial plants becomes saturated at about 20% of full sunlight. Thus, if nutrients and other limiting factors are sufficient to foster growth of vegetation so that light is reduced below this level, then it will be limiting (as it frequently is, for instance, to seedlings on the forest floor). Strategies for light interception, particularly those related to growth form, are therefore important (Horn 1971; see also Chapter 5 and Givnish 1986b).

The foregoing consolidates to the general statement that energy is obviously not invariably limiting, but neither is anything else. Providing its shortcomings are recognized, energy remains the best common denominator for comparisons.

The second limitation is that to be broadly applicable, optimal foraging theory must be combined with optimal digestion theory because, as formulated, it deals only with acquisition and not with conversion of the resource (Figure 3.6). Digestion efficiency concerns the amount of resource (energy) extracted per unit of food ingested (Chapter 2 in Sibly and Calow 1986; as they point out, a better definition would account for the amount of energy expended in processing that food). There is considerable research on different assimilation strategies, such as gut design and retention times to maximize the *net* rate of obtaining energy (Sibly and Calow 1986; Penry and Jumars

1987). Thus, foraging *and* digestive strategies should jointly maximize net energy return. Framed in this fashion, the concept is complete (search → capture → conversion) and can be applied to all organisms, despite confusion arising from the semantic inconsistencies in use of the term efficiency. Finally, to take the sequence to completion, I emphasize that foraging and digestion cannot operate independently of reproduction (Figure 3.6). Reproductive activities influence and are in turn influenced by the vigor of the soma. Nutrients acquired by a bear or a squirrel in the fall will be stored as potential energy and metabolized later for rearing offspring. Likewise, fungi typically acquire glucose from a culture medium early in the growth cycle, convert it to macromolecules such as mucilage or cell wall, and later catabolize the polymers to provide energy for sporulation.

An incomplete approximation to the ecologists' digestion terminology is the biochemists' and bacteriologists' use of the word 'efficiency' with respect to the degree to which nutrients are oxidized to completion (Pardee 1961; Krebs 1981). Bacteria and fungi may trade off high extraction of energy for high numbers of progeny. Pardee (1961) proposed that bacterial metabolism is geared to maximize growth rate, not complete extraction or the 'efficiency' of substrate utilization (energy production). As an example he cited the aerobic growth of *Escherichia coli* on glucose. Under these conditions glucose is only partly catabolized to acetate. If acetate conversion to CO_2 and H_2O were complete (which it is when glucose is exhausted), more energy is obtained. However, glucose enables faster growth than does acetate. This is because while ATP is being generated in catabolic processes it is being used in biosynthesis of monomers (amino acids, fatty acids, nucleotides). The amount used depends greatly on the substrate (Gottschalk 1986). It is small with glucose as a substrate because conversion of the precursors (phosphoenolpyruvate, pyruvate) to the monomers is connected with a gain of ATP, whereas a loss occurs when the same compounds are produced from acetate. For *E. coli*, 1 g of cells synthesized from glucose requires 34.8 mmol ATP compared with 99.5 mmol from acetate (Stouthamer in Gottschalk 1986, p.103). This is really just a variation on the theme of maximizing versus optimizing, or absolute gain versus net gain. Selection favors the net of benefits over costs. The analog would be to say that it is not cost efficient for an animal to design a longer intestine.

Krebs (1981) argues that making optimal use of resources is not as critical for microorganisms as it is for macroorganisms because for the latter starvation leads to death, whereas microbes under nutrient stress can shift easily to a dormant phase. The evolutionary advantage of spore formation is particularly obvious in those habitats where drought or heat kill nonsporeformers. While this observation is true for the fungi and to a lesser extent for the bacteria (few of which form spores), it needs to be added that dormancy per se is not a productive venture for a microbe and can at best lead only to survival in place or distribution in space (Chapter 7). However, time spent in suspended animation is lost to growth and reproduction and hence to spreading one's genes in the gene pool. Largely quiescent individuals would be selected out of the microbial milieu unless they counterbalanced prolonged dormant states with rapid growth, reproduction, and effective dispersal characteristics when transient favorable conditions arose. Microbial strategies are discussed when I consider the implications of size (Chapter 4).

Bacteria and to a large extent fungi appear to accomplish by metabolic versatility (see **Versatility**, below) what higher animals do by mobility and behavior, and plants by growth pattern. To illustrate, the search strategy for foraging animals involves first locating a profitable habitat patch and then modifying their tactics to capitalize on abundant prey. Thrushes explore for worms on a lawn by running a short distance, pausing to look about, turning, and making another run. Once food is located, the birds turn more sharply and frequently; the result of this activity is a concentrated search pattern over the zone of high food density (Figure 3.7; Smith 1974). Motile single cells (protozoa and some bacteria) can move short distances in random walk fashion (Figure 3.8). Although crude compared with animal search strategy, this improves their chances of finding resources over Brownian movement alone. Photo- (Pfennig 1989) and chemotactic (Adler 1976; Koshland 1979) responses exist (Figure 3.8 and Carlile 1980a). Thus, bacterial cells move to a food source as grizzly bears move to garbage cans in Yellowstone National Park! Moreover, bacteria were doing this for about 3 billion years before their distant furry relatives evolved! Other microbes such as the fungi move in effect by different forms of filamentous growth (Chapter 5) or by plasmodial mobility. Other than this, microbial movement is limited essentially to passive dispersal. I take up microbes and plants within the context of how growth form relates to resource acquisition in Chapter 5.

To summarize, provided that the limitations of energy as a common currency are recognized, optimal foraging and digestion theory can serve as a useful conceptual framework (illustrated by examples in the remainder of this chapter and in Chapters 4 and 5). *The growth form of an organism and its behavior are compromise responses to many selection pressures, of which energy acquisition and allocation is only one.* The means by which very different organisms forage are not homologous but they are analogous (Table 3.2). For instance, animal ecologists refer to 'foraging', to the 'catchability of prey', to 'handling time', to 'feeding efficiency', and so forth because the life forms that they study usually move about. In spinning a tensile web to trap its prey, the orb-weaving spider is behaving fundamentally like an aquatic filter-feeder, which must expose a prey-capturing surface to water currents. In essence, plant ecologists deal with the same issue but discuss, for instance, leaf orientation and optimal leaf area indices. Microbial ecologists consider enzyme kinetics and catabolite repression. Pathologists and epidemiologists deal with

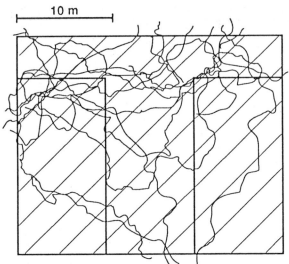

(a) Low density: food is distributed
     throughout all areas

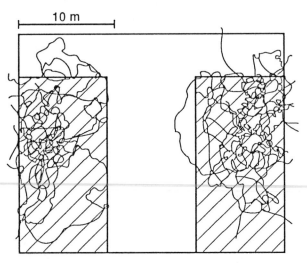

(b) High density: food is distributed
     in specific areas only

Figure 3.7   The patterns (track maps) of European thrushes searching for food on the ground. (a) In low-density areas artificial prey were distributed at 0.064 prey/m^2 throughout the grid area. (b) In high-density areas, prey were distributed at 0.299 prey/m^2 only in the two small rectangular grids. Where prey density is low, a less intensive search results because the birds turn less sharply and infrequently on the meadow, compared with the high-density areas. Redrawn from Smith (1974).

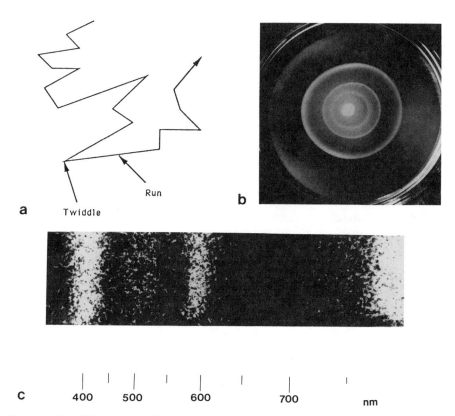

Figure 3.8 (a) Random walk of the motile bacterium *Escherichia coli* in a uniform chemical gradient. Movement consists of brief 'runs' in a particular direction, followed by 'twiddles' during which reorientation occurs. (b) If the gradient is nonuniform, directed movement can take place either towards (attractant) or away (repellent) from the source. In this example, bacteria inoculated as a drop in the center of a petri dish have moved outwards as three successive rings, each following a specific nutrient which has become locally depleted. Photograph courtesy of Julius Adler. (c) Aggregation of individual cells of the photosynthetic bacterium *Thiospirillum jenense* at wavelengths of light at which its pigments absorb. Photo courtesy of Norbert Pfennig.

host range, age or sex of suscept, tissues invaded. Semantics and detail aside, it all amounts to the same thing to the foraging creature.

Versatility   Although it is common dogma in microbiology that most microbes are 'versatile', what is meant by the term is often unclear. Certainly bacteria, for instance, can be found almost everywhere, from hot springs, to the ocean floor, to Antarctic ice sheets. Although this has been taken to mean that bacteria as a *group* of organisms are versatile (in this case with regard to

Table 3.2 Optimal foraging theory and optimal digestion theory broadly construed: Some analogous components for animals, plants, and microbes

Organism	Terminology	Measures of foraging efficiency
Animals	Prey size: digestibility; gut length; wait or pursuit; territory size; hunt individually or in packs; foraging time; filtering rate	Prey caught per unit time; net Kcal acquired per unit time
Plants	Chromatic adaptation; photosynthetic pathways; growth habit; shoot and root architecture; leaf area indices; substitutable ions	Percent solar radiation trapped
Microbes	Metabolic versatility; partial or complete oxidation; growth habit; photo- or chemotaxis; host range and tissue specificity; proto- or auxotrophy	Maximum specific rate of food acquisition (q_s^{max}); efficiency of food conversion to biomass (Y_s); ability to acquire food at very low food densities (K_s)

extremes of physical habitat), it says nothing about versatility of the individual, which is the relevant context here.

From the standpoint of the organism, versatility can be used in two contexts. First, we say that someone who can do *many things*—wrestle, play the saxophone, speak German and French as well as English, debate, and write short stories—is versatile. Notice that by this definition versatility is increased by extending the range of activities undertaken, but is *not* altered either by the degree of complexity of a specific feat, or by the competency with which a task is executed. Second, a person who masters many different situations quickly and with apparent ease is said to be versatile. Unlike the first definition, the emphasis in this latter case is on *speed of response* to new conditions and less on the spectrum of accomplishments. The constraint is time. After all, it can be argued, many things can be accomplished given sufficient time; the truly versatile are those who adjust quickly. Finally, since there are no absolute standards for versatility it is implicit that the term, however defined, is used in a comparative or relative sense.

Because the above interpretations emphasize different attributes, versatility can be assessed variously, leading to markedly different conclusions. As examples, some heterotrophic microbes will be compared with plants and animals. 'Metabolic' is used with reference to anabolic and catabolic pathways (in terms of their number and regulation) and to the range of substances synthesized or degraded.

Evaluated by the first criterion given above (being able to do many things), both microbes and macroorganisms are highly versatile metabolically with respect to nutrition. It cannot be said whether microbes are more or less so than the higher organisms because all the metabolites, end-products, and metabolic pathways for even a single organism are not yet known. The potential metabolic complexity of even a small organism is illustrated by *E. coli* which is about one-five hundredth the size of a plant or animal cell. This is the best understood and probably the most studied of all organisms. Known to date are about 80% of the metabolic pathways and one-third of the gene products in some biochemical depth, and about 10% of the genome has been sequenced (Neidhardt 1987, pp.1-6; see also Chapter 4 in Watson et al. 1987).

Most bacteria and fungi synthesize the organic compounds they need for cellular processes and hence will grow if almost any relatively simple carbon source is available. (Some members of both groups are fastidious and require growth factors such as vitamins, amino acids, or purines and pyrimidines in small amounts.) *E. coli* has extraordinary biosynthetic versatility, being able to synthesize all its cellular components from just glucose and minerals. This means that a pathway, separate at least in part, exists for each of the 20 amino acids synthesized. In the absence of a sugar, any amino acid can serve as a sole carbon source. (However, this bacterium cannot use cellulose, starch, or PCBs as a carbon source!) Since degradative pathways are distinct from biosynthetic pathways, this means that there are also numerous pathways for amino acid degradation to release nitrogen and carbon for biosynthesis. Degradation pathways, most of which are fairly specific, exist also for lipids, carbohydrates, and other compounds. For instance, if acetate is to be used it must flow into the glyoxylate pathway which regenerates oxaloacetate for each turn of the tricarboxylic acid (Krebs) cycle. These degradative reactions produce key precursor metabolites (plus ATP and reducing power in the form of NADH) through central pathways that are common to all species of bacteria. *E. coli* and other bacteria also have peripheral metabolic pathways which may be unique to the species, or function only if the bacterium grows on compounds that are not part of the central pathway. *E. coli* has more than 75 such pathways consisting of at least 200-300 different enzymatic reactions (Watson et al. 1987, p.116). The number could be expected to exceed this in certain bacteria such as the pseudomonads, which grow on diverse molecules (see **Generalists and specialists**, p. 88).

Plants are even more versatile than heterotrophic bacteria and fungi. Starting with only CO_2 rather than a sugar as a carbon source, and using light instead of chemical energy, they manufacture an immense array of primary and secondary compounds. Animals, on the other hand, lack the biochemical machinery to synthesize all their requirements from a simple molecule such as glucose. Vertebrates, for example, can make only half of the

basic 20 amino acids; the others must be supplied in the diet. The complex
nutritional requirements of animals are also evident in culture, where 13
amino acids, 8 vitamins, and various undefined growth factors supplied by
horse serum, in addition to glucose and inorganic salts, are needed in order
to grow human (HeLa) cells (Eagle 1955).

To summarize, by the first definition, nutritional versatility in the case
of plants and the bacteria or fungi essentially entails the capacity to synthesize
all macromolecules from a few, occasionally a single, relatively simple, inorganic or organic compounds. (This does not mean that in nature they do
not degrade complex substrates such as lignin or cellulose or hydrocarbons,
but rather that they possess the ability, as demonstrated in cultural studies,
to construct all they need from simple raw materials.) In contrast, nutrition
for animals generally involves degrading complex food sources, the component units of which are rearranged into new metabolites. While in terms of
the ability to do many things this may seem to imply that animals are less
versatile than microbes, it does not. Few microbes synthesize specialized cell
types or complex molecules akin to antibodies, eye pigment, neurotransmit-
ters, or any of the vast assortment of other chemicals and structures characteristic at least of the complex metazoans. These gene products are often
encoded by many spatially separate genes. Such products interact with others
in a precisely timed sequence of events to coordinate assembly and function
of cells, tissues, organs, and ultimately the entire organism. Thus, of the
genetic coordination mechanisms functional at the levels of 1) biochemical
pathway; 2) interaction between gene products; and 3) sequence in timing of
gene activity, the latter two regulatory processes are relatively more important
in the macroorganisms (Stebbins 1968). This issue is developed in Chapter 7.

The second interpretation of versatility, discussed at the beginning of
this section (speed of response to new biochemical [nutrient] conditions), is
more amenable to analysis. Bacteria and to a lesser extent fungi have the
remarkable and apparently unique ability to respond rapidly to different nutrient resources by altering metabolic pathways. Thus, what is impressive
about the bacteria is not just the numbers of alternative central and peripheral
pathways within a single cell, but that entire peripheral pathways can be
rapidly switched on and off. The set of enzymes needed for a particular route
is present or absent as a unit. Details of how these pathways and intermediates
change are available (e.g., see Doelle 1975; Baumberg 1981; Gottschalk 1986).
I shall consider briefly the ecological implications and the mechanistic basis
for the phenomenon.

Prompt response to ambient nutrient conditions undoubtedly reflects the
rapidly fluctuating environmental conditions to which microbes are exposed
(Chapter 7). These range from excess of a particular substrate, to growth-
limiting levels or, most commonly, to situations where the organism 'sees'
different but functionally similar substrates (e.g., glucose and acetate supply
both energy and carbon) in varying concentrations. Enzymes useful at one

moment could be useless or detrimental (wasteful) at the next. From the microbe's standpoint, the premium is on being able to avoid synthesizing catabolic enzymes if it is currently well nourished, while being able to adjust quickly to new resources if the need arises. In general, when resources are in high concentration (not growth-limiting) the response of bacteria is one of sequential utilization, with the nutrient supporting the highest growth rate being used preferentially from the mixture (Harder and Dijkhuizen 1982; Harder et al. 1984). This phenomenon is known as *diauxie* and supports Pardee's (1961) contention noted earlier that in bacteria the enzyme machinery adjusts itself to permit as rapid growth as possible. If substrate concentrations drop to growth-limiting levels, the various nutrients are used simultaneously.

Despite some similarities, there are important differences between gene regulation in prokaryotes and eukaryotes. Detailed discussion of these is not feasible or relevant here (for details see Watson et al. 1987; Ptashne 1989), but it is pertinent to emphasize specific aspects that contribute to the versatility of bacteria. While the eukaryote cell may respond to environmental cues (e.g., hormones, endorphins) in similar fashion, the pattern of gene expression in contrast is determined progressively during ontogeny. Once in motion it is stably inherited and relatively fixed. So, while a bacterial cell (or the entire clone) can respond rapidly to changes in its present environment (Chapter 7), the eukaryote cell is constrained by past conditions—an exceptional situation for bacteria.

In bacteria, fast changes in gene expression are possible because of the kinds of control mechanisms that operate at both the transcriptional and translational levels. With respect to the former, new mRNA molecules are produced continuously. There is virtually complete replacement of protein templates every several minutes. The process is facilitated by DNA supercoiling, which makes the initiation of DNA or RNA synthesis easier. In contrast, the DNA of eukaryotes is bound by histones and most appears not to be supercoiled. The resulting complex, chromatin, must first be decondensed so that the nucleotides are accessible for transcription. The genomic blueprint is also located in the nucleus, away from the site of translation. Finally, bacteria exhibit 'gene clustering', whereby enzymes in a particular pathway may be encoded by adjacent genes. This allows for coordinated regulation of the genes which are transcribed as one polycistronic mRNA, and sequentially translated by ribosomes into each of the proteins. Probably the classic example is the architecture of the *lac* region in *E. coli*, which encodes three genes involved in the metabolism of lactose (Chapter 16 in Watson et al. 1987). The three structural genes are grouped as an operational unit or operon which is transcribed into a single mRNA molecule. The essence of the mechanism is that the availability of external food molecules (lactose in this case), in conjunction with the energy status of the cell as mediated through cyclic AMP, controls the rate of synthesis of the small inducible enzymes by regulating the synthesis of particular mRNA templates. The situation is different for more complex organisms, where enzymes are coded individually and extensive RNA splicing is involved in the formation of messenger. As an example of comparative speed, translation proceeds about eight times faster in bacteria than in mammals (reviewed in Koch 1971). Transcription is also rapid: the average rate is 48 nucleotides per second, or three-fold the step time for incorporation of an amino acid.

Details of gene expression for microorganisms other than the bacteria remain poorly known because, with few notable exceptions such as *Saccharomyces, Neurospora*, and the nematode *Caenorhabditis*, they are not favored genetic models. Being eukaryotes presumably they are regulated more like the higher eukaryotes than the prokaryotes. Two examples illustrate the differences between *Saccharomyces* and bacteria. First, although the reaction sequence for biosynthesis of most of the amino acids is essentially identical in the two organisms, regulation of genes that code for the enzymes is different (Arndt et al. 1987). If bacteria are starved for a particular amino acid, transcription is enhanced of only those genes in the relevant pathway, whereas in yeasts and many other fungi, genes are transcribed for several unrelated amino acid pathways. Second, when amino acids are present in the medium, bacteria completely stop transcription of the corresponding genes, while yeasts continue to express high levels of gene transcription (basal level control). These excess amino acids are stored in the vacuole, an organelle which bacteria lack, where they provide an endogenous supply for a couple of cell generations in the absence of amino acid biosynthesis (Jones and Fink 1983; Arndt et al. 1987).

In overview, ecological comparisons based on versatility defined as the ability to do many things are currently a dead end because all of the metabolites for even a single life form are not yet known. Extrapolations can be made based on the coding capacity of the haploid DNA molecular weight of an organism, but these only define at best a theoretical upper limit. When versatility is defined as speed of response to new nutrient conditions, bacteria (and possibly the fungi) are more versatile than macroorganisms because they can rapidly alter entire metabolic pathways. Bacteria can grow at a high rate without excess enzyme 'baggage' (Chapter 7), on a preferred nutrient source, under substrate-abundant conditions, while retaining the ability to adjust quickly under nutrient limitation by using multiple substrates concurrently.

Generalists and specialists   Koala bears eat *Eucalyptus* leaves. The diet of Panda bears is exclusively bamboo. Rabbits, on the other hand, are largely indiscriminate grazers, and Virginia opossums and humans are omnivores par excellence. The same dietary spectrum applies among microbes. *Sulfolobus*, discussed previously, is a specialist at oxidizing elemental sulfur. *Methylococcus* lives on methane (Smith and Hoare 1977). In contrast, *Pseudomonas putida* can grow on any one of at least 77 different compounds as a sole source of carbon (de Jong in Kluyver 1931; see also Clarke 1982). *Sulfolobus* is the koala bear of the microbial world, and *P. putida* is the opossum.

The adjectives 'generalist' and 'specialist' can refer either to the range of habitats occupied by an organism or to the breadth of food sources it uses. Additionally, feeding may be specialized with respect to type of food yet generalized with respect to *source* (as in only seeds or nectar being consumed but from many species). Plant parasitic nematodes, for instance, commonly have several hosts yet they invade only specific organs (usually the roots) and typically feed only within a particular zone in those organs (e.g., as exo- or endoparasites near the tip). Habitat range and dietary breadth of organisms are related: Use of several types of food implies at least the ability to occupy more environments than does a restricted diet. Also, dietary differences can often be inferred from spatial separation, as in the case of birds feeding on different insect communities in the air, within the foliar canopy, on tree trunks, or in the litter. It is the dietary and only indirectly the habitat connotation of the terms generalist and specialist that will be pursued here.

The generalist/specialist issue relates to versatility as follows. In being able to do many things a generalist is versatile (versatility definition one; see above). A generalist is not necessarily versatile, however, in the sense of speed of response to new conditions (versatility definition two; see above). While a specialist cannot be versatile according to the first definition, it can be in terms of speed of response. For instance, a microbe may be restricted or fastidious in its range of utilizable substrates yet highly responsive to fluc- tuations in the density of those nutrients it does use (Chapter 5). This em- phasizes the point made earlier that versatility is only a relative concept and additionally that, however one chooses to define it, the context must be made clear.

What are the costs and benefits of generalist or specialist feeding? Prey items for the generalist are easier to find, search times are shorter, and con- sequently starvation is less likely because fluctuations in abundance of food types tend to cancel out on balance. Less nutritional, possibly toxic, or un- palatable items can be avoided by 'resource switching' as the consumer adjusts and 'optimizes' foraging tactics from one environment to another. Thus, where prey require long search times or are compositionally inadequate (nutritionally poor; toxic), natural selection should favor increasing diet breadth. Of course it follows that for each item by which the menu of the generalist is increased, one or more competitors that also use the food are also potentially added (Chapter 3 in Begon et al. 1986).

A means of decreasing competitive pressure is to become expert at per- forming one activity. This is demonstrated elegantly by the work of Brian (1952) and Heinrich (1976) on the foraging dexterity, acquired by learning, of individual bumblebees. The bee requires considerable skill to collect pollen and nectar, particularly when flowers are morphologically complex. Because any given bee colony generally contains individuals specializing on different types of flowers, it is able to extract energy from diverse floral sources (Figures 3.9 and 3.10). This strategy of high specialization by the channelling of time and energy into the most efficient extraction of resources is an effective alternative to one of repelling competitors by fighting. The bumblebees also illustrate a general principle that where food sources exert evolutionary pressure (as shown in this instance by their structural complexity) which demands a specialized response by the forager (here it is behavioral), a restricted diet will tend to be favored by natural selection. Other forms of response are physiological, morphological, or pathological specialization (see the case study later). Though the mechanistic interpretation of pathological specialization remains to be clarified (gene products of host and parasite identified), undoubtedly it resides in specific physiological and possibly also morphological adjustment of the partners.

A corollary to optimal foraging theory is that generalists should be less efficient at harvesting or perhaps using any particular resource than specialists. It is evident intuitively that the ability to do many things (e.g., harvest numerous food types in various ways) involves compromising high achievement (e.g., efficiency) in each. This trade-off is supported in theory, and there is evidence from both macroorganisms and microbes that it exists in nature.

The corollary implies that organisms feeding as specialists but which are otherwise similar to generalists should be competitively superior in that narrow portion of the resource spectrum in which they excel. This has been demonstrated in models by assigning prey with specific characteristics such as a particular size, predators which vary only in harvesting efficiency as a function of prey size, and by making various simplifying assumptions such as that of a homogeneous environment. (Environmental fluctuations complicate the issue and may allow for coexistence [see e.g., van Gemerden 1974]. This matter is taken up in Chapter 7.) MacArthur and Levins (1964, 1967) modelled the theoretical performance of ant-eating lizards with jaws of different sizes in situations where first, the food source consisted of an ant colony with equal numbers of 3- and 5-mm ants and second, where the ants were 2 and 8 mm long. Briefly, the very general result is that if the sizes of prey in a mixed population do not diverge too far beyond the optimal efficiency of the generalist (e.g., the phenotype that could best exploit 4-mm ants in the first situation), it will outperform the specialists on a particular size. If the differential in prey size extends beyond a certain point (as for the second case), however, the specialists on the specific sizes will outcompete the generalist. It is interesting to note in passing that this model based on *size* discrepancies is directly analogous to influencing, by varying substrate *chemistry*, the competition between specialist and generalist (the latter called "versatile" by Harder and Dijkhuizen 1982) bacteria on mixtures of inorganic and organic carbon compounds. Either type may win, depending on the ratio of the respective nutrient (Gottschal et al. 1979; Laanbroek et al. 1979).

A similar conclusion is reached by assuming a heterogeneous (patchy)

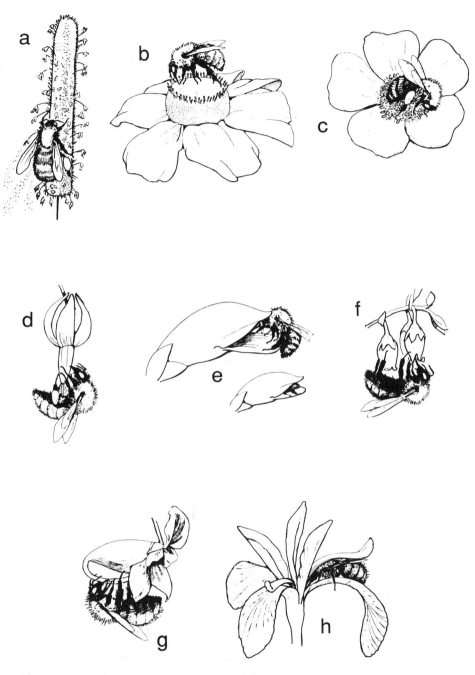

Figure 3.9 Differences in foraging posture of individual bumblebees (*Bombus vagans*) specializing to collect pollen (a-f) and/or nectar (b, e-h) from different kinds of flowers. From Heinrich (1976).

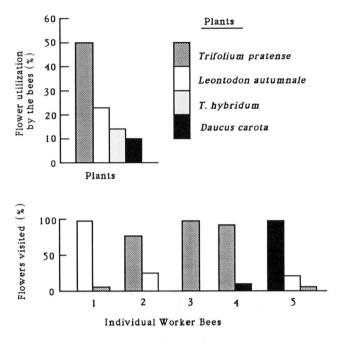

Figure 3.10   Specialization on flowers by bumblebee workers. (Top) Percentage of the four intermingled, concurrently blooming flowers in a hayfield used by the bees as a group, suggesting floral species-preferences. (Bottom) Percentage of flowers visited by five individual bees, suggesting independent, individual preferences are superimposed on species-preferences. Based on data for *Bombus vagans* from Heinrich (1976).

environment and asking at what point a jack-of-all-trades will outcompete patch specialists. MacArthur and Pianka (1966) predict that this will be when the generalist makes up in reduced travelling time between patches for what it loses in lower hunting efficiency. By being able to feed in more patches and consequently being less preoccupied with travelling between suitable habitats, it will have more hunting time available and will harvest food to a density below which the specialist can compete.

Although circumstantial evidence may abound, rigorous data to support such predictions from studies of macroorganisms are generally lacking because experiments are fraught with difficulties: To test the hypothesis of a trade-off unambiguously would require knowing in detail the feeding characteristics of the supposedly competing populations and that the generalist and specialist were otherwise effectively identical. Clearly, species do best in certain habitats and there is abundant evidence for niche shifts or competitive displacement (e.g., Krebs and Davies 1984; Grant 1986). The reasons are usually complex, however, and may or may not involve feeding behavior, of which the generalist/specialist attribute is only one consideration.

For instance, studies by Werner and his colleagues over several years on competition in sunfish show that the bluegill (*Lepomis macrochirus*) and the green sunfish (*L. cyanellus*) have similar food habits (e.g., Werner 1977; Werner and Hall 1977). The green sunfish makes broader use of the food size spectrum, probably because its larger mouth and other morphological features enable it to handle large prey. The bluegill feeds on smaller items more efficiently. When the distribution of the two species overlaps, as is commonly the case, the bluegill is forced to open water where it eats small, abundant zooplankton, whereas the green sunfish feeds predominantly on prey associated with vegetation bordering the ponds. While this work to an extent involves generalized versus specialized feeding, it is complicated by the fact that neither species is really a specialist relative to the other, and by behavioral patterns (homing and aggressiveness in the green sunfish; tendency to school in the bluegill) which also affect the displacement. For similar reasons the large body of data on birds, illustrated by detailed studies over decades on Darwin's finches (summarized by Grant 1986), is equally difficult to interpret conclusively within the generalist/specialist framework.

In some cases, different forms of the same species can be compared. This situation offers greater precision and begins to approach what can be done with the microbe systems discussed below. Meyer (1989) studied the feeding behavior of two morphs of a cichlid fish, *Cichlasoma citrinellum*, from Nicaragua. The molariform morph has a different jaw structure from the papilliform morph. This allows it to feed on harder shelled prey, such as snails, as well as on soft prey, preferred by both morphs. It thus has an alternative food source available when needed, which the papilliform type lacks. The papilliform morphs, however, feed more efficiently than do their counterparts on soft prey (as measured by handling time). The proportion of molariform morphs in the population appears to be correlated with the abundance of snails in the Nicaraguan lakes.

The generalist/specialist issue can be approached rigorously, if somewhat artificially, in microbial systems. First, point mutations can be made and the resulting auxotrophic mutants (unable to synthesize a particular nutrient) can be tested competitively under controlled conditions against their fully functional (prototrophic) parents. This must be done in the presence of the specific, required nutritional factor; the reciprocal experiment comparing competitiveness in the absence of the nutrient cannot be done because the auxotroph will not grow. The hypothesis is that the metabolically more simple 'specialist' will have a selective advantage because it does not carry the additional biosynthetic steps of what is in effect the 'generalist'. This requires certain assumptions discussed below. Data from numerous experimental systems involving mutants support the assertion (e.g., Zamenhof and Eichhorn 1967, below). Second, competition experiments between various generalist and specialist microbe strains or species lead to similar conclusions (e.g., Gottschal et al. 1979; Laanbroek et al. 1979). Indeed, the weakness of the evidence

A Case Study: The Gene-for-Gene Interaction and Parasites as Specialists

Nowhere are specialized responses better illustrated than in the co-evolution of parasites and their hosts. For some plant pathogens in particular, a precise correspondence between genes governing host resistance and those controlling parasite virulence has been well documented (e.g., Flor 1956; Person 1966; Vanderplank 1986; see also Wolfe and Caten 1987). As a result of this gene-for-gene relationship, each virulence gene in the parasite population is matched by a resistance gene in the host population. This occurs because 1) under conditions of disease pressure, resistant plants will be at a competitive advantage over their susceptible counterparts, and 2) a single new gene for resistance can be overcome by a single new gene for virulence in the pathogen and vice versa. Concurrently, as resistant hosts displace those that are susceptible, fitness of the now avirulent parasite declines, so any individuals with virulence genes will be selected. In other words, for this system, specialization at the level of the individual gene has been documented.

One manifestation of the interaction is what has often been called an evolutionary 'arms race' between parasite and host, most evident in the 'boom-and-bust' cycles of many cultivars of major crop plants. Thus, as for transferable drug resistance (Chapter 2), we see here with coevolution a phenomenon both of much basic ecological interest and major practical implications, in this instance with respect to breeding strategies for crop plants. The phenomenon, however, is not restricted to agronomic situations (Burdon 1987).

This is not to imply that every host-parasite association involves a gene-for-gene phenomenon. Nevertheless, close association and resulting tightly knit interactions can obviously coevolve. On one hand, the host is fighting for survival; on the other is the parasite which depends for its existence on the host. This intimacy could not develop to the same degree for free-living organisms and prompted Price (1980, p.19; see also Trager 1986; Price 1990; Esch et al. 1990) to comment that "parasites represent the extreme in specialized resource exploitation". A specialized lifestyle is further promoted by the small size of parasites, the poor mobility of many phases in their life cycles and, for some such as the insects, the use of specific cues to locate food items (Price 1977, 1980). The coevolutionary lifestyle is a dramatic example that evolution generally results in the narrowing of options. In this instance, the partners "each drive the other into an ever deepening rut of specialization" (Harper 1982, p.17; see also

Huffaker 1964), constraining what each can do. Philosophically, as Harper (1982) notes, this is in contrast to viewing (optimistically!) such relationships as a precise fit between organism and environment (Chapter 7).

By virtue of the products of its virulence genes, an individual germinating spore of the flax rust fungus is specialized for collecting nutrients. This is directly analogous to the individual bumblebee that is a specialized harvester of pollen from the wild rose (see text).

from microbiology to date lies not in documenting the phenomenon, but in extrapolating its possible role to nature.

Zamenhof and Eichhorn (1967; see also Atwood et al. 1951) compared various nutritionally deficient mutants and fully functional (revertant) strains of *Bacillus subtilis* in continuous (chemostat) culture in liquid media. One experiment demonstrated that a histidine-requiring mutant (his⁻) was competitively superior in dual culture to its histidine-nonrequiring, spontaneous revertant (his⁺) (Figure 3.11). Two starting ratios of the strains were used; curves of the decline in ratio of his⁺/his⁻ over time were parallel, indicating that the auxotroph his⁻ intrinsically had a selective advantage over his⁺ and that the effect was not due to some specific inhibitor. In experiments with other mutants the authors went on to show, first, that dispensing with an earlier rather than a later biosynthetic step gave the auxotrophic carrier a selective advantage and, second, that a derepressed strain producing the final metabolite (tryptophan) in quantity was at a selective disadvantage compared with the normal repressed strain.

The most obvious interpretation of the work is that the prototroph continues to synthesize at least some of the metabolite in question, thereby incurring energy costs, despite the availability of the chemical in the medium. These functions are completely blocked in the auxotroph. Evidently the feedback inhibition and gene repression mechanisms are incomplete in the prototroph. The mutants studied by Zamenhof and Eichhorn contained point mutations, and the economy lies only in synthesis of the metabolite—that is, the structural gene, mRNA, and in some cases the inactive enzyme are still manufactured. If, instead, deletion mutants were produced resulting in shortening of the chromosome, additional savings (e.g., faster replication; fewer nucleotide building blocks) should accrue, resulting in even greater selective advantage (Zamenhof and Eichhorn 1967). The authors go on to speculate that it may have been through energetic savings that parasitism evolved, once nutritious 'media' in the form of macroorganisms became available to support fastidious variants of fully functional, free-living microbes.

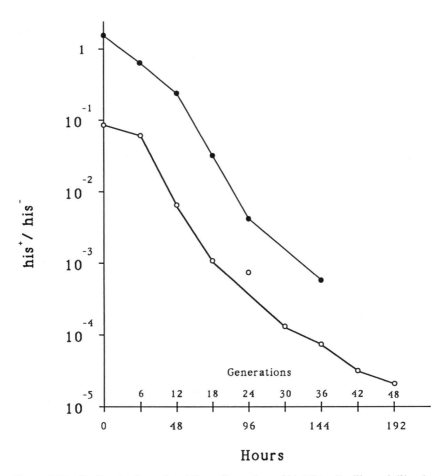

Figure 3.11   Decline in the ratio of the cell number of his⁺/his⁻ *Bacillus subtilis* when the two strains were inoculated together in broth in a chemostat and incubated for 192 hours (48 generations). The two curves represent two different starting ratios. From Zamenhof and Eichhorn (1967). Reprinted by permission from Nature, vol. 216, p.63. © 1967. MacMillan Magazines Ltd.

There is a direct biotechnology counterpart to the above example. A common production problem is that bacteria genetically engineered to produce a particular desired protein (e.g., human growth hormone) are at a selective disadvantage in continuous culture because of the imposed gene function. Unless the gene is kept switched off by manipulating inducer or temperature levels until the desired growth stage is reached for expression, the original recombinant is displaced by revertant wild-type cells (Helling and Lomax 1978; Dykhuizen and Hartl 1983).

Where and how can generalists and specialists operate to their respective advantage? Within phylogenetic limits, organisms can change phenotypically and genotypically to meet their nutrient (and other) needs in particular environments. One manifestation of this is the grouping of foragers as generalists or specialists. Generalists such as the opossum or *Pseudomonas putida* are 'versatile' (in the sense of doing many things) in dietary intake and can respond to scarcity of food (nutrient limitation) by resource switching (mixed substrate utilization). Other things being equal, they will be favored in environments or under conditions where prey items are diverse and search times are high (nutrients growth-limiting due to high turnover or low concentration; mixed substrate utilization). Specialists such as koala bears, *Methylococcus* spp., and auxotrophs, being more efficient harvesters of a particular resource spectrum, will be favored where prey items tend to be similar and where prey occur locally in abundance (nutrients not growth-limiting; single substrate abundant or mixed substrate dominated by one component).

3.5   **Summary**

In an evolutionary context the three major nutritional questions posed by every organism are "What do I eat?", "How do I get the food before someone else does?", and "How do I use the resulting energy and molecular building blocks so as to effectively propagate my genes?". Restated in ecological terms, these three issues involve resource categories, resource acquisition, and allocation of metabolized resources, respectively.

Resource categories can be illustrated by considering the sources of energy and of carbon building blocks for an organism. ATP is the major carrier of biologically usable energy in all organisms and there are only two basic mechanisms for its generation, electron transport phosphorylation or substrate-level phosphorylation. Energy is obtained directly from the sun by phototrophs (phototrophic bacteria, algae, plants), in which case light energy is converted by electron transport phosphorylation into the high energy phosphate bonds of ATP. In most organisms (chemotrophs), ATP is generated from reduced inorganic compounds or from organic compounds by electron transport phosphorylation or substrate-level phosphorylation. Carbon is acquired either directly from CO_2 (autotrophs) or from organic compounds (heterotrophs). Although the resulting energy/carbon permutations form several potential resource categories, most living forms, in terms of species or biomass, use either light energy to fix CO_2 for their biosynthetic needs (photoautotrophs = C-autotrophs, e.g., plants), or derive both their energy (i.e., in this case electron donors) and carbon from organic molecules (chemoheterotrophs = C-heterotrophs, e.g., animals and most microbes). Chemolithotrophs (chemotrophs using an inorganic electron donor), which evidently exist only as microorganisms, are quantitatively insignificant. However, they play an essential, global role in biogeochemical cycles.

The most obvious result of energy/food relationships within the living world is the trophic structure of ecosystems. This is usually depicted as a grazer (herbivore) food chain based on phototrophs associated with a decomposer (saprovore) chain based on dead organic matter. Microbes play a role in both, particularly the latter. Apart from the microscopic plankton in aquatic systems, however, they are generally not explicitly recognized as part of grazer systems. Hence their various ecological roles tend to be underestimated. This is mainly because of the zoological bias of limnologists. Likewise, conventional depictions do not accommodate food chains based on other sources of energy input, such as that fixed chemosynthetically by lithotrophic bacteria. These microbes may ultimately support, for example, substantial oases of benthic invertebrates around deep-sea hydrothermal vents.

The sources of energy and carbon also set broad limits on the distribution of living things, as in the restriction of aquatic plants, algae, and photosynthetic microbes to the photic zone, and of obligate parasites to their hosts.

As originally devised, optimal foraging theory was essentially a cost/benefit analysis in energetic terms developed to interpret the foraging behavior of vertebrates. It can be construed broadly, merged with optimal digestion theory, and applied conceptually to all organisms. Thus, as shall be shown, bacteria and fungi appear to do largely by metabolic versatility what animals accomplish by mobility and behavior, and plants and other sessile organisms by growth form. Growth form and behavior are, however, compromise responses to many selection pressures, of which energy acquisition and allocation is only one.

From the standpoint of the individual and with respect to nutrition, versatility can be defined either as the ability to do many things or to respond rapidly to new conditions. As judged by the first definition it cannot be said definitively whether microbes are more or less versatile than macroorganisms because all of the metabolites for even a single life form are as yet unknown. Estimations can be made based on the coding capacity of the haploid DNA of an organism, but these define only a theoretical upper limit. Although the metabolic pathways in a given cell type of a multicellular eukaryotic organism may be comparatively few, in aggregate for the individual they could well exceed those for a microbe, given the range of cell types, the subcellular complexity, and the diversity of chemicals which can be produced. When versatility is construed as speed of response, bacteria (and possibly some other microbes such as the fungi) appear to be more versatile metabolically than macroorganisms because they can rapidly switch entire metabolic pathways.

Organisms are either generalists or specialists with respect to dietary range. For the generalist macroorganism with diverse prey items or the generalist microbe able to use many substrates, 'food' is easier to find, substitutable resources can be alternated, search times are shorter, and starvation is less likely. When food sources exert evolutionary pressure by their structural, behavioral, or physiological complexity, demanding a specialized response, the most likely outcome is restriction in dietary breadth. This may be seen, for example, among individual bumblebees specializing as foragers on a particular flower type, by the coevolution of parasites with their hosts, and by specialist microbial strains that will not grow in the absence of a specific energy source. More generally, all specializations present opportunities (optimization at doing certain things) but also impose constraints (fewer options). Evolution generally moves organisms towards increasing specialization, thereby narrowing options and limiting what they can do.

A corollary to optimal foraging theory is that generalists should be less efficient in acquiring or using any particular resource than specialists. This is accepted intuitively and it has been documented with theoretical models. The supporting evidence from field studies is, however, largely circumstantial and tenuous: There is difficulty in isolating the variable of interest (feeding range) with experimental systems of macroorganisms and, for microorganisms, in extrapolating the results of competition experiments (e.g., of facultative lithotrophic generalists versus obligate heterotrophic specialists; or auxotrophs versus prototrophs) under laboratory conditions to nature.

3.6   Suggested Additional Reading

Alberts, B. et al. 1989. Molecular biology of the cell, 2nd ed. Garland, N.Y. A good overview of cellular biochemistry, structure, and genetics, including comparisons of prokaryotes and eukaryotes.

MacArthur, R.H. and E.R. Pianka. 1966. On optimal use of a patchy environment. Amer. Nat. 100: 603-609. The starting point for considering optimal foraging theory.

O'Brien, W.J. et al. 1990. Search strategies of foraging animals. Amer. Sci. 78: 152-160. The search strategies of most (unitary) animals involves a pattern of starts and stops (saltatory search) which repositions the hunter; at the extremes the other strategies are a 'cruise search' (hawk) or 'ambush search' (lion).

Pardee, A.B. 1961. Response of enzyme synthesis and activity to environment. Sympos. Soc. Gen. Microbiol. 11: 19-40. The thesis that bacteria have evolved to multiply as rapidly as possible.

Schopf, J.W. 1978. The evolution of the earliest cells. Sci. Amer. 239: 85-103. Aspects of bacterial metabolism, the origin of eukaryotes, the interaction of organism and environment in evolution.

4

Size

*What is a microorganism? There is no simple answer to this question.
The word 'microorganism' is not the name of a group of related
organisms, as are the words 'plants' or 'invertebrates' or 'frogs'. The use
of the word does, however, indicate that there is something special about
small organisms; we use no special word to denote large animals or
medium-sized ones.*

—W. R. Sistrom, 1969, p. 1

4.1 Introduction

The most remarkable feature of any assemblage of organisms is the difference
among species in size and architecture. Excluding entities such as viruses that
reproduce but are nonliving, the size range spans 21 orders of magnitude
(Figure 4.1), from mycoplasmas at about 10^{-13} grams to blue whales which
exceed 10^8 grams. The blue whale, incidentally, is the largest animal ever
known. It is almost twice as big as the largest dinosaur, *Brachiosaurus*, which
probably weighed about 85 tons (ca. 8.5×10^7 grams), and is equivalent in
mass to about 50 elephants at approximately 3 tons (3×10^6 grams) each.

Two observations on Figure 4.1 are worth highlighting. Notice first that
implicitly the sizes given are for an *adult* form, which is the life stage we
usually associate with an organism. (With respect to size, the specific stage
of the life cycle matters relatively little for a bacterium but a lot—more than
an order of magnitude—for, say, a juvenile versus an adult elephant, and
several orders if the mature form is compared with the zygote or seed or
fetus!). Instead, however, comparisons can be made by integrating the size
of the organism through its entire life cycle (Bonner 1965). For all sexually
reproducing species there is a point in the life cycle where the organism
consists of a single cell—the zygote—and there are as well all the intervening
sizes and shapes to maturity. (The counterpart in many asexually reproducing
species is the unicellular spore.) This point will be developed in the next
chapter.

101

Figure 4.1   The size spectrum of living
things. Mass in grams on a logarithmic scale.
From McMahon and Bonner (1983, p.4).

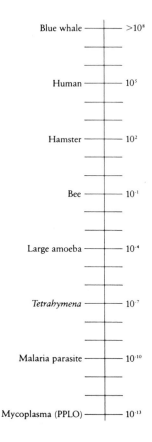

Second, note that the scale in Figure 4.1 is not readily applicable to modular organisms—such as bracken fern covering a hillside, given as an example in Chapter 1. Such creatures, as genetic individuals, are indefinite in size. This aspect will also be pursued in Chapter 5, but it is worth commenting here that Oinonen (1967) observed clones of bracken fern (*Pteridium aquilinum*) extending almost 500 meters across that were estimated to be 700 years old, and suggested that others may reach 1,400 years. Clones of sea anemone may approximate several hundred meters in area and "probably persist for many decades" (Sebens 1983, p.441). While as a result of environmental onslaught these organisms may not come to reach the biomass of a blue whale, they can theoretically weigh more, and they do eventually occupy large areas. It appears that though they lack homeostasis and the advantages it confers (Section 4.5), clonal organisms escape many of the problems of large size, such as having to support a massive bulk. Many such organisms grow prostrate rather than in upright fashion, they tend to get large simply by the iteration of countless small parts, and often spread themselves about by fragmentation (Chapter 5).

4.2 Constraints on Natural Selection: Phylogenetic, Ontogenetic, and Allometric

The three major, interrelated constraints which limit the ability of natural selection to shape organisms are phylogenetic (taxon-related), ontogenetic (development-related), and allometric (size-related). Unlike the trade-off concept, which underlies most of the examples in this book, evolutionary constraints are absolute and cannot be moved to-and-fro by changing an opposing selection force (Stearns 1977, 1982). Thus, one way to look at any organism is "as a mosaic of relatively new adaptations embedded in a framework of relatively old constraints" (Stearns 1982, p.249).

The constraints placed on a species by its evolutionary pedigree are *phylogenetic*. Organisms evolve in lineages. What is available to a grasshopper is not available to a human. A specific pattern of development for each is both characteristic and similar to that of ancestral forms. Humans, unlike millipedes, have only two legs and lack the power of regenerating amputated appendages. Starfish and their relatives all show five-point symmetry. All plants are modular in growth form; all higher animals are unitary. In every instance, limitations as well as opportunities are bestowed on the organism by its birthright. So, given a particular design, certain things are impossible or, though possible are strongly selected against, or both (Stearns 1983).

Selection is also limited in what it can do at any point in the life cycle by what has gone before in the developmental program of an organism (Bonner 1982b; see especially pp.1–16). Such limitations are termed *ontogenetic or developmental*. Changes in the timing of events in the life cycle (heterochrony) may occur (events speeded up, as in larvae which are sexually mature; slowed down, as in development of the large brain in humans), but critical phases or structures cannot be eliminated. For instance, the basic sequence of stages in the cell cycle is established, as is the pattern of events, in embryogenesis. This is true even when functionally useless vestigial structures (gill arches in vertebrates; appendices in humans; tails in birds and mammals) appear in the fetus or adult. The structures persist as a consequence of engineering size increase by the building block method (Dobzhansky 1956; Chapter 4 in Bonner 1988). What is important is not the particular item so much as the overall process and the end product.

Allometric limitations relating to size include changes in chemistry, physiology, and morphology. Size differences and the attendant allometric relationships can be examined during the course of development of an organism (ontogenetic comparisons) or across taxa at arbitrarily selected stages (phylogenetic comparisons) in an historical or contemporary context. I consider in this chapter the extent to which microorganisms and macroorganisms see the world differently and the ecological implications of this size differential. Such considerations must allow for not only differences of the conventional allometric sort applicable to organisms of similar geometry, but acknowledge also the pronounced differences in shape as well as size of microorganisms versus macroorganisms. There are no spherical cows to compare with coccoid bacteria! Even if there were, the types of environments and the scale on which the organisms would interact with these environments would be entirely different.

4.3   Why Are There Macroorganisms?

Life appears to have begun about 3.5 billion years ago, some one billion years after the earth appeared. Bacteria-like organisms were the first living creatures in the world and for at least two billion years were the *only* living forms (Schopf 1983; Chapter 3 in Raven and Johnson 1986). All phyla can be traced to a Universal Ancestor, but the exact pathways are unknown and contentious. This life form did not survive, though evidently had properties in common with the eukaryotes as well as with the two other major groups, the Archaebacteria and the Eubacteria. Based on a phylogenetic tree derived from similarities in 16S rRNA, it has been suggested (Pace et al. 1986; Woese 1987) that archaebacteria are as different from eubacteria as either group is from the eukaryotes (the 'Three Kingdom' or more correctly 'Three Domain' Hypothesis [Woese et al. 1990]; for comments on molecular versus morphological data in the inference of phylogenies see below). Archaebacteria are distinct from eubacteria in having (like eukaryotes) introns in rRNA and tRNA. Unlike eubacteria, they also commonly display genomic rearrangements; branched-instead of straight-chain fatty acids, and ether-linked instead of ester-linked lipids; they have no peptidoglycans in their cell walls.

Certainly very early forms of life were various kinds of lithotrophic archaebacteria (a physiologically diverse group including extreme halophiles, thermophilic sulfur metabolizers, and methanogens). As noted in the previous chapter, lithotrophs can obtain all their energy by oxidizing some kind of inorganic electron donor and generate ATP by electron transport phosphorylation. The first kinds of these lithotrophs were probably highly thermophilic anaerobes which got their energy by reducing sulfur (Woese 1987). They were likely followed by the the obligately anaerobic methane-producing bacteria. Using electrons, probably from H_2, the methanogens reduced CO_2 as the terminal electron acceptor to CH_4, also producing from CO_2 their cell carbon. The eubacteria, a diverse assemblage of phototrophs and chemotrophs, may have appeared before or after the archaebacteria. The photosynthesizers included the green lithotrophic sulfur bacteria, which used light energy together with the dissolved H_2S in the primeval sea as a source of electrons (rather than water) to generate ATP by anaerobic photosynthesis. They were probably followed by the cyanobacteria (blue-green algae), which utilized oxygen-forming photosynthesis. The nonphotosynthetic forms comprised the chemolithoautotrophs and chemoorganoheterotrophs, which got their energy and carbon from organic molecules (cf. Table 3.1, Chapter 3).

The eukaryotes as we know them today appear to have originated 1.3 to 1.5 billion years ago (Vidal 1984), though some earlier eukaryotic lines evidently predate this period (Woese 1987). This is the estimated time that the hydrosphere became aerobic, hence it is when an oxygen-utilizing mitochondrion could have appeared (Yang et al. 1985). For about 700 million years probably all eukaryotes were unicellular microorganisms. According to conventional phylogenetic schemes, such early forms gave rise independently to several multicellular lines (Figure 4.2) culminating in the major groups that are now assigned to the fungal, plant, and animal kingdoms. These, along with the bacteria and protists, comprise the 'Five Kingdom Hypothesis' (Bonner 1965, 1988; Margulis and Schwartz 1988; summarized in Chapter 3 in Raven and Johnson 1986). Mitochondria and chloroplasts of the eukaryotes probably originated as bacterial endosymbionts (reviewed by Gray 1983; see also Palmer 1985). (The contentious issue of the 'best' approach to the study of genealogy as it relates to the Three Kingdom Hypothesis versus the Five Kingdom Hypothesis cannot be explored here. Suffice it to say that *both* morphological and molecular evidence should be used to infer phylogenetic relationships [reviewed by Hillis 1987; Sytsma 1990].)

As noted above, details surrounding the connection between the eukaryotes and prokaryotes are unclear and controversial, as are those pertaining to the origin of life and the early phylogenetic changes. Woese (1987) argues that we have a simplistic and misleading preconception of these evolutionary events colored by dogma about: 1) the prokaryote-eukaryote distinction (prokaryotes were defined on the basis of *lacking* something, which is no reason to assume that they are related); 2) the 'Oparin ocean scenario' (primitive oceans were an anoxic nutrient 'soup' which leads to the false conclusion that the first organisms were anaerobic *heterotrophs*); and 3) Darwin's "warm little pond" image (the picture conveyed is one of hospitable 'warmth', probably not intended by Darwin; thermophilic adaptations; anthropomorphism). Whatever its origin, the bacterial lineage is extremely old and has numerous representatives! Judged by their ancient ancestry and their vast numbers (whether quantified as individual cells or as biomass), the bacteria have been eminently successful. In any flourishing tide pool, there are vastly more bacterial genomes than those of all other organisms combined.

Why, then, did macroorganisms evolve? Fortunately, this question can be pursued independently of controversy about genealogical relationships! Natural selection for increased size is most easily interpreted in terms of obtaining food, of avoiding being used as food by someone else, or for dispersal. There are two ways of becoming large—by increasing cell size or

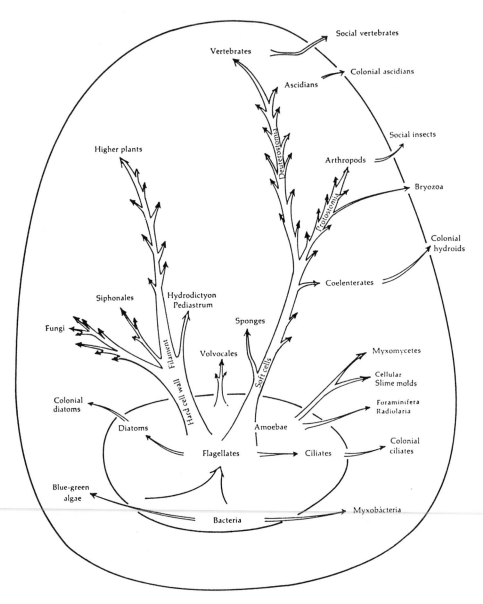

Figure 4.2 Bonner's (1974, p.81) hypothetical scheme for the independent development from unicellular forms (innermost ellipse) of multicellular organisms (large ovoid), and colonial and social organisms (shown outside the ovoid). Reprinted by permission of the publishers from *On Development: The Biology of Form* by John Tyler Bonner, Cambridge, Mass.: Harvard University Press, © 1974 by the President and Fellows of Harvard College.

number (or both). Plants do relatively more of the former and animals do more of the latter. Originally, larger variants of unicells may have been favored because they were less likely to be eaten or because they were more effective predators than their smaller counterparts. As discussed below, there is an upper limit on cell size beyond which efficient metabolism becomes impossible. Multicellularity confers the advantages of large size by the building block approach. This method resolves metabolic concerns by delimiting the size of the individual blocks, and allows for regularized genetic oversight of the developing soma by mitosis (Chapter 4 in Bonner 1988).

The progression towards a 'multicellular' (broadly defined) organism was accomplished either by the retention of daughter cells following fission, by the formation of cross walls within an expanding multinucleate body, or by aggregation of cells (Figure 4.3). This innovation fostered different kinds of cooperation among the component units, as in the the pooling of enzymes from a consortium of organisms to digest a substrate recalcitrant to attack by an individual cell. Indeed an analogous form of mutualism is evident today in the aggregated mass of hyphae termed 'infection cushions' in fungi such

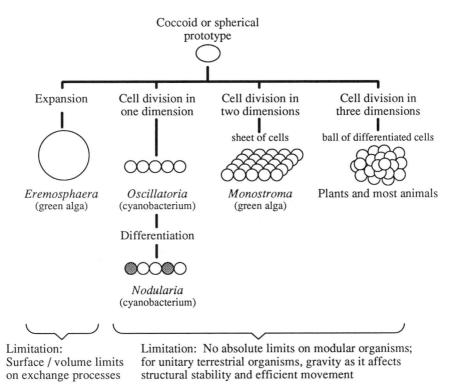

Figure 4.3 Four simple examples of how increase in size could have evolved.

as *Rhizoctonia*. These structures provide sufficient localized inoculum potential to enable the breaching of host defenses. One can envisage an evolutionary progression from unicellular individuals, to a *colony* (clonal aggregation) of such individuals of characteristic but variable appearance, to a true multi-cellular functional unit in which cell autonomy and competition among the component cells is suppressed (Carlile 1979). The success of the multicellular experiment among prokaryotes is evident in a range of forms—from the rel-atively unorganized mass of individual bacterial cells termed 'microcolonies' common in nature (e.g., Gibbons and van Houte 1975; Savage 1977), to the more organized cell aggregations characteristic of the swarming myxobacteria, which grow, feed, and eventually form fruiting bodies as cooperative units of cells (Figure 4.4 and Shapiro 1988), and the many filamentous cyanobac-teria (e.g., *Anabaena*, *Aphanizomenon*).

Evolution of the eukaryotic cell was a major event in the development of macroorganisms. The details of how this novel cell type arose remain unclear and there are no known intermediate types in the fossil record. Al-though the differences between prokaryotic and eukaryotic cells are many

Figure 4.4 Mature fruiting bodies of the myxobacterium *Chondromyces crocatus* show-ing elevated, multiple sporangia, or cysts, each of which contains resting cells embed-ded in slime. Photograph courtesy of Pat Grilione. From Grilione and Pangborn (1975).

(for a good general synopsis see Chapters 5 and 6 in Raven and Johnson 1986), the significant overall advance was probably the compartmentation of the eukaryotic cytoplasm which allowed localization of cellular activities, together with an oxidative metabolism (in common with some prokaryotes) that improved energy extraction beyond that obtainable by fermentation. These activities were fostered by acquisition of endosymbionts capable of photosynthesis and respiration. Eukaryotic cells are on average 1000-fold larger in volume than prokaryotic cells (Bonner 1974, p.79). Even a unicellular eukaryote carries more DNA and in general more structural and genetic information than does a prokaryote. A stunning array of morphological forms is apparent even among the protozoa and single-celled algae.

The eukaryote design must have been a good building block for multicellular architecture, for it seems more than coincidental that the cells of all the more complex organisms are eukaryotic. Apart from establishing a lineage (the unicellular protists) that could obtain food by engulfing it, the primitive eukaryote gave rise to: 1) multicellular, soft-celled, heterotrophic forms that digested food internally (primitive invertebrates); 2) multicellular, hard-walled forms that digested food externally (the fungi); and 3) multicellular forms that, by symbiosis with bacteria, came to manufacture their own food (green algae and the higher plants). (Prokaryotes approach this multicellular, differentiated condition only in the complex structures of some of the fruiting myxobacteria such as *Chondromyces*, Figure 4.4.) The way was now opened for cell differentiation (Figure 4.5) and the radiation of life forms (cf. Figure 4.2).

One of the earliest and most important developments was probably a boundary layer (the epithelium) separating the interior of an organism from the exterior. The culmination of cell specialization is demonstrated by the exquisite division of labor among the some 200 cell types of a vertebrate. The driving force behind this series of innovative events was that first the larger cell and then the multicellular macroorganism, by virtue of their size, were able to exploit some circumstances (environments, resources) better than could the microbe. This does not imply that there were pre-existing niches waiting to be filled. Rather, the changing environment presented certain opportunities as well as constraints (Chapter 7). The new forms, through growth and activity, shaped their surroundings and in turn evolved in ways that their progenitors could not. For instance, one consequence of size is speed of locomotion: the fastest eukaryotes move about 10 times faster than the swiftest prokaryotes (Bonner 1974, p.79; see comments in Section 4.6 about *relative* speed). This is of obvious value in catching prey or escaping from a predator.

Overall, there has been an increase in size during evolution in the sense that among (aclonal) plants and animals there has been an increase in upper size limits through geological time. However, several qualifications must be made. I am considering the grand picture of the maximum sizes of organisms at different geological periods. Larger species tend to appear later within a

Eukaryote cells

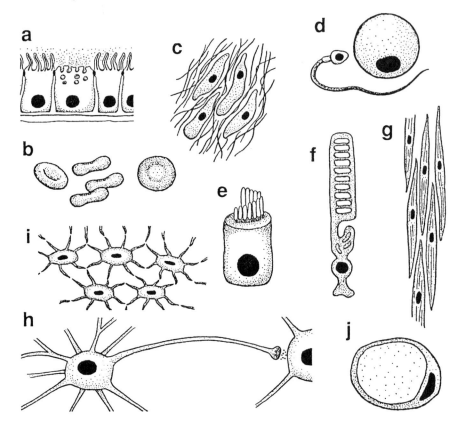

Prokaryote cells

Figure 4.5 Architecture of some eukaryotic and prokaryotic cells (not drawn to scale). (Top) A few of the more than 200 cell types in the human body: (a) ciliated and a secretory epithelial cell; (b) erythrocytes; (c) fibroblasts; (d) sperm and egg cells; (e) sensory hair cell of inner ear; (f) rod cell of retina; (g) smooth muscle cell; (h) nerve cell (neuron); (i) bone cells (osteoblasts) surrounded by calcium; (j) adipose cell. Redrawn from Alberts et al. (1989, pp.24-25). (Bottom) Basic cell types of the bacteria. From Brock and Madigan (1988, p.12). *Biology of Microorganisms,* 5th ed. © 1988. Reprinted by permission of Prentice Hall, Englewood Cliffs, NJ.

group's phylogeny. This does not mean that the paleontological record is without periods of size decrease (Kurten 1959), and it has been shown that the rate of change of size varies inversely with the duration over which it is studied (Gingerich 1983). That most animal groups evolve towards large size is expressed as Cope's Rule, in honor of the paleontologist Edward Drinker Cope who first described the phenomenon (Rensch 1960; Stanley 1973). A comparable phylogenetic increase in size has been demonstrated for various invertebrates including the corals, echinoderms, mollusks, and brachiopods (Newell 1949).

However, natural selection does not necessarily favor large size. Organisms such as the birds, amphibians, and rotifers are smaller than their ancestors, and at least some of the unicellular fungi are evidently derived secondarily from multicellular lines (Bonner 1968; Stanley 1973). The reptiles of today are clearly much smaller than their dinosaur relatives of another geological period; likewise, the modern club mosses (lycopods) are a few centimeters high compared with their ancestors, many meters in height, which proliferated in the luxuriant Carboniferous coal forests of 300 million years ago. Of course, the progenitors of both lineages *were* small. For many taxa, Stanley (1973) has shown that while the maximum size does increase with evolution, the median and minimum have not changed to any extent. For any population there will be an optimal size. Whether evolution moves the population upwards or downwards in size depends on whether the mean is above or below this optimum. The trend to larger size is evidently not because of any *intrinsic* advantage of being large per se, but rather because of the tendency for animal groups to originate at a small size relative to their optima (Stanley 1973). With increasing size comes increasing structural complexity and specialization for support and locomotion (Section 4.5). This constrains the larger forms from acting as evolutionary ancestors for new lineages.

The evolutionary sequence to increased size and complexity runs in tandem with phylogenetic increase in nuclear DNA content. For instance, the amount of haploid DNA in mammalian cells is, to an order of magnitude, about 10^3-fold greater than that of bacteria (Figure 4.6; see also Chapter 11 in Raff and Kaufman 1983; Watson et al. 1987, pp.621-622). These values are approximate and the correspondence between DNA content and complexity is not perfect—some amphibians have 25 times more DNA per cell than do mammals. Nevertheless, the pattern must be more than coincidental and suggests that, in general, gene amplification played a major role in phylogenetic change (Stebbins 1968). Within phyla, however, there is no correlation between DNA content and complexity. Stebbins attributes this apparent anomaly to the possibility that the origin of new phyla and the associated requirement for new cells and organs necessitates an increase in different enzyme systems. In contrast, evolution within a phylum may be more a matter of integration of function or alteration of conformation, processes which could be accomplished by mutation and recombination of existing genes.

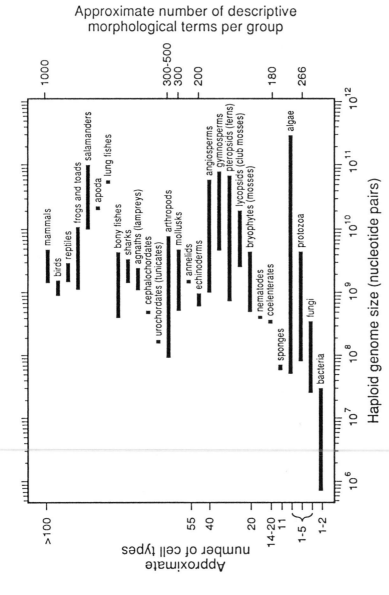

Figure 4.6 Complexity shows a very general increase with the amount of DNA in haploid genomes (C-value). There is considerable variation and the haploid DNA content of simple organisms can exceed that of more complex organisms. Complexity here is based primarily on number of cell types (left y-axis; cf. Figure 4.8) and secondarily by a useful though much cruder index, the approximate number of descriptive terms used by taxonomists to describe each group (right y-axis; cf. Schopf et al. 1975). That the latter index is an imperfect descriptor is indicated by the larger number of terms for the protozoa (266) than the sponges, coelenterates, and plants (180–200). Redrawn from Raff and Kaufman (1983, p.314).

4.4   On Seeing the World As an Elephant or a Mycoplasma

Lower limits and upper limits   Physical and chemical laws set the lower and upper constraints on life. The unicellular prokaryote (or the individual eukaryotic cell) has to be large enough to accommodate its genetic and metabolic equipment: Bacterial DNA molecules in aggregate can be on the order of 1,000-times the length of the bacterium (Cairns 1963) and accordingly must exist in a compact form within the cell. Polymerases need to have access to the genome. Ribosomes must have space to produce proteins. At the same time, the cellular volume (varying as the cube of the radius) cannot exceed a critical maximum if the surface area (varying as the square of the radius) available for exchange of nutrients and waste products is not to be exceeded. It has long been speculated (reviewed by Gould 1966) that enlargement of the protists is constrained by increasing demands placed by size on surface-dependent functions, and that progressively decreasing surface:volume ratios may initiate cell division. Thus, almost all cells have volumes between 1 and 1,000 μm^3. Apparent exceptions can be explained by specialized function, such as the large water-containing vacuoles of many plant cells, or food storage in the case of the ostrich egg, which reaches 10^{15} μm^3. In general, cell size has had little if any direct effect on organism size or complexity.

The upper limit on size of multicellular organisms seems less rigorously demarcated than the lower limit. While there are advantages to larger size (discussed later), mechanical problems ensue if an animal becomes too large. Huxley (1958, p.24) has said, "It is impossible to construct an efficient terrestrial animal much larger than an elephant". However, the matter really hinges on growth *form*, that is, whether the organism is of unitary or modular design (Chapter 5). Modular life forms have, by iteration, the capacity to add components indefinitely. Organizing a soma in this fashion allows an organism to increase biomass without transgressing critical morphological limits (Hughes and Cancino 1985; see also Chapter 5).

The organism as geometrician   Before going on to contrast specific organisms, I pause to consider the role of geometry and certain simple mathematical relationships in such comparisons. *Geometrically similar* bodies, that is, those in which all corresponding linear dimensions are related in the same constant proportions, are said to be *isometric* (Chapter 2 in Schmidt-Nielsen 1984). For instance, the surface areas of two cubes vary as the square of their lengths and the volumes as the cube of their lengths. Surface areas are related to the 2/3 power of volume. In other words, in isometric forms, surface area relative to volume decreases with increasing size.

However, as developed in later examples, macroorganisms are rarely isometric, though they may be superficially similar in appearance. The term *allometric or nonisometric scaling* denotes the regular changes in certain proportions (Chapter 2 in Schmidt-Nielsen 1984). In general, body size relationships take the following form:

$$Y = aW^b$$

where Y is the characteristic to be predicted, W is the body mass, and '*a*' and '*b*' are constants, derived empirically. This equation describes a straight line with a slope of '*b*'. Note, then, that '*b*' defines the general nature of the relationship as follows (Lindstedt and Swain 1988): *When '*b*' is near 1.0, Y varies as a fixed per cent of body mass, that is isometrically.* Examples include lung, gut, and blood volume, all of which increase in direct proportion to increase in body size. *When '*b*'>1.0,* the increase in Y becomes relatively greater as size increases. Thus, bones, tree trunks, and shells (discussed below) get bigger with larger organisms: The skeleton in a shrew is about 4% of the animal's body mass but accounts for 25% of the body mass of an elephant (Lindstedt and Swain 1988). *When '*b*' is between 0 and 1,* Y increases fractionally for each unit of increase in body mass. Hence, small animals have a higher weight-specific metabolism than do larger animals (Section 4.6). *When '*b*' is negative,* Y is maximized in small organisms. For various biological rates (heart, respiration) the exponent is close to $-1/4$. Translated into biological reality, this means that while the heart of a shrew can exceed 1,000 beats per minute, that of an elephant beats at only about 25-30 times per minute.

The greatest constraints on increasing size are gravity and decreasing area-to-volume ratios (Thompson 1961; Gould 1966). While the oceans provide buoyancy for whales, gravity conditions the forms and actions of all terrestrial macroorganisms. The giant sequoia tree has hard cell walls and massive arrays of supporting fibers and hence does not collapse under its own weight. Almost all animals have some sort of skeleton (hydrostatic, earthworms; exoskeleton, arthropods; endoskeleton, vertebrates). Bones in animals and stems in plants must become disproportionately thick with increasing size of the species if additional mass is to be accommodated. The wings of birds are designed to provide lift and to be strong yet light in weight. Because weight varies as volume while strength of a device to support it varies as cross-sectional area, engineering considerations mean that to remain strong the arthropod exoskeleton must be thicker in larger animals. But beyond a certain point the shell would be cumbersome and would decrease survival. It is probably because of this structural constraint that insects were not able to evolve into forms able to occupy the niche now filled by passerine birds (Stebbins 1968).

Analogously, macroorganisms similar in appearance undergo a change in shape with increasing size because efficiency in many biological (exchange) processes requires *proportionate*, rather than ever diminishing changes in surface areas with increasing volume which an isometric relationship would dictate, as seen above. However, elastic properties, (resistance to loading; see below) remain similar. So selection on size typically also affects shape (Bonner and Horn 1982).

Another size-related consideration for macroorganisms is heat loss in the homeothermic animals. A human, for example, consumes about 1/50 of its weight per day, a mouse ½ its weight. The design of a warm-blooded animal much smaller than a mouse becomes impossible because such an organism could not maintain a constant body temperature (Chapter 2 in Thompson 1961). Peters (1983, p.33) calculated that small mammals degrade about 10 times as much chemical energy per unit time as would an equivalent mass of large mammals. Haldane (1956) observed that there are no reptiles or amphibians in arctic regions, and that the smallest mammal in Spitzbergen (island in the Artic Ocean) is the fox. In contrast, the constraint at the upper extreme pertains to heat dissipation (as evidenced by lethargy and various heat exchange devices or behaviors of various kinds to promote cooling) because the larger the animal the greater its heat production relative to heat loss. Heat production for the mouse and heat dissipation for the elephant are both a consequence of the fact that heat-generating capacity varies as the cube of the linear dimension while loss depends on surface area, which varies as the square.

A related area:volume problem concerns efficient exchange of metabolites such as nutrients and various gases and waste products. Haldane (1956, p.954) said that "comparative anatomy is largely the story of the struggle to increase surface in proportion to volume". How is this achieved? The three evolutionary 'solutions' (Gould 1966) have been: 1) a differential increase in surfaces among the more advanced animals (e.g., fish gills; lung alveoli; intestinal villi); 2) a change in shape by flattening or attenuation without structural elaboration (e.g., the tapeworms); and 3) the incorporation of inactive organic matter within the soma (e.g., jelly in the case of Coelenterates; wood fibers in plants). To his list of solutions could be added the efficient partitioning of surface area:volume for the capture of resources—organisms as diverse as the fungi, sponges, plants, and corals all show essentially this same feature (Chapter 5).

In contrast, microorganisms largely escape the problems of size and strength. Despite large differences in the size of fruiting bodies of the cellular slime molds, gravity has little effect on the relative diameter of their supporting stalks. Weight in this case has probably been negligible as an evolutionary consideration (Bonner 1982a), the stalks being geometrically similar, unlike the stalk of the relatively much larger mushroom discussed above or of a growing tree, which becomes disproportionately thicker as it elongates. Hence, Bonner and Horn (1982) have said that small organisms tend to have geometric similarity; large organisms have elastic similarity. The former organisms scale essentially as the function of one axis, that is, the diameter is directly proportional to the length ($d \propto l$); the latter vary as two axes such that $d \propto l^{3/2}$ (McMahon 1973; McMahon and Bonner 1983). How this 3/2 relationship arose during evolution of large organisms is explored elsewhere (Chapter 4 in Bonner 1988) and is one example of how selection for size in macroorganisms also affects shape (for general comments on engineering as it relates to organisms, see Wainwright et al. 1982).

By definition a microorganism is small and its environment will obviously be small in absolute terms relative to that of a macroorganism. This is clear for unicellular microbes such as bacteria, yeasts, the protists, and many small invertebrates. It is less straightforward for filamentous microbes, which technically are microscopic only by virtue of their narrow cross-section. A filamentous shape confers tremendous surface area, which is desirable whenever there is the need to increase the amount of environment an individual is in contact with for surface-related activities such as absorption or excretion; analogous examples include root hairs, capillaries, and the intestinal villi. Notice, however, that if the component strands of the mycelial network of a fungus become aggregated, the organism becomes macroscopic, as happens when fruiting structures (mushrooms), mycelial sheets, or root-like rhizomorphs develop. Mushrooms grow on a wet lawn within a few hours, as the fungus moves through size and related changes in shape and gravity effects very quickly. So, size changes can often occur abruptly. Nutrient signals are the trigger the cellular slime mold uses to change from a disaggregated state of solitary, grazing amoebae to the social organism comprised of about 100,000 cells.

Microbes are governed by the forces of diffusion, surface tension, viscosity, and Brownian motion (McMahon and Bonner 1983). Theirs is a world of molecular phenomena not noticed by macroorganisms any more than bacteria notice gravity. Even exclusively terrestrial microbes, including the protists and nematodes, are usually associated with liquid in some form. This may be mucilage or other secretions of their own making; soil capillary water; the interiors of plants and animals; or boundary layer films of various origins. Free water is required almost invariably for such activities as mobility, growth, and reproduction. Propagules are frequently released into or must escape through a liquid film. Surface tension can even be a major factor for the smaller macroorganisms. Haldane (1956) observed that whereas a human emerging from a bath carries only a thin film of water weighing approximately one pound, a wet mouse has to carry about its own weight of water and a wet fly is in very serious trouble, "an insect going for a drink is in as great a danger as a man leaning out over a precipice in search of food" (p.953).

With few exceptions, small organisms and the propagules of most organisms assume simple, often spherical or subspherical shapes. Such forms are exceedingly rare in the external morphology of macroorganisms. Allen (1977) argues that the critical size is 10 μm. Below this limit, organisms essentially become spheroids and overcome species-specific problems by internal means; above it form can be used as part of a survival strategy. Constancy of external body form masking the evolution of a diverse biochemistry has been called "The Volkswagen Syndrome" (e.g., Schopf 1983, p.361).

External simplicity was once thought to be governed purely by surface tension (Thompson 1961). It is in part, but more recent evidence emphasizes the role played by internal structural proteins located at the cell periphery (McMahon and Bonner 1983). Of course the question still remains as to why the sphere should be a good design. One reason is that a curved (oblate) shape means that the container, whether a cell wall or the steel skin of a storage tank (Figure 4.7), must withstand tensile but not bending stresses. Furthermore, because the covering is under the same tension per unit length throughout, there is no region more likely to rupture than any other (Chapter 3 in Thompson 1961; Chapter 6 in McMahon and Bonner 1983). Of all possible designs, a true sphere encloses the greatest volume with the least surface area. Hence, a spherical shell is robust, spacious, and economical. It seems a logical starting point for elaborations on the architecture of organisms, including the cylindrical form, also common among microbes.

Movement in a fluid   The Reynolds number (*Re*) is a dimensionless or relative velocity related to movement in a fluid, expressing the ratio of inertial forces (numerator) to viscous forces (denominator) as follows:

$$Re = \frac{\rho l v}{\mu}$$

where Re = Reynolds number
  ρ = fluid density, g/cm^3
  l = characteristic organism length, cm
  v = characteristic organism speed, cm/sec
  u = fluid viscosity, g/cm sec

McMahon and Bonner (1983, see their Chapters 5 and 6) discuss some interesting implications of Reynolds numbers, particularly as they apply to the propulsion of microorganisms. For the movement of small organisms (i.e., with short lengths and speed), viscous forces (density and frictional components) dominate, and Re is low. Conversely, for large organisms, inertial forces dominate and Re is high. The blue whale, because of its huge size and relatively fast speed, swims at a Re of about 10^8, the porpoise of 10^5, while that for a moving bacterium is 10^{-6}.

Inertial forces of water, used to advantage by fish for movement, are of little consequence at a low Re. A microscopic whale would get nowhere by its mode of propulsion. Conversely, a bacterium or protozoan cannot swim like a whale. Their motion is dominated by the viscosity component, which also means that, unlike fish, they have essentially no glide distance once their propulsive engines have stopped. Furthermore, at a low Re, swimming is

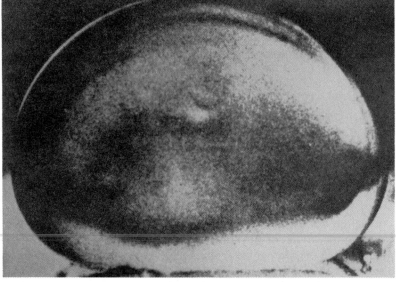

Figure 4.7 Oblate shape of storage tanks (top) confers stability when they are filled with a heavy liquid. The design is similar to the form assumed by a liquid drop (bottom) on a solid surface due to surface tension and gravity. From McMahan and Bonner (1983, p.215).

completely reversible. In theory, if a microorganism moved ahead and then back by exactly the same number of propulsive movements, not only would it return to the identical spot but all the displaced water molecules would also return to their original places. Finally and most fascinating of all are the implications for the design of the propulsive equipment (McMahon and Bonner 1983). The problem, because of the reversibility of movement at low *Re*, is that if microbes had to move by solid 'oars' they would go nowhere (forward on the thrust, followed by an equal distance back on the return stroke). The 'solution' is provided by flexible oars in the form of flagella or closely related but structurally distinct projections called cilia. The cilium, for example, is held straight out on the power stroke but collapses parallel to the body for the return, which reduces drag. (It is noteworthy as an aside that although ciliated organisms range in size over two orders of magnitude, in general cilia length and frequency of beating remain approximately constant. This means that most ciliated organisms swim at about the same speed of about 1 mm/ sec or approximately 10-fold that of a flagellated bacterium.)

To summarize, physical constraints imposed on an organism vary with its size. The elephant with a huge bulk supported by a massive skeletal system experiences a world dominated by gravity and thermodynamics (Section 4.6, later). The tiny mycoplasma, without even a cell wall for support, knows nothing about gravity or homeostasis, and exists in a world governed by fluid dynamics and diffusion phenomena. Such physical limitations are probably as important in evolution as they are in civil engineering (Lindstedt and Swain (1988). These constraints influence shape—shapes possible for one organism are not options for another—means of locomotion, speed, and many other related features which are also discussed further in Section 4.5. There is probably an optimal size associated with each type of activity (Haldane 1956; Pirie 1973).

4.5   Some Correlates of Size

Complexity   To function, a large organization depends on a division of labor among its component parts. This is intuitively true and is evident whether we compare a model airplane with the space shuttle, a village council with a national parliament, a water flea with a rhinoceros, or a prokaryote with a eukaryote. Bonner (1988) discusses the association between size and biological complexity and only a few of his many interesting points can be noted below. An increase in size has of necessity been matched by an increase in complexity. This is because of demands for efficiency (locomotion, metabolism, circulation, support) and by allometric considerations dictated by the changing relationship between linear dimensions and volume, discussed in Section 4.4.

Complexity of organisms is defined in terms of numbers of interconnected parts, in practice best considered as number of cell types (which usually have different, specialized functions and hence a division of labor is established; Chapter 5 in Bonner 1988). There is a direct correlation between number of cell types and size (Figure 4.8), although for either a given size or number

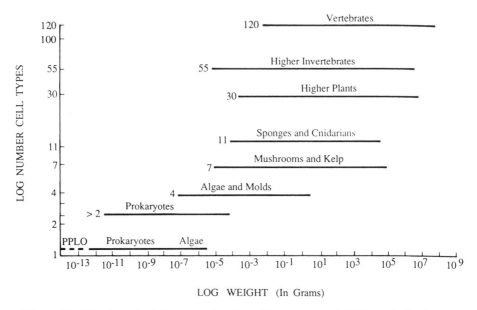

Figure 4.8   Number of cell types as a function of increasing weight. Complexity (number of cell *types*) generally increases with size (weight or approximate number of cells) of micro- and macroorganisms. PPLO, pleuropneumonia-like organisms, are now more commonly called mycoplasmas. "Algae" (at the bottom) designates the small cyanobacteria and the green alga *Protococcus*. The prokaryotes appearing as more than two cell types (second line) include the large cyanobacteria and sporeforming bacteria. From Bonner (1988, p.123).

of cell types there is considerable variation in the other parameter. Natural selection can act on size, shape, or complexity independently, yet a change in one influences the others (Bonner 1988, p.226). Bonner notes that selection acting on complexity is probably more important than that acting on size because increase in number of cell types opens the way for a large increase in size. In contrast, while size alone is somewhat plastic, its upward movement in the absence of an associated increase in complexity is limited by losses in efficiency.

One of the implications of the size-complexity issue is that a large organism is locked into an intricate developmental pattern, each step of which is predetermined by what has gone before (see comments, Section 4.2, on ontogenetic constraints). Microorganisms, being small and less complex in number of cell types (Figure 4.8), escape this complicated ontogeny.

A body plan built on specialized, interdependent subunits offers the benefits of high efficiency at particular tasks (e.g., sight, taste, translocation, support, defense) at the cost of vulnerability, often death, if an intricate system fails. Failure of the system can occur, for instance, when tetanus exotoxin binds specifically to one of the lipids of human nerve synapses in the central nervous system; when T-4 lymphocyte helper cells are killed by the AIDS virus; or when propagules of the fungus causing Dutch elm disease block xylem vessels. Organs can fail also for hereditary and environmental reasons. By having fewer and less complicated parts, microbes lack the advantages but avoid the shortcomings of life based on a complex blueprint. The colonial form of existence for certain clonal organisms (Chapter 19 in Wilson 1975) is perhaps a compromise between the two alternatives but, judging from the relatively smaller number of species that have adopted it, seems neither as 'successful' as multicellularity with specialized function on the one hand nor unicellularity on the other.

Bulk, within limits, confers tolerance of environmental vicissitudes. Extremely large animals require correspondingly large quantities of food and water and hence are as vulnerable as entire organisms to fluctuations that smaller animals could tolerate. (This is true despite the fact that large organisms can survive longer on stored energy than can small organisms; see Section 4.6 and Peters 1983). But, whether the macroorganism is relatively large or small, its component cells are buffered, accomplishing in aggregate what an isolated cell could not. The cells within a plant or animal soma live in an inner world sheltered from external influences. In contrast, microbes, especially those that are unicellular, are vulnerable to environmental changes because of their direct exposure to the elements. The response of the individual is to adapt rapidly by phenotypic changes, as in adjusting to oscillations in the type and concentration of nutrients (Chapters 3 and 7).

Chronological versus physiological time   Increase in size can be viewed as an integral of steps through the life cycle; it follows that more steps (hence usually more absolute or chronological time) are required to produce a larger than a smaller organism (Yarwood 1956; Bonner 1965). This in turn means that the generation time (Figure 4.9), or the time to reach sexual maturity (more broadly, to produce offspring whether by clonal or aclonal means), is usually longer (as must be life span in general terms for larger organisms). Generation times range from an order of a few minutes for bacteria growing under favorable conditions, to a few hours for protozoa, a few days for the house fly, about 20 years for humans, and 60 years for the giant sequoia.

There is also another time scale, one which has a physiological basis and is size-dependent. Hill (1950) speculated that all physiological events in an organism might be set by a clock which ran according to body size. Thus, small animals have a fast pace of life but do not live long. On balance, both the shrew and the elephant experience about the same number of physiological events or actions per life span—insofar as life span can be estimated meaningfully (Chapter 6; see also Chapter 12 in Schmidt-Nielsen 1984). Almost all biological times in birds and mammals (e.g., muscle contraction, blood circulation, respiratory cycle, cardiac cycle) vary with the same body-mass

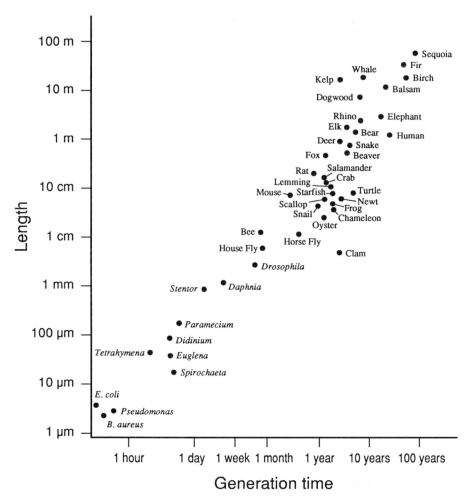

Figure 4.9 Lengths of micro- and macroorganisms versus their generation times. Redrawn from Bonner (1965, p.17).

exponent (mean = 0.24; Lindstedt and Calder 1981), regardless of the tissues, organs, or systems involved. The relationship also holds for a range of life history or ecological times, for example, life span, time to reproductive maturity, time for population doubling. It follows that the *ratios* of any two biological times (e.g., respiration cycle:heart cycle, or time to reproductive maturity:blood circulation time) versus body size plot as a horizontal line (Lindstedt and Swain 1988). In other words, just as the number of minutes per hour is constant in absolute time, so are biological activities per unit of physiological time: All mammals have 4–5 heartbeats *per breath*, and use about the same number of calories per unit weight *per lifetime*. Because physiological time accounts for body mass, hence how each organism sees the world, in a way that absolute time cannot, it may be the most appropriate criterion for interspecific comparisons (Lindstedt and Swain 1988; Lindstedt and Calder 1981). However, physiological time would appear to have little value in comparisons involving modular organisms (including microorganisms) because of their indeterminate, iterative mode of growth (Chapter 5), other than possibly at the level of the module (i.e., for the physiological rather than the genetic individual).

Density relationships   Typically there are more species of small than of large animals (May 1978, 1988). This generality is subject to several qualifications, among them that the evidence is for terrestrial, aclonal life forms; that many small organisms remain unsampled and unidentified; and that conventional taxonomic criteria for the smallest of organisms are highly controversial and may be without real biological meaning (Cowan 1962). Within the length range of 10 to 10^4 mm, the number of species (S) of terrestrial animals varies with length (L) by the relationship $S \propto L^{-2}$ or with weight (W) as $S \propto W^{-0.67}$ (Peters 1983, p.179). This means that for each 10-fold reduction in length, a 100-fold increase in the number of species would be expected (Figure 4.10). The relationship does not appear to hold for organisms whose body length is less than 5 to 10 mm. Bonner (see his Chapter 5, 1988) has extended this plot to all life forms, from bacteria to the largest macroorganisms (Figure 4.11). Why there is a reduced number of species at both ends is unknown, but probably relates in part to the caveats noted above pertaining to life form, sampling, and taxonomy. Analysis (May 1978; Dial and Marzluff 1988) of relationships within related assemblages (e.g., families and orders) suggests that it is not the smallest organisms that are the most species-diverse, but those which average 38% larger than the smallest. In other words, the small-to-medium sized taxa may be the most numerous.

Population density tends to vary inversely with organism size. Brock (1966, p.112) reports that a given amount of nutrient can support 10^9 small bacteria, 10^7 yeasts, 10^5 amoebae, or 10^3 paramecia, and that densities in nature follow similar trends. Among macroorganisms there is also good evidence that the population density of aclonal animals decreases with body size (Figure 4.12, and Chapter 10 in Peters 1983). So, if we look at different species of animals, there will be many more per unit area of the smaller than of the larger. Damuth (1981) has expressed the relationship as follows:

$$\text{Population density} \propto (\text{body mass})^{-0.75}$$

An analogous relationship seems to hold where interspecific competition has been studied in plants and is commonly expressed as the "$-3/2$ thinning

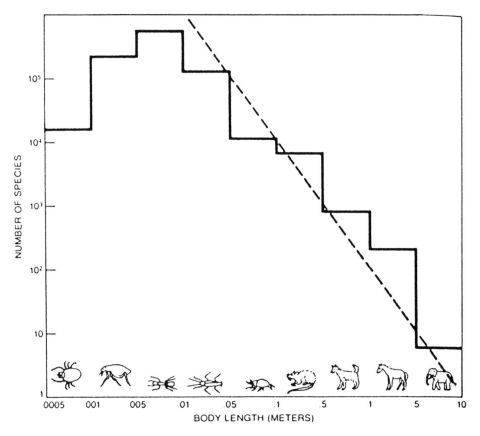

Figure 4.10 Numbers of species of terrestrial animals versus the typical body length of the component individuals (log/log scale). Dashed line shows general trend for organisms longer than 0.01 meters. From "The Evolution of Ecological Systems" by R. M. May. Scientific American 239(3) p.175. © 1978 by Scientific American Inc. All rights reserved.

law" (Yoda et al. 1963) because density (d) is related to mean weight per plant (w) by the equation

$$w = cd^{-3/2}$$

where c is a constant. This means that in a stand containing mature plants, the mass of tissue per unit area varies as the $-3/2$ power of the number of plants per unit area. Thus, plants sown (as a cohort of a single species) at different densities and harvested over time, approach, and then follow a self-thinning line with slope of $-3/2$. The slope reflects the relationship that in any particular population sown at a density sufficient to show self-thinning, if followed over adequate time, total biomass (yield) increases faster than

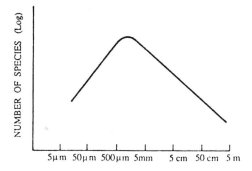

Figure 4.11   Postulated relationship between the number of species of all organisms as a function of the typical length of the constituent individuals (log/log scale). From Bonner (1988, p.106).

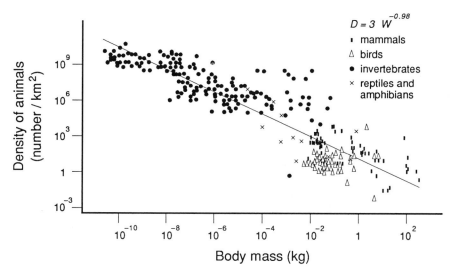

Figure 4.12   Relationship between animal body size and density. Redrawn from Peters (1983, p.169), "The Ecological Implications of Body Size", Cambridge University Press. © 1983.

density decreases (plotted as changing density of survivors versus total weight per plant over time). Why plants should thin according to this precise relationship remains unclear (e.g., see Givnish 1986a; see also criticism by Lonsdale 1990), as does the basis for species-size relations among animals, other than that both must ultimately reflect in some way energy relationships in fluctuating environments. With respect to plants, Hardwick (1987) argued that the relationship follows directly from a 'core-skin' hypothesis which postulates that plants function in effect as a thin, active skin (two-dimensional leaf plus phloem and stem cortical tissues) over an inactive core (three-dimensional xylem).

4.6 Some Ecological Consequences of Size

Benefits and costs   Because generation time increases with size, the individual may be killed before reaching sexual maturity and hence leave no descendants. To function the individual depends on the integrity of specialized, interrelated cell types, the failure of any one of which could have lethal consequences. Adaptation to change is slower in large organisms as is the ability to colonize new habitats. Peters (see his Chapter 8, 1983) examined this aspect by assuming that a catastrophe had depressed populations of micro- and macroorganisms to an arbitrarily low (1 g·km⁻²) density. He asked how long it would take for the biomass to return to 100 kg·km⁻², assuming each species increased at its maximum intrinsic growth rate (r_{max}). The time ranged from on the order of days for the bacteria (an overestimate of recovery potential in nature) to one century for the large vertebrates (Figure 4.13). Possibly the tendency for body size to increase during evolution is offset by

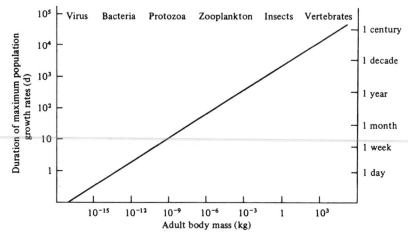

Figure 4.13 Influence of body size on *potential* duration of *maximum* (exponential) population growth rates (r_{max}) and thus on relative colonizing potential of micro- and macroorganisms. For bacteria this is on the order of days; for insects, it is on the order of years (right y-axis) required to reestablish an arbitrary population biomass of 100 kg·km⁻² (average density rounded to the nearest order of magnitude of many animal populations) from a negligible starting density of 1 g·km⁻². Redrawn from Peters (1983, p.138), "The Ecological Implications of Body Size", Cambridge University Press. © 1983.

higher extinction rates among larger organisms (Chapter 9 in Stanley 1979; for caveats and details see Pimm et al. 1988). Why a species goes extinct is a complex question. An answer involves the longevity of the organism, its generation time, population density, and capacity for geographic dispersal— all of which are size related—not to mention other factors such as competitive pressure, environmental change, and parasitism.

Worded differently, Peters' recovery time example means that although the microbes respond faster, the duration for which they are able to grow potentially at a maximum rate is orders of magnitude less than that for the largest vertebrates. This is because of the short generation times of microbes and hence the relatively early onset of crowding effects. The implications for temporal scale in biology are far-reaching, as Peters notes (see also Allen 1977). The apparent stability of the larger macroorganisms may be simply because they cannot fluctuate appreciably during the time course chosen for their observation. A wildlife ecologist studying moose, lions, or elephants may make observations at monthly or yearly intervals for decades. At the other extreme, observations on bacteria or phytoplankton kinetics typically occur over hours, possibly weeks, perhaps for a month or a growing season, depending on the experiment. Choice of the appropriate interval is critical, particularly when interpreting microbial succession on natural substrata (Swift 1982). Finally, concepts such as stability or community equilibrium may well be an illusion: they depend on the frame of reference, the scope of the community under scrutiny, the length of life of the organisms concerned, the frequency of the observations, and the duration of the study. It may be that our concept of time is only relevant when the parameters by which we measure it (sun/moon; night/day; seasons) are also responded to by the organism of interest.

Small organisms have an adaptive advantage in certain habitats. For example, size imposes constraints on access to a resource. The size of the opening to a small burrow or shelter may exclude a large predator. Sometimes organisms adapt to space restrictions by conformational changes. Vascular wilt fungi produce a small form of spore which facilitates movement within the conducting vessels of their living hosts. Also, because of space limitations, such spores are produced in situ by budding rather than on conidiophores (see the case study on the following page). Many systemic fungal pathogens of humans and animals are dimorphic (Romano 1966; Szanislo 1985), alternating between a single-celled, yeast-like, parasitic state and a saprophytic mycelial form.

Kubiena (1932, 1938) made fascinating observations pertaining to the size of fungi in soil. Some organisms were limited in occurrence to large spaces because they were too big to develop in the smaller pores. The fruiting bodies of *Cunninghamella* occurred only in spaces of diameter exceeding 600 μm. Conidiophores of *Botrytis* and sporangia of *Mucor* and *Rhizopus* were similarly restricted. The sporangiophores (spore-bearing stalks) of *Rhizopus*

A Case Study: When It Helps to Be Small

The plant pathogenic fungi which cause vascular wilt diseases (primarily *Fusarium, Verticillium,* and *Ceratocystis*) and certain wilt-inducing bacteria are unique because they remain within the xylem tissue (and specifically for the most part within the water-conducting elements or xylem vessels) until the late stages of disease development. Although details of life cycles and pathogenesis vary with each system, there is a common pattern. All these fungi must be able to grow within a relatively narrow, segmented pipeline with rough walls, interspersed with obstacles, through which is continuously flowing a dilute nutrient solution. In such an environment the production of large spores packed with nutrients would not be advantageous. Rather, the premium is on numbers of propagules. Success or failure hinges on the ability to colonize rapidly and thereby avoid being delimited by resistant responses of the host. How is this accomplished?

It seems more than coincidental that all wilt-inducing fungi are dimorphic (Puhalla and Bell 1981), that is, capable of growing either in single-cell, yeast-like fashion (as bud cells or microconidia) or as hyphae. The obvious adaptive value of the yeast phase in the xylem vessel is that fungal biomass is channelled directly into many small cells which move passively upward in the xylem fluid to systemically colonize the host. Numbers may be at least as high as 6,870 conidia per ml of xylem fluid (for *Verticillium albo-atrum* in the hop plant; Sewell and Wilson 1964). The xylem vessels throughout the length of the plant are open to water, but not to spores which become trapped by the reticulate walls or at the ends of the individual vessel elements. The common strategy of the wilt fungi is to release into the sap stream abundant spores which travel for a distance before becoming entrained; they then germinate quickly, the fungus grows through the obstruction, another round of yeast-like cells is released, and the sequence is repeated. Thus ensues a pitched battle between the host—which attempts to restrict the invasion by various mechanical and biochemical mechanisms, and the pathogen—which must advance rapidly if it is to escape the counterattack. It does so in successive waves of propagule release followed by growth through obstacles, much like an infantry battalion advances to secure successive beachheads. The difference between success for the invader (systemic colonization; ultimately more genes contributed to the gene pool) and failure (restricted colonization; pathogen delimited by host and ultimately displaced by more virulent biotype in the population) depends crucially on the *speed* of colonization of the vessel elements as mediated by the production of small propagules. For additional information, see Mace et al. (1981), Pegg (1985), and Beckman (1987).

nodosus, normally produced upright, were bent and occasionally spirally curled, molded to fit the space available. Where pores were sufficiently large, soil insects, mites, protozoa, and springtails were found, along with the above molds and a larger fungus, *Humicola*. Likewise, Swedmark (1964) found the smallest representatives of several phyla in the interstices of marine sand. The larger forms among them were frequently threadlike. Comparable size-distribution limits of the filaments of the gliding sulfur bacteria in marine sediment (Jorgensen 1977) and for the percolation of bacteria and fungal zoospores through soil (Wilkinson et al. 1981) have been reported.

In aquatic systems, smallness may also be advantageous for phytoplankton by promoting suspension in the photic zone, by increasing the process of light absorption itself and, because of the increasing surface area:volume ratio with decreasing size, by fostering nutrient uptake (Sournia 1982; Smith et al. 1985; reviewed by Fogg 1986 and Raven 1986b). Small size, irregular shape, and extensive vacuolation increase the area of plasmalemma per unit cytoplasm; these charges provide more sites for nutrient transport into the cell, and tend to reduce the sinking rate. Cells in the size range of picoplankton (broadly speaking, planktonic organisms between 0.2-2.0 μm; see Fogg for caveats and details) sink at rates that are so slow as to be immeasurable (Takahashi and Bienfang 1983). Phytoplankton sink faster when they are larger, but if turbulence is adequate, they will cycle rapidly through the whole water column, scrubbing nutrients as they go. Theoretical calculations show that while a spherical cell of 10 μm in diameter may sink at about 25 cm per day, the rate for a 1-μm cell is 2.5 mm per day (Fogg 1986). Further reduction in size below 1 μm appears to have no additional benefit on sinking rate or light absorption. Increase in nutrient flux to cells below this limit is evidently in excess of any requirement. This example also illustrates that surface area:volume relationships are especially important for essentially sessile organisms because it directly affects how they forage (Chapter 5).

Being small can also be hazardous. The same surface area:volume relationship that was beneficial for solute exchange in picoplankton could be disadvantageous, for instance in the aerial dissemination of microbes or the microscopic propagules of macroorganisms: Small structures not only contain less food reserves, but are relatively more exposed to desiccation. Savile (1971a), commenting on the larger spore size of plant rust fungi from the mesic areas of the USA and the Mediterranean, observed that increased size aids water economy. Analogously, among the Californian flora there is a positive correlation between seed weight and dryness of habitat (Baker 1972). Larger propagule size may not only reduce desiccation in transport but also provides for a relatively larger root system so that the young seedling obtains water quickly (see Section 4.7).

What are the advantages of large size? Some of these have been noted in passing. Among animals, locomotory stamina increases with size (Hill 1950). Within limits, speed increases with size for all the life forms (up to the largest representatives of each), regardless of the form of locomotion (McMahon and Bonner 1983; Peters 1983). Unlike the case in water, where the largest are the fastest, on land there is a trade-off between size and speed because of increasing weight with volume. The case for swimming is shown in Table 4.1. It is interesting to observe, however, that if swimming speed is expressed in body lengths per unit time (i.e., relative to the amount of new environment sampled by the organism) rather than in absolute terms, a bacterium explores at the same rate as, say, a dolphin. Large species consume more energy per unit time and distance, but *specific* costs (i.e., per unit weight) on either basis decrease.

Large size can deter predation. A mature, healthy elephant is too large for any predator; puffer fish deter attackers by inflating themselves with water and projecting their spines. Weapons of defense or attack (e.g., horns or antlers in animals; spines or thorns in plants) increase with increasing allometry. Being larger than one's rivals may offer a competitive advantage and can extend the size range of available food items. Size differences also mean that individuals of the same or related species can use the same type of food in the same habitat concurrently. Diamond (1973) observed eight species (two genera) of fruit pigeons in the New Guinea rain forest. The size differences among them were associated with the fruit sizes they could eat and the branch sizes on which they tended to perch. Up to four consecutive members in the species size sequence were found in any given tree. Trees with larger fruit supported on larger branches attracted larger pigeons. The smaller species ate the smaller fruit and concentrated on the smaller, peripheral branches. Birds commonly foraged out along a branch until it bent under their weight. This situation is generally interpreted as an example both of competition related to size and as a partitioning of food resources. However, the evidence is purely observational (see important caveats on such field observations by Begon et al. 1986, pp.270-273; see also Hairston 1990). The smaller birds

Table 4.1   Length and swimming speed of representative micro- and macroorganisms[1]

Species	Length	Swimming speed Rate (cm·sec⁻¹)	Body lengths per sec
Bacillus subtilis	2.5 μm	1.5×10^{-3}	6
Spirillum volutans	13.0 μm	1.1×10^{-2}	9
Paramecium sp.	220.0 μm	1.0×10^{-1}	5
Pleuronectes platessa (plaice;larval)	7.6 mm	6.4	8
Leuciscus leuciscus (European dace)	15.0 cm	175	12
Pygoscelis adeliae (Adelie penguin)	75.0 cm	380	5
Delphinus delphis (common dolphin)	2.2 m	1,030	5

[1]Modified after McMahon and Bonner (1983).

could not cope efficiently with the larger food and additionally were driven from the fruit by their larger competitors. While the larger species could eat the smaller fruit, they tended not to because the slender branches did not support their weight. So, there can be selection simultaneously for large and small variants of essentially the same taxon (another better known example is Darwin's finches, which vary in beak size, hence food items, among closely related species).

Observations on competition of *Hydrobia* mud snails in Denmark (Fenchel 1975; Fenchel and Kofoed 1976) are analogous to those of the fruit pigeons. *H. ulvae* and *H. ventrosa* can live apart or together. When apart, individuals of the two species are similar in size and consume similarly sized food particles. When their ranges overlap, *H. ulvae* is larger; it predominantly eats larger food than *H. ventrosa* and food over a size range expanded at the upper limits. This example is of interest to ecologists because it provides circumstantial evidence for character displacement (morphological variation in the presence of competition). It is noteworthy in our context because of an apparent advantage conferred by larger size, in this instance an extended feeding range.

Metabolic issues   Peters (1983, see especially his Chapter 3) reviewed the literature and discussed at length the metabolic consequences associated with a particular size. Only a few of his many interesting points can be summarized here. A note on methodological differences should be made first. For higher animals, the power required to just maintain life is estimated from some minimum, so-called *basal* metabolic rate; this figure is supplemented by estimates of additional power required for each type of activity. For other taxa, such as the microbes, the term *standard* metabolic rate replaces basal rate, implying that the data are obtained under standard but not necessarily minimal conditions.

Peters has produced useful comparisons for unicellular organisms, poikilotherms, and homeotherms. The first of these is that the relationship between metabolism and body mass is similar for the three groups. Consider again the general equation, $Y = aW^b$. As expected, metabolic rate (in watts) increases with size (mass in kg): within each class, large organisms respire at higher rates than do smaller organisms. Also, because the value of a above is highest for homeotherms, declining in turn for poikilotherms, and then for unicells, metabolic rate for a hypothetical 1-kg organism in each group declines comparably. As Peters remarked, one ecological implication is that the relative demands of the three types of organisms on their bodies and on the environment must decline in similar fashion. Consequently, homeotherms need high resource levels and have to be relatively efficient in resource utilization.

If power production is expressed instead in *specific* terms (watts·kg^{-1}), the rate of energy consumption decreases with body size. This means that within each of the three groups the maintenance cost for large organisms is less than that for the same amount of smaller organisms. Another way to look at the same issue is to ask what the maximum amount of biomass is that could be supported per unit of energy supply (kg·watt⁻¹). Thus, it turns out that the same amount of energy could support about 30 times more poikilotherms (and still more unicells) than homeotherms. Within each class, a greater biomass of the larger than the smaller organisms could be sustained per unit of energy flow.

Crude estimates can be made of turnover time, that is the time needed to metabolize an amount of energy equivalent to the energetic content of tissues (Peters 1983, pp.33-37). Again, among organisms hypothetically of equivalent weight, turnover times are shortest for homeotherms, and progressively longer for poikilotherms and unicells in which energy is mobilized more slowly. Within each metabolic group, energetic reserves of the larger organisms last longer than those of the smaller. Consequently, large forms can survive longer on their energy reserves and as such are less dependent on a reliable food supply. Smaller species or juvenile forms have been the first to disappear when food has been restricted in competition experiments under controlled conditions (Goulden and Hornig 1980).

Although changes in size from small to large or vice versa may carry adaptive benefit for the organism, it is worth noting in passing Gould's (1966) observation that this does not necessarily apply to specific *structures* which must be above a certain minimum size to function at all. There are examples both among the micro- and macroorganisms. To insure effective spore dispersal, toadstools must be sufficiently high above the ground. The stalks (stipes) of the larger forms need not exceed this length and hence on a biomass basis are proportionately smaller than those of the small species (Ingold 1946). Among animals, the size of rods and cones of the eye does not vary with organism size, but is evidently set by optical properties (Haldane 1956; Thompson 1961, pp.34-35).

On balance it seems that for every proposal in favor of large size, a counter proposal based on its disadvantages can be put forward. Summarizing the situation, Peters (1983, p.193; see also Barbault 1988) said "arguments for the adaptive advantage of large size are [thus] distinctly flat". Obviously both large and small species are 'successful' in that both remain extant. Which size is 'best' has no real meaning. Nevertheless, the process of examining specific constraints and opportunities presented by any particular size is productive in that it helps to clarify how organisms live from the perspective of those organisms, rather than one imposed by human perception. Allen (1977) has made this point more imaginatively: "Much as Victorian children were viewed merely as young adults, microscopic algae may be mistaken for plants that happen to be small".

4.7   Size and Life History Theory

Size of dispersal units   Salisbury (1942) developed a theory about seed size in flowering plants which was extended by Garrett (1973) to plant pathogenic fungi. The idea in Garrett's words is that "the average size of a reproductive propagule is determined by the nutritional needs for establishment of a new young individual of the species in its typical habitat". Salisbury's thesis was that seed size is determined by the length of time during which a seedling must be self-supporting before it can supply its own needs by photosynthesis (for alternative hypotheses on seed size see Thompson and Rabinowitz 1989). Salisbury's generalization was based on data on seed and fruit production by 240 species of British flowering plants. He found that the spectrum in seed size could be related to habitat type: species with the smallest seeds were characteristic of open habitats; at the other extreme were the heaviest seeds from shade-adapted woodland flora. By evolutionary adjustment of seed size to habitat type, any plant species could efficiently allocate resources, providing the appropriate level of reserve in each instance.

Garrett (1973) followed this up with examples from *Botrytis* and *Fusarium* illustrating the principle that the level of endogenous reserves in fungal spores appears to be adjusted to take advantage of supplementary exogenous nutrients supplied in plant exudates. *B. fabae* produces spores about nine times the size (and with proportionately more nutrient reserves) than those of *B. cinerea*. The former is able to overcome the resistance of healthy, vigorous leaves. In contrast, *B. cinerea* cannot infect healthy leaves, except under extreme conditions of inoculum pressure or in the presence of additional nutrients (e.g., as may be leached from pollen deposited on leaf surfaces). This fungus is well known to be a weak or 'wound' parasite. Its typical infection courts are wounded, senescent, or chlorotic leaves, floral organs, overripe fruit—all of which are predisposed to infection by virtue of nutrient leakage or reduced resistance or both. So Haldane's (1956) maxim "on being the right size" for *Botrytis* means one of two strategies: The organism could produce numerous small propagules of limited infectivity and rely on nutrients from a debilitated host to compensate for their small nutrient reserve (analogous to smaller but more abundant seeds characteristic of plants from open habitats). Alternatively, fewer but large, well-supplied spores could be produced, capable of overcoming highly resistant host organs. The benefit would be a propagule of higher infectivity, less subject to certain external conditions; the cost is reduced output. Both strategies are evidently successful, but it is interesting that *B. cinerea* is the more abundant and widespread of the two species (Garrett 1973).

Salisbury (1942) extended his seed size hypothesis to encompass vegetative propagules such as rhizomes and stolons, insofar as these were another means of nutritional support from the parent plant. Garrett (1973) recognized the striking visual and functional parallel presented by certain structures of root-infecting fungi such as species of *Fomes, Armillaria,* and *Phymatotrichum.* Mycelial strands and rhizomorphs differ in detail but all are basically subterranean, macroscopic (often several mm in diameter), multistranded 'telephone cables' of hyphae that may extend for dozens of meters. In *Armillaria,* clones extending at least 450 meters have been documented (Anderson et al. 1979), and the maximum size of a single clonal population remains unknown. All such strands or rhizomorphs function in translocation of nutrients from a food base to the growing apex, ultimately to the point of infection. Why does an essentially microscopic organism consisting of fine mycelial threads suddenly 'choose' to expend resources to produce such a massive, elaborate structure? Unlike the roots of herbaceous plants, which are relatively vulnerable to infection by single spores, those of undamaged, woody hosts resist invasion. By aggregating multiple hyphae in a rhizomorph, the pathogen can breach the defenses of a mature tree that would be impenetrable by a single hypha.

Harper (1977, pp.672-673) makes a most interesting extension of Salisbury's story as it relates to the very small seeds of certain symbiotic plants, including the orchids, the saprophytic *Monotropa,* and the parasite *Orobanche.* Unlike some of the larger-seeded symbionts (mistletoes and dodder which must penetrate bark or grow if they are to colonize a host), these small-seeded plants are assured of an external food immediately upon germination. In evolutionary terms, this removed the need for a food reserve, and natural selection acted to increase seed number rather than size. The species have acted opportunistically, seizing on "an alternative mode of embryo nutrition to reduce seed size to tiny dried bags of DNA and expand their reproductive capacity to a new limit" (p.673). This is in contrast to the conventional interpretation that they "need" to produce more seeds to "find" a host—an explanation which, Harper points out, is wrong.

To what extent can the foregoing adaptive arguments on propagule size be generalized? Overall, both adaptation and constraint appear to play a role. Among animals, egg weight may not be correlated with offspring-fitness parameters (Wiklund et al. 1987) and may be due purely to allometric scaling factors. Among bird species, egg size varies directly, though not proportionately (exponent is about 3/4), with body size (Chapter 3 in Calder 1984), suggesting a major influence of allometry. There are some exceptions, however, the most spectacular of which is the kiwi. This chicken-sized bird produces eggs about five-fold larger than those a chicken would lay. Evidently this is because the absence of predators in New Zealand removed selection pressure for smaller eggs and hence shorter incubation periods (Calder 1984, pp.335-338).

Microorganisms are not necessarily *r*-strategists   Salisbury's and Garrett's hypotheses prompt the more general question of how size may influence survival strategies, including reproductive tactics. Do microorganisms behave fundamentally differently from macroorganisms? Are the life history patterns displayed by organisms of different sizes a consequence of selection that acts primarily on *size* or is the primary selection for a particular *reproductive rate* which in turn affects size? The original and most general concept that deals with this issue is r- and K-selection (MacArthur and Wilson 1967; reviewed by Horn 1978; Boyce 1984; Andrews and Harris 1986).

Consideration of r- and K-selection requires a brief introduction to population growth which is taken up further in Chapter 5. The basic idea is that the per capita or specific rate of population increase for *any* organism, $(1/X)$ dX/dt, can be expressed as the difference between the per capita birth rate, b, and the per capita death rate, d:

$$\frac{1}{X}\frac{dX}{dt} = b-d$$

The terms r and K are derived from the logistic equations for population growth, which are merely an expansion of the above relationship into the various components for birth and death. These equations specify that under uncrowded conditions the per capita rate of increase, r, is maximized and with increased crowding (see below) a decline in the per capita rate of increase occurs until the population density equilibrates at its upper asymptote or carrying capacity, K. The simplest mathematical expression for a change from an r- to a K-condition is a linear decline (Figure 4.14) in the specific rate of change with increasing population density (from a maximum rate, r, at zero density) to zero (at a maximum population density, K) (MacArthur 1972, pp.226-230). This becomes the differential form of the Verhulst-Pearl or logistic equation when expressed as:

$$\frac{1}{X}\frac{dX}{dt} = y = r - \frac{r}{K}X$$

where

$\dfrac{1}{X}\dfrac{dX}{dt} = y =$ specific rate of population change (unspecified conditions)

$\quad r =$ per capita rate of increase (uncrowded conditions)

$\quad K =$ carrying capacity of the environment

$\quad X =$ population density (specified as N for numbers or M for biomass)

From the equation and Figure 4.14 it can be seen that at low X, y will be dominated by r; as X approaches K, y will be dominated by K.

The logistic expression thus describes population growth in a limited environment. Such growth curves are well known and appear in some form in every introductory ecology and microbiology book as linear (log-transformed) or sigmoid plots of change in population density versus time. The simplifying assumptions of the logistic expression have long been acknowl-

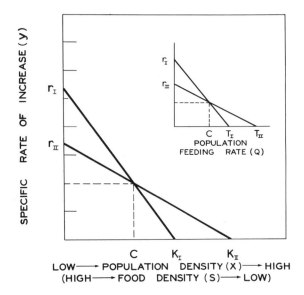

Figure 4.14 Comparative specific rates of increase for two microorganisms or two macroorganisms (I and II) versus population density or population feeding rate (inset). At low population densities organism I outcompetes II because r_I exceeds r_{II}. At high densities organism II outcompetes I because K_{II} exceeds K_I. Organism I will prevail when the environment fluctuates enough to keep populations growing at a density less than the competitive cross-over point C (r-selection); organism II will win in environments stable enough to permit populations to remain at population densities greater than C (K-selection). The inset shows the analogous population feeding rate (Smith equation); T is the rate of food consumption at equilibrium. From Andrews and Harris (1986; modified after MacArthur 1972, and Pielou 1969, respectively).

edged. However, the point is rarely made that the equation is descriptive, not explanatory. Its insurmountable defect is that the *mechanistic* basis of the decline in the specific rate of increase with increasing population density is not explicitly recognized (Andrews and Harris 1986). I shall discuss briefly some implications for macro- and microbial ecology. In particular, the implications of 'crowding' should be fully understood because, especially in macroecology, they are often interpreted narrowly to involve only competition for resources. Crowding implies all factors that change in density-dependent fashion, including parasitism, predation, and—especially in microbiology—production of toxic metabolites.

It is apparent, then, that the terms r and K as used now describe in a comparative sense either *conditions* (which in aggregate constitute particular environments or selection regimens) or, less formally, *species*. Uncrowded environments (r-selection) are usually unpredictable and transitory. Colonists that have a high r (in microbial semantics, a high μ_{max}) should do best because this characteristic favors discovery of the habitat, reproduction, and dissemination before the habitat either disappears or becomes crowded. Conversely, organisms that can compete well in a habitat that is either stable or predictably seasonal should be favored under crowded conditions (*K*-selection) where there is typically a shortage of resources (Figure 4.14). Although both *r* and *K* are subject to evolutionary adjustment, the ecological dogma is that a high r occurs at the expense of a low K. In other words, an individual cannot maximize both parameters (Wilson and Bossert 1971, pp. 111-112; Roughgarden 1971).

The central question is what life history traits are associated with high *r*- or *K*-selecting environments. *r*-selected individuals are predicted to be smaller in size, to mature earlier, and to have more and smaller progeny. In contrast, *K*-selected individuals should be larger, show delayed reproduction, allocate resources more to growth (size) for competitive advantage, and hence have fewer but larger offspring. Pianka (1970) recognized several correlates (Table 4.2; however, see criticism by Boyce 1984). On balance it is clear from the table that small organisms, microbes in particular, would be shifted toward

Table 4.2   Some correlates of *r*- and *K*-selection[1]

Attribute	*r*-Selection	*K*-Selection
Climate	Variable and/or unpredictable: uncertain	Fairly constant and/or predictable: more certain
Mortality	Often catastrophic; nondirected; density-independent	More directed; density-dependent
Population	Variable, nonequilibrium, usually well below carrying capacity; unsaturated communities or portions thereof; recolonization annually	Fairly constant, equilibrium; at or near carrying capacity; saturated communities; no recolonization
Competition	Variable, often lax	Usually keen
Selection favors	Rapid development; high r_{max}; early reproduction; small body size	Slower development; greater competitive ability; lower resource thresholds; delayed reproduction; larger bodies
Length of life	Short; usually less than 1 year	Longer; usually more than 1 year
Leads to	Productivity	Efficiency

[1]Abbreviated after Pianka (1970). For caveats and criticism, see Boyce (1984).

the *r*-end of an *r-K* spectrum. It is important to recognize, however, that microorganisms themselves can be relatively *r*- or *K*-selected (Table 4.3; see also Swift 1976; Andrews and Rouse 1982; Andrews and Harris 1986; Andrews 1991).

While *r*- and *K*-comparisons across widely different taxa can be informative in a general sense, several limitations need to be kept in mind. First, the concept of reproductive value (Chapter 6 and Horn 1978), familiar to macroecologists in life history models, is foreign to microbiologists. This is a measure of contemporary reproductive output and residual reproductive value; for macroorganisms it entails life table statistics based on the likelihood of survival and reproduction for specific age classes (juveniles versus adults). A particular life history exhibited by an organism should be one that has the greatest overall reproductive value. Although the *r/K* model assumes that environmental fluctuations affect all age classes equally, competing models ('bet-hedging'; Schaffer 1974) do not. The issue is moot in microbial ecology because of the short generation times and the plasticity and totipotency of

Table 4.3 Some life history features of *r*- or *K*-selected microorganisms[1]

	Organism	
Trait	*r*-strategist	*K*-strategist
Longevity of growth phase	Short	Long
Rate of growth under uncrowded conditions	High	Low
Relative food allocation during transition from uncrowded to crowded conditions	Shift from growth and maintenance to reproduction (spores)	Growth and maintenance
Population density dynamics under crowded conditions	High population density of resting biomass; high initial density compensates for death loss	High equilibrium population density of highly competitive, efficient, growing biomass; growth replacement compensates for death loss
Response to enrichment	Fast growth after variable lag	Slow growth after variable lag
Mortality	Often catastrophic; density-independent	Variable
Migratory tendency	High	Variable

[1]From Andrews and Harris (1986).

microorganisms—which preclude any analog to adolescence or pre- and pos-treproductive classes used in plant and animal biology.

Second is the distinction between modular and unitary organisms taken up in Chapter 5. By growing in modular fashion, the genetic individual or genet can increase indefinitely in size by adding modules (as in flowers, branches, or leaves of a tree; hyphal tips of a fungus). Population growth for these organisms may thus be exponential and, unlike the case for unitary organisms, is not necessarily delayed by postponing reproduction. The predictions from r/K theory for modular versus unitary organisms may be quite different (Sackville Hamilton et al. 1987). These are taken up in Chapter 5.

Third, comparisons are best made among organisms on the same trophic level because whether resources or some other mechanism acts to limit organisms may depend on their position in the trophic web (Wilbur et al. 1974).

Finally, when organisms are compared, body size should be considered to determine whether natural selection influences reproductive output independently of size. In other words, does selection act primarily on size (Stearns 1983), with a corresponding intrinsic rate of increase (r_{max}) following as a *consequence*? For engineering reasons noted above related to complexity, it takes less time to construct a small than a large organism. Small organisms mature and breed earlier than large organisms, hence the correlation between smaller body size and a higher r_{max}. Or, alternatively, is selection mainly for a particular r_{max}, which then *dictates* a given size because of the relationship between the two parameters (Ross 1988)? Ross examined reproductive patterns in 58 primate species and found generally that after size differences are factored out, species inhabiting unpredictable, r- environments have a high r_{max}. Selection has evidently acted directly to increase r of these species rather than indirectly by decreasing body weight. There was no evidence, however, that either the raw or relative r_{max} values were as predicted for species in predictable environments. This anomaly remains to be reconciled but may by explained by imprecise classification schemes for the respective environments (Ross 1988). Such formal comparisons have not been made among the microbes. However, because microorganisms of similar size behave very differently in different environments (Andrews and Harris 1986) and have distinctive intrinsic growth rates (Brock 1966, p.95), it seems that selection has been primarily for an r- or K-type strategy rather than indirectly as an unavoidable consequence of size.

That the logistic equation underpinning r- and K-selection has no mechanistic basis has undoubtedly contributed to current disenchantment with the theory by many ecologists. The complexity of most plant and animal systems means that it is difficult if not impossible in population dynamics to delve deeper than gross phenotypic expressions (birth, death, growth). Microbial systems, particularly bacteria, can provide for a causative interpretation of the effect of crowding. Experiments can be conducted under controlled as well as uncontrolled conditions with organisms that are relatively well characterized in terms of their genetics, growth rates and efficiencies, nutritional requirements, and metabolic pathways. This is illustrated briefly as follows.

For microbes, the decline in specific rate of increase usually reflects food limitation, toxic stress (for equations see Pirt, 1975), or predation/parasitism (for equations see Williams, 1980). Of these, crowding with respect to food is probably the key limiting factor in most ecosystems, and will be used here to show how microbial systems can allow identification and quantification of a representative parameter of interest. Mechanistic equations for growth phase bacterial metabolism and the transition from growth phase to vegetative resting phase metabolism exist (Pirt 1975). These can provide an interpretative basis for the competitive outcome depicted in Figure 4.14 by illustrating the underlying relationships such as that between the specific rates of food consumption (acquisition) or specific rates of growth (individuals or biomass production) versus food density for the competing organisms. Components of these and similar equations constitute refined correlates for r- or K-selected organisms (Table 4.3). These include refinements pertaining to the productivity (unit output of numbers or biomass per unit time) versus efficiency (unit output per unit of energy input) trade-off (Tables 4.2 and 4.3, and Pianka 1970). Experimentally, techniques are available in principle for measuring such components, although practical limitations such as the detection of limiting nutrient concentrations, or discrimination among growing, metabolically active, resting, and dormant biomass exist (Andrews and Harris 1986).

In overview, the r/K theory is an excellent example of an ecological concept with universal application. The theory addresses two population parameters common to all living things; hence there is no reason why analogies should be constrained, subject to the caveats noted above. It is simplistic and misleading to argue that macro- and microorganisms are K- and r-strategists, respectively; a continuum can be discerned within both groups. Modifications of the r/K theme have been developed (reviewed in Southwood 1988; Andrews 1991) to accommodate various aspects of the environment, such as 'harshness', and some of the extensions focus on particular groups of organisms such as plants or fungi. While many organismal traits may not be directly r- or K-related, it is still informative to consider all environments and organisms from an r/K perspective, especially when a mechanistic interpretation can be given.

To what extent can comparisons and analogies be made on the basis of size? These fall into at least three categories. First and possibly the least relevant in our context, organisms within the same genus or species can be compared. Studies such as those described earlier of the pigeon species in New Guinea or the snails in Scandinavia provide evidence on the role of size in niche shifts and competitive displacement. Even at the population level the influence of size is apparent in larger males having larger harems, and in fecundity increasing with size of the female (more eggs, animals; more seeds, plants). Size differences within a species also result from sexual selection: males can be very large (Alaskan fur seal) or very small (some spiders) relative to the females. So, fairly substantial size differences can occur even at the species level. Second, microorganisms, individually or as a group, can be compared with macroorganisms. While it is clear from the above arguments that the two classes of organism must see the world differently due to size differences *alone*, there is little more that can be said because of the confounding correlates of size, including generation time, energy relationships, complexity, shape, and longevity. These complications include but transcend the standard allometric relationships governing organisms of similar shape. Hence, microbes and macroorganisms must be compared as the sum of their traits. Third, general statements about size can be made independently of taxon. These include the role of physiological time in species comparisons and adaptation versus constraint in shaping the size of propagules.

4.8 Summary

Change in shape associated with increase or decrease in size is termed allometry. The consequences to the organism include changes in chemistry, physiology, and morphology, and are evident both across taxa and during the development of any individual. Comparisons of micro- with macroorganisms must allow not only for these conventional allometric changes, but recognize also the gross differences in shape of representatives of the two groups. Both factors profoundly affect the environments sampled by the respective types of organisms. Allometric considerations are one of three interrelated types of constraints on the phenotype of organisms. The others are phylogenetic (taxon-related) and ontogenetic (development-related).

The bacteria were the first living forms and for at least 2 billion years, or half the age of the world, were the only living forms. Two major events underpinned the evolution of large organisms. The first, oxygen-generating photosynthesis, enabled more energy to be extracted from foodstuffs by the use of oxygen as an electron acceptor (aerobic respiratory metabolism instead of anaerobic fermentation). The second was emergence of the eukaryotic cell. The unicellular eukaryotic microorganisms probably gave rise independently to several multicellular lines culminating in life forms classified traditionally as members of the fungal, plant, and animal kingdoms. Recent molecular evidence confirms that all biota appear to be 'rooted' to a common ancestral state, and suggests that organisms fall within three major domains which supercede the kingdom taxon: Bacteria (eubacteria), Archaea (archaebacteria), and Eucarya (eukaryotes). An early stage in the evolution of multicellular organisms was the colonial life form (in the sense of cluster of clonal organisms, not a social aggregation as in seabird colony). The colonial form could have arisen in various ways, most simply by daughter cells remaining attached after cell division.

Because all the cells of a multicellular organism contain essentially the same DNA, the evolutionary trend to large, multicellular, macroorganisms must have depended on the ability of eukaryotic cells to *express* their hereditary information differently and to function as a cooperative unit. The driving force behind this series of events was that macroorganisms could exploit resources and environments in different ways from microorganisms.

Overall there has been, up to a point, an increase in size during evolution: Macroorganisms are derived from microorganisms and, within a class, larger species tend to appear later in a group's phylogeny. However, natural selection does not necessarily favor large size. For some taxa, while the maximum size does increase with evolution, the median and minimum do not; also, some organisms are smaller than their ancestors.

Physical and chemical laws presumably set the lower and upper limits on life. A cell must be larger than a minimum size to accommodate its genetic and metabolic machinery. Microorganisms exist in world governed by the forces of diffusion, surface tension, viscosity, Brownian movement, and Reynolds numbers. For macroorganisms, at the other extreme, the greatest constraint on large size is gravity, which conditions the forms and actions of all macroorganisms. The upper limit on size is less rigorously drawn than the lower limit. It is related to growth form and habitat: Many clonal plants grow indefinitely in a horizontal rather than vertical direction; blue whales are larger than elephants and are partially buoyant in their aquatic medium.

If major taxonomic groups are compared, large species are usually more complex (i.e., they have more cell types), have longer generation times, and in general are longer lived in absolute time than smaller species. Chronological time is size-independent, but physiological time ("rate of living") is dictated by the size of an organism. Size confers a stable internal physiology (homeostatic ability), diverse architecture, speed, competitive ability, predator deterrence, and certain metabolic advantages such as a decreased rate of energy expenditure per unit mass. Larger organisms require more resources, however, and they adjust slowly to change, are slower colonizers, and evidently have higher extinction rates. Both large and small organisms are 'successful' in that both remain extant.

The terms r and K as used now describe in a comparative sense either *conditions* or, less formally, *species*. Uncrowded environments (r-selection) are usually unpredictable and transitory. Colonists that have a high r (in microbial semantics a high μ_{max}) should do best because this characteristic favors discovery of the habitat, reproduction, and dissemination before the habitat either disappears or becomes crowded. Conversely, organisms that can compete well in a habitat that is either stable or predictably seasonal should be favored under crowded conditions (K-selection) where there is typically a shortage of resources. Size and correlates of size mean that macroorganisms and microorganisms are relatively K- and r-selected, respectively, within the r/K scheme as it has come to be depicted. Within each group, however, r-

and *K*-strategists can be identified. Microbial systems, particularly bacteria, provide a means for causative interpretation of the effect of crowding, hence a mechanistic basis, on which the theory of *r*- and *K*-selection is based.

At least three kinds of useful ecological comparisons can be made with respect to size: 1) Large and small forms within a species or genus can be compared for competitive success and behavior. 2) The forces acting differentially on macro- and microorganisms can be identified and the implications noted. However, because size is invariably correlated with other phenotypic traits, the two groups cannot be compared meaningfully on the basis of size alone. 3) General principles pertaining to size can be made irrespective of taxon. An example is the Salisbury-Garrett theory that the average size of a reproductive propagule of a species reflects the nutritional needs for a developing new individual in its usual habitat.

4.9 Suggested Additional Reading

Bonner, J.T. 1988. The evolution of complexity by means of natural selection. Princeton Univ. Press, Princeton, N.J. An excellent, stimulating synthesis on why there has been a progressive increase in size and complexity from bacteria to the plants and animals.

McMahon, T.A. and J.T. Bonner. 1983. On size and life. Scientific American Books, N.Y. An interesting book for the general reader about the effects of size on plants and animals.

Peters, R.H. 1983. The ecological implications of body size. Cambridge University Press, Cambridge, U.K. A review of the literature with incisive comments, from a quantitative standpoint.

Thompson, D'A. W. 1961. On growth and form. (Abridged edition edited by J.T. Bonner.) Cambridge University Press, Cambridge, U.K. The benchmark of excellence on the analysis of form.

Valentine, J.W. 1978. The evolution of multicellular plants and animals. Sci. Amer. 239(3): 104-117. Traces the development of multicellular organisms from their unicellular progenitors, emphasizing the fossil record.

5

Growth and
Growth Form

*Geometry is the obvious framework upon which nature works to keep her
scale in "designing." She relates things to each other and to the whole,
while meantime she gives to your eye most subtle, mysterious and
apparently spontaneous irregularity in effects.*

—Frank Lloyd Wright, 1953, p. 53

5.1 Introduction

Growth form, that is the shape and mode of construction of an organism,
together with size, set fundamental limits on the biology of living things (the
so-called phylogenetic and allometric constraints, respectively; Chapter 4).
Similarity of form in nature occurs at two levels: First, all organisms are either
basically unitary or modular in construction. Being designed according to one
blueprint or the other carries numerous implications, the most important of
which are evolutionary consequences pertaining to fecundity and the transfer
and expression of genetic variation. Second, there is a geometric commonality
among unrelated entities, for example between tributaries and circulatory
systems; and among spirally shaped shells, horns, and certain inclusion bodies
in virus-infected cells. Does this reflect merely coincidence—just as the log
normal distribution may describe simply chance effects of random variables
acting on large and diverse collections of objects? Or, if all possible permutations of form are imagined, is there a biological reason why some patterns
are more common than others? What shapes of a fungus or a tree are the
most fit, most likely to occur in particular circumstances? I examine in this
chapter these two levels of the replication of form.

5.2   Unitary and Modular Organisms: An Overview

The unitary design   Unitary organisms, represented by most mobile animals, follow sequential life cycle phases predictably and their number of appendages is fixed early in ontogeny. Their growth is noniterative (not based on a repeated multicellular unit of construction), and determinate (of strictly limited duration). They display generalized (systemic) senescence, and are generally unbranched. Reproductive value (see Section 6.3 of Chapter 6) increases with age to some peak and then declines. The genetic individual (genet) and the physiological or numerical individual are the same entity, and are repeated only at the start of each new life cycle.

The most important evolutionary implication of a unitary design is that the genetic individual and the physiological individual are one. In other words, if the genetic individual through the zygote-to-zygote life cycle is defined as a *genet* (see Chapters 1 and 2; also Kays and Harper 1974; Harper 1977, pp.26-29), and the basic unit of growth or construction broadly as a *module*, then for unitary organisms the ratio of genets:modules is 1:1. The concept that the product from one sexual event is a single unit or genetic 'individual', regardless of the number of divisions of that product, or whether the units stay together (as in a clonal herb) or separate (as in a *Lemna* clone), dates back at least to Thomas Henry Huxley's (1852) observations on coelenterates.

Many animals are of unitary growth form. Operationally, with few exceptions, this can be equated with mobility. Such animals follow sequential life cycle phases in a predictable fashion and their number of appendages is fixed early during embryonic development. Harper (1977, pp.515-516) has illustrated the direct correspondence between the genetic and numerical individual by observing that a count of rabbits measures the number of genotypes and also gives an estimate of biomass. Conversely, he notes that within a factor of about 10 it is possible to estimate roughly the number of individuals if the biomass is known. Because form is determinate in unitary organisms (as is cell number within relatively strict limits), one could determine the number of rabbits by counting their legs and dividing by four or their ears and dividing by two. Thus, in unitary organisms "the zygote develops to a determinate structure that is repeated only when a new life cycle is started from a single-celled stage, usually a zygote" (Harper et al. 1986, p.3). As will become apparent below, these relationships do not apply for modular organisms.

The validity of the notion that the genet exists as a distinct genetic entity from zygote-to-zygote hinges on the premise that the zygote (followed by its developing products) is *the* unit of variation (Harper 1977, p.27), faithfully and exclusively displaying mutational and recombinational events. While clear and unquestionably useful conceptually, the notion has become clouded by increasing recognition (reviewed in Chapter 2) of variation in suborganismal replication (e.g., of intracellular organelles; transposable elements, including retroviruses), of which some variation is heritable in some organisms (Reanney 1976; Shapiro 1983; Buss 1985). Also, environmentally induced heritable changes that mimic Lamarckian effects have been reported for several varieties of flax and tobacco (Perkins et al. 1971; Durrant 1971, 1981; see also Silander 1985; Cullis 1987) and are analogous to those reported for the immune system in animals (Steele 1981). In other words, *heritable* genetic change can potentially occur horizontally, among organisms in the same generation, as well as by the conventional vertical means, that is, by sexual reproduction. The best conventional examples are the unusual genomic systems of bacteria and fungi, where it has long been known that the 'physiological individual' (see below) can change genetically without fertilization. A second consideration is that what constitutes a genet in practice for many organisms is frequently impractical or impossible to determine as the genealogy of all individuals in a population cannot be observed continuously (Jackson 1985).

The modular design Contrasting with unitary life forms, in modular organisms the zygote produces a unit of construction (module) which is iterated indefinitely, giving progressively more modules which in aggregate constitute the genetic individual. A typical, easily visualized module is the leaf together with its axillary bud and internode of a stem. The number of modules—whether they be leaves or roots on a tree, hyphae of a fungal colony, or polyps on a coral—is indeterminate. The ratio of genets:modules is accordingly 1:many. The major contrasting features between unitary and modular organisms are summarized in Table 5.1. As will be developed in the following sections, whether an organism is unitary and mobile, or modular and sessile, probably has greater biological significance than its size (microorganism versus macroorganism).

Modules may remain attached and contribute substantially to the organism's architecture (as in trees and corals, much of which often consists of accumulated dead modules). Alternatively, they can operate as separate physiological units or ramets (see below). Recall from Chapter 1 that the ramet is the 'physiological individual', functional on its own if severed (strawberries; buttercups) or sloughed off (lichens; corals; the floating aquatic plants *Lemna* and *Salvinia*) from the parent. Thus, modular organisms frequently grow *clonally*, that is, by the formation of functional individuals of identical or nearly identical genetic composition (Jackson et al. 1985). Modular growth is thus analogous to but distinct from clonal growth. Organisms that reproduce asexually are clonal and often, but not invariably, modular. There are some clonal unitary creatures (mobile animals which reproduce parthenogenetically or apomictically, for example, aphids, rotifers, lizards, earthworms).

It is common in botany and zoology for 'the individual' to be taken as a particular unit of modular construction (Harper and Bell 1979). These in-

Table 5.1 Some major attributes of unitary and modular organisms[1]

Attribute	Unitary organisms	Modular organisms
Branching	Generally nonbranched	Generally branched
Mobility	Mobile; active	Nonmotile[2]; passive
Germ plasm	Segregated from soma	Not segregated
Development[3]	Typically preformistic	Typically somatic embryogenesis
Growth pattern	Noniterative; determinate	Iterative; indeterminate
Internal age structure	Absent	Present
Reproductive value	Increases with age, then decreases; generalized senescence	Increases; senescence delayed or absent; directed at module
Role of environment in development	Relatively minor	Relatively major especially among sessile forms
Examples	Rabbits; birds; humans; vertebrates typically	Plants; hydroids; corals; colonial ascidians; invertebrates typically; fungi; bacteria

[1]From Harper et al. (1986) and other sources.
[2]Juvenile or dispersal phases mobile. Many bacteria and protists are somewhat mobile (motile). Phytoplankters may move vertically in a water column by buoyancy mechanisms; clonal plants may 'move' horizontally on land by varying lateral patterns of local senescence and growth.
[3]Pertains to degree to which embryonic cells are irreversibly determined. Preformistic = all cell lineages so determined in early ontogeny; somatic embryogenesis = organisms capable of regenerating new individual from some cells at any life cycle stage (cells totipotent or pluripotent). See Buss (1983).

dividuals constituting members of a clonal population have been termed *ramets* (Chapter 1) and are modules at a level of organization capable of separate existence. Operationally, they are typically the countable units in the pasture or the coral reef or in the petri dish. Finally, clonal organisms may be *colonial* or noncolonial (solitary, often dispersed over large areas) in growth habit. The term colony is used broadly here to mean the tendency of units related by descent to clump. These units (modules in the case of modular organisms) may be physically linked together (as in rhizomes and the basal or root suckers of many clonal plants) or not linked (clumps of *Lemna*; bacterial cells; aphids).

For clonal organisms the issue of what constitutes an individual cannot be sharply defined. This defiance of arbitrarily imposed order is refreshing, but it has nevertheless spawned considerable ambiguity, debate, and sematic difficulties (van Valen 1978). While proceedings from two major symposia to date (Jackson et al. 1985; Harper et al. 1986) provide exciting new perspectives, the gulf between approaches and terminology of the botanists and zoologists is quite evident. Even more striking is that apart from one paper by Trinci and Cutter (1986), the microbial world is omitted entirely from the synthesis. Why this is the case seems to be more a matter of disciplinary isolation than irrelevance.

The confusion exists primarily because of two levels of complication. First, among clonal macroorganisms, the physiologically functional individual (ramet) does not correspond to the genetic individual (genet). The ratio of genets:ramets is 1:many. The illustration above for plants applies as well to any of the several clonal invertebrates. Among corals (reviewed by Rosen 1986), logical arguments could be made depending on a particular context for the individual to be represented by the polyp, by numerous polyps together forming a colony, by the zooid, or even by an entire reef. Second, and superimposed on the first, is the genetically mosaic nature of bacteria and many fungi. In these microbes, the genet itself is frequently discontinuous: The organism through its life cycle is in a state of genetic flux, comprising both many ramets *and* many genets (Chapter 2).

5.3   Fungi As Modular Organisms

Knowledge of hyphal growth patterns has been been obtained almost exclusively from laboratory culture of representative fungi (and prokaryotic actinomycetes) on solid nutrient media. From these relatively uniform and highly controlled conditions the following general pattern emerges.

Hyphae initiated from a germinating spore grow radially outwards, branch, occasionally fuse, and show mutual avoidance reactions (Butler 1966, 1984; Gregory 1984; Oliver and Trinci 1985; Trinci and Cutter 1986). As long as growth remains unrestricted, the ratio between total hyphal length and number of branches eventually becomes constant for any particular strain, a value referred to as the *hyphal growth unit* (Plomey 1959; Trinci 1984; Trinci and Cutter 1986). (This phenomenon, incidentally, is characteristic of branching systems in general and is not unique to fungi; see Section 5.7.) Thus, under ideal conditions the mycelium can be regarded as resulting from duplication of this hypothetical growth unit comprising a hyphal tip and an associated growing mass of constant size. Conceptually, this is a convenient vegetative module, as is the leaf unit of plants. Most fungi (and actinomycetes) go on to produce other forms of vegetative (asexual spores) and sexual modules, just as plants develop vegetative modules which are later alternated with or replaced by sexual modules such as flowers, stamens, and carpels. Superficially similar branching patterns are evident among other sessile organisms such as the stoloniferous colonies of the marine hydroid *Hydractinia*

(Buss 1986). Details of fungal growth forms and the case that fungi are modular organisms are discussed in Andrews (1991).

While in its diffuse feeding mode the branching mycelium is conspicuously modular in construction, it can differentiate to function in two other roles, namely survival and dispersal. In both cases, hyphae that were formerly kept separate by some repulsive mechanism come together under control to produce aggregated structures (Figure 5.1). These forms often have architectures unlike anything else on earth. The frequently complex and massive fruiting bodies (basidiocarps) of the basidiomycetes represent the extreme in such morphogenesis. In such cases the mycelium performs a structural (and transport) function. It plays no role in nutrient uptake, serving only to provide an exposed surface for spore dissemination. In the order Agaricales (the 'agarics', including the common toadstools and mushrooms), the basidiocarps are fleshy, transient, and rely mainly on turgor pressure for support. In contrast, fruiting bodies of the bracket fungi (Aphyllophorales) are large, and may be woody and perennial (*Phellinus, Ganoderma*). In this sense they are analogous

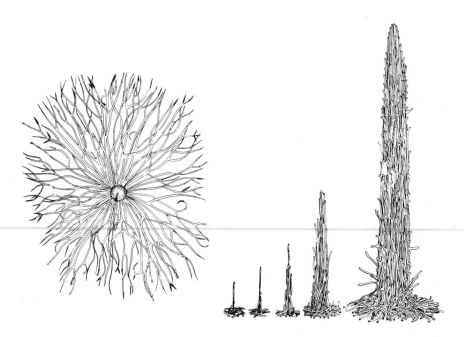

Figure 5.1   Fungi as modular organisms. Left, *Achlya*, a water mold, growing in diffuse fashion, allocating biomass primarily to foraging (from Bonner 1974, p.97; redrawn from J.R. Raper). Right, *Pterula gracilis*, a basidiomycete, growing as an aggregate unit, allocating biomass primarily to reproduction. From Bonner (1974, p.96; redrawn from E.J.H. Corner). Reprinted by permission of the publishers from *On Development: The Biology of Form* by John Tyler Bonner, Cambridge, Mass., Harvard University Press, © 1974 by the President and Fellows of Harvard College.

to trees in accumulating structural dead biomass. Regardless, the important point with respect to growth form is that the fungi seem to exploit two major patterns: a branched, foraging mode and one that is organized and constrained for reproductive or other purposes. That substantial biomass can be diverted away from resource acquisition to survival or reproductive structures attests to the importance of dormancy and relatively long-range dispersal of new genets (as opposed to localized growth by mycelial extension) in the life histories of these fungi (see also Chapter 7 and Andrews 1991).

5.4 Bacteria As Modular Organisms

Bacteria range in complexity from unicellular rods and cocci (the typical bacteria) of more or less spherical shape to multicellular, branched (the actinomycetes) or unbranched (e.g., *Leucothrix*) filaments. Some filamentous forms may show gliding motility or be enclosed by a sheath. The myxobacteria (noted in Chapter 4) are typically unicellular rods that aggregate and produce ornate fruiting bodies, usually containing specialized resting cells. Despite occasional complex morphology or developmental cycles, all bacterial types can be resolved essentially into growth forms based on either the unicell or a filamentous unit.

Bacteria may exist in nature as single cells, but apparently this is a rare event (Shapiro 1985). 'Microcolonies' (aggregations visible by light microscopy) form when cell division is not followed by dispersal. Winogradsky (1949) examined such cellular consortia on soil particles and termed them "families". They have since been observed in many other habitats including the surfaces of plants and animals; the inner walls of the colon; dental plaque; and as suspended particulate matter in lakes and oceans. The particular form of a microcolony is affected by the way a cell divides and the nature of its surface. Macroscopically visible 'macrocolonies' (or simply 'colonies' in bacterial semantics) of the sort typifying growth in a petri dish probably rarely occur in nature (Carlile 1980; Pfennig 1984). Conspicuous exceptions are the massive growths in sulfur hot springs or occasionally in stagnant water (e.g., Pfennig 1989).

Because bacteria tend to form microcolonies, for the most part they function ecologically as multicells. The disadvantage of this state of affairs for the bacterium is largely one of competition. Depending on the circumstances, the advantages include enhanced return from concerted activity (e.g., migration or enzyme synthesis) or the resistance of pathogenic microbes to host defense mechanisms (phagocytosis; antibodies) and of free-living species to abiotic factors (desiccation; UV light; mechanical erosion). Since genetic recombination is a relatively rare event (Chapter 2), cells in a microcolony will initially be clonal and will tend to remain so, apart from progressive change due to mutations. As cell division continues, the aggregation will fragment from time to time under the influence of various dispersal mechanisms. The bacterial equivalent of the genet is thereby dispersed, and the asexual process is directly analogous to the shedding of plantlets by *Lemna* or *Salvinia*.

The module of a microcolony of unicells is thus the individual bacterial cell. *The bacterial cell qualifies as the module, whereas the cell of unitary organisms does not, because it is iterated indefinitely.* Although cell number in unitary organisms changes (e.g., in response to disease; by regular turnover as in replacement of blood and epidermal cells), the number is defined within fairly tight limits by intrinsic genetic and developmental factors. In contrast, the cell number of a bacterial genet is limited in practice by extrinsic factors, for instance, by whether or how soon mutation/selection or recombination occurs (see Chapter 2). By convention, the term module in eukaryotes is reserved for multicellular structures; while its use in the bacterial context is thus strictly a departure, the relationship of a single cell to the bacterial clone with that of, say, a multicellular *Salvinia* plantlet to the *Salvinia* clone is analogous.

5.5   Some Implications of Being Modular

Modular organisms, especially those that are clonal, tend to be highly plastic in size and fecundity (Harper and Bell 1979) because the number of modules can change readily by birth and death processes. This high variability tends not to extend to the level of the individual module. Indeed it is the consistency in size and shape of such modules as flowers and leaves that makes them reliable taxonomic features. Plasticity at the ramet (functional individual) level is another reason why counts of individuals in themselves convey little information. Individual plants of the same age can vary substantially (Chew and Chew 1965), for instance, by 50,000-fold in weight or reproductive potential (Harper 1977, pp.25-26). To the extent that age correlates with size it can give some measure of fecundity (as in many trees), which is typically size-related (Downs and McQuilkin 1944).

The iterative property affects organism shape as well as size because modules are commonly retained even after their death, as in the dead branches of a tree or the dead zooids in a coral reef. Of course modules may eventually be lost, for instance selectively, by the self-pruning of large plants. This has been interpreted in relation to loss of photosynthetic structures and support tissue that are no longer contributing positively to the net energy and carbon balance or, if so, not justifying their allocation of recyclable nutrients such as N, P, and K (Raven 1986a). The related issue for clonal organisms is the extent to which ramets may 'cooperate' for the benefit of the genet. This is an interesting but open question. In theory, ramets could be spaced at distances to reduce competition among them, but the evidence is mixed (Harper 1985).

To what degree any overall control may be lost when a clone fragments is also intriguing, but unanswered. Small colonies of the ectroproct (soft coral) *Alcyonium* evidently have higher rates of prey capture per polyp than single, large colonies (McFadden 1986). There are obvious and interesting implications for food acquisition by other modular organisms (Chapter 8 and Andrews 1991).

Growth form of sessile modular organisms is either predominantly vertical or horizontal. For plants the competitive strategy may be to stack modules vertically thereby shading one's neighbors. Alternatively, and for modular organisms generally, it may be to capture resources by lateral expansion and perhaps in so doing avoid some of the mechanical limitations of size (see Chapter 4; also Watkinson and White 1985). In rhizomatous or stoloniferous plants (or for fungi that invade new territories by rhizomorphs and related structures) new root (or diffuse hyphal) systems are typically produced. The intervening segments may rot or break under the influence of water currents or from mishap (e.g., under the hooves of grazing sheep) so that the original zygote becomes represented by physiologically independent individuals (ramets). Ultimately the number of modules produced—whether they remain attached or become separate—will affect the number of progeny produced, in turn the number of descendants, and hence fitness. In some circumstances, fitness is evidently increased by retention of the intact genet; in others, by a dissociated genet (Harper 1981a).

Iteration means that different parts of the same genet are affected by different environments and consequently are subjected to different selection pressures (Harper 1985; Harper et al. 1986). Progeny are produced from those portions of the genet that are most successful. These might be the segments that have adequate nutrition and avoid being eaten by predators or, in the case of a pathogenic fungus, those clonal spores which contact a suitable host. Thus, for modular organisms, the testing of a particular gene combination is achieved by *growth* whereas for unitary organisms it is achieved in large part by *motility* within the population. Architecture supplants movement. Phrased differently, growth *is* movement for sessile organisms. (Strictly speaking [Harper 1977, p.27; and Section 2.3] the term growth includes what is often called asexual or vegetative 'reproduction'. Reproduction implies the formation of a new genetic entity which in turn forms a new individual by replicating ['reproducing'] its encoded genetic information. Most other authors [Buss 1985] distinguish growth from reproduction arbitrarily, based on the degree to which the clonal unit is dispersable; thus, new plantlets or polyps would constitute growth; production of gemmules or spores would be asexual reproduction. Bacteriologists tend to use the terms interchangeably; mycologists use them in the sense of Buss [1985].)

Because modular organisms occupy relatively fixed positions, interactions with neighbors develop over time and can be expected to be relatively strong. Modules capture resources (nutrients; light) which sooner or later tend to become locally exhausted creating a *resource depletion zone* (RDZ) (Figure 5.2). The size of a RDZ and, in fact, if such a zone develops at all, depends on whether the solute is absorbed relatively faster than water (Nye and Tinker 1977, pp.129-142). If this is the case, then the concentration at the module

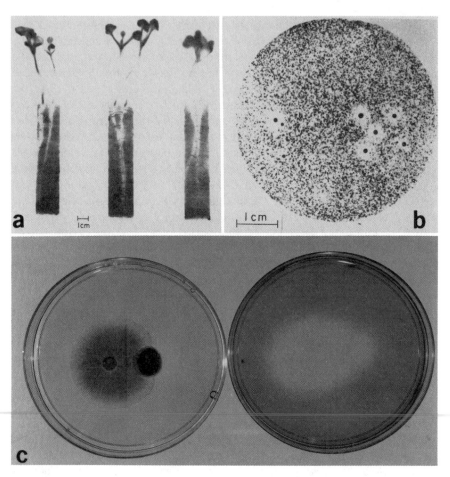

Figure 5.2   Resource depletion zones (RDZs) in soil (a) labelled with [33]P (around roots of rape, from Bhat and Nye 1973), and (b) with [35]SO$_4$ (around onion roots, from Nye and Tinker 1977, p.141, based on Sanders 1971). (c) Phosphate RDZs in agar medium around 4-day-old colonies of (right colony) *Athelia bombacina* and (left colony) *Aureobasidium pullulans* growing on potato dextrose agar. The agar was overlain by a dialysis membrane prior to inoculation and incubation. The membrane with colonies was then stripped off the plate and both the membrane and medium were stained separately. Note staining of fungus (left, denoting phosphate uptake) and the corresponding phosphate-depleted zone in the stained medium (right). From Andrews (1991).

surface (e.g., the root) will be lowered. In contrast, if water is taken up faster, the solute may accumulate at the root surface and diffuse away. (It should also be remembered that root zones, and presumably also those of certain other modules such as polyps of coelenterates, can actually be zones of enrichment for nutrients such as amino acids and carbohydrates which are exuded by the host.) Thus, the onset and dimensions of RDZs depend on how rapidly the resource is used by the module and resupplied by the medium. There has been considerable theoretical discussion of optimal growth form and search patterns to capture resources. These are developed later in Section 5.6.

Germ cells are sequestered in unitary organisms. This is one mechanism to preserve genetic individuality, but segregation of totipotent cells precludes the ability to produce ramets (Buss 1983; 1985). Modular organisms do not separate soma from germ plasm which is in effect articulated as the organism grows, for example, in the meristems of plants. Many modular organisms remain capable of clonal replication from certain tissues at any life stage (somatic embryogenesis; Buss 1985). Consequently, somatic mutation has potentially a significant role in their evolution, a prospect which is fascinating but poorly documented to date (Chapter 2).

Somatic polymorphism (Chapter 7) occurs in modular organisms in that modules of a genet are born over time and are frequently specialized for a particular role such as being leaves, flowers, stamens, or carpels (Harper and Bell 1979). For example, the form and arrangement of leaves may be quite different on juvenile and mature branches of *Eucalyptus* (Harper 1977, p.707), and the aerial and submerged leaves of aquatic plants such as water milfoil (*Myriophyllum* spp.) are strikingly different in morphology. There are at least two important consequences of somatic polymorphism: The first is on the number of modules allocated to a particular reproductive or vegetative role. Second, the diversity in form that results from this allocation pattern also influences the behavior of associated organisms. Insofar as parasites and herbivores are concerned, flowers are quite different from leaves or carpels or roots. Rather than the entire genet being directly affected (as is the case with unitary organisms), the potentially damaging activity is concentrated on the module which may be destroyed and replaced without great consequence to the organism. Here is yet another example of the issue of trade-offs—the modular organism must 'decide' what is the 'best' balance of functionally different modules, each of which will be a drain on resources.

Age-specific fecundity, important in the population biology of most animals, means less for modular organisms because the number of reproductive units increases exponentially with age and this exponential increase can theoretically last indefinitely (Harper and Bell 1979; Watkinson and White 1985; Harper et al. 1986). Death occurs at the level of the module, rather than being total, which is the case in unitary organisms. As long as the birth rate of modules exceeds their death rate (and barring somatic mutation that affects the gametes, Chapter 2), the genet is (theoretically) immortal. Under such circumstances, Hamilton's (1966) postulate that senescence is an inevitable consequence of natural selection does not apply (see Chapter 6).

Extremely long-lived genets have been recorded, particularly among clonal plants: quaking aspen (*Populus tremuloides*) has been estimated at 10,000 years (Kemperman and Barnes 1976); creosote bush (*Larrea tridentata*) in the Mojave Desert of California at approximately 11,000 years (Vasek 1980); and box huckleberry (*Gaylussacia brachycerium*) at about 13,000 years (Wherry 1972). Watkinson and White (1985) suggest that among clonal plants the causes of death of old genets include fire, disease, and competition. Fragmentation, dormancy, and dispersal mechanisms would all act to reduce the risk of death to entire genets. Lichens are among the slowest growing of organisms. Extrapolations based on the calculated rate of radial expansion of crustose forms in the Arctic have provided crude estimates of the age of some thalli at 4,500 years (Beschel 1958; for information on growth rate studies see Ahmadjian and Hale 1973; Hale 1983). According to Hale (p.81), "lichens surely have a maximum thallus size and a finite life span". If this is so, it must be meant in a probabilistic sense (see Section 6.3 in Chapter 6), rather than in the sense of a finite size and lifespan as those terms are applied to unitary organisms. A given thallus may senesce and die, but the genet continues indefinitely by virtue of vegetative (asexual) reproduction through the dispersal of diaspores (lichenized structures containing both the algal and fungal components). Fungal clones can also be long-lived. Dickman and Cook (1989) found several genets of the wood-rotting fungus *Phellinus* [*Poria*] *weirii* estimated to be older than 1,000 years in mountain hemlock (*Tsuga mertensiana*) forests of Oregon. The fungus spreads outward from a focus of infection primarily by root-to-root contact. Sibling ramets of the various genets survive forest fires which periodically destroy large areas of the stands.

Finally, it remains unclear to what extent life history differences between modular and unitary organisms arise from this fundamental biological distinction alone or because of associated attributes. For example, within the modular category, organisms can be either clonal or aclonal (and as noted above, some unitary animals are clonal). Is the clonal/aclonal division more important biologically than the modular/unitary distinction? How important is branching as a correlate of modularity and what constraints or opportunities does it impose on the organism? Clonal modular organisms may be essentially branched (fungi; some lichens; many corals) or unbranched (other corals; bacteria; the crustose forms of various taxa such as lichens and bryozoans). Future studies that compare the modular and unitary designs need to address the implications of these correlates. Attempts to interpret life history concepts as they pertain to clonal modular organisms as opposed to unitary creatures have recently been made (Hughes and Cancino 1985; Sackville Hamilton et al. 1987). Differences in predictions may be quite profound. For example, Sackville Hamilton et al. show by modelling that while traditional theory associates greater reproduction with *r*-selection, in a modular, clonal species with guerrilla genets, *r*-selection in general favors nonreproductive clonal growth.

Growth of modular and unitary organisms   For modular organisms there are two levels of population structure in a community (Chapter 1 in Harper 1977): The genets, formally equivalent to the original zygotes (called N in animal population dynamics) and the quantity of modules (called η in plant population dynamics), a variable number of which occurs *per genet*. In theory both N and $N \cdot \eta$ (i.e., the number of modules in a given *population*) can be quantified. For aclonal modular organisms such as annual plants and most trees, N can be counted easily. The difficulty occurs with clonally spreading perennials (e.g., Harberd 1967) or among certain microorganisms and invertebrates where intermingling, sporulating, or fragmenting clones necessitate either a marker or mapping system to separate genets from ramets. Individuals tallied where possible as genets are meaningful to the geneticist or evolutionist interested in the genetic variation of a population. The production agriculturalist or population biologist or microbiologist is more interested in the number of ramets, that is, functional individuals, which gives a rough idea of biomass.

The most obvious difference, then, between a modular organism and a unitary organism with respect to growth is that growth of the former is a *population* event. There is an increase in number of modules, whether these be multicellular units of construction (e.g., leaves) or units of clonal growth capable of separate existence (i.e., ramets such as 'countable' individuals of bracken fern; bacterial cells). This is what Harper meant (1977, pp.20-29) in viewing "the plant as a population of parts" which could be varied in response to environmental conditions (see also White 1985). The genet of modular organisms thus has considerable morphological flexibility and can accumulate large biomass, often rapidly. Growth (enlargement) of the *module* is formally equivalent to growth (enlargement) of the entire soma (a single deer, a single bear, etc.) in unitary organisms, and this is distinct from population growth, which is the aggregate expression of each unit of construction. Hence, while each cell of the bacterial or yeast clone increases in mass more or less arithmetically prior to fission or budding, the population of such cells increases logarithmically. Similarly, for the fungi, increase in germ tube length is initially exponential, then linear; however, because new branches are formed continuously the germling as a whole continues to expand exponentially, both in terms of total length of mycelium and number of branches.

Where modular organisms are also clonal, growth of the genet can be especially rapid (and is coupled with regenerative capacity following partial losses; Hughes and Cancino 1985; Sebens 1987). Thus, under favorable conditions, the number of ramets may increase (more poplar trees), or the number of modules per ramet (more leaves per poplar), or both. Conversely, 'degrowth' or shrinking can occur at either or both levels. For colonial clones, such as bryozoans and bacteria, there may be larger colonies, a larger number of colonies, or both. The prodigious multiplication rates of bacteria are the extreme example. Here, the capacity for clonal growth, coupled with the efficiencies of small size (Chapter 4), are reflected in extremely short generation times (doublings on the order of minutes under optimal conditions). However, even among bacteria, intrinsic (maximal) growth rates vary greatly (Brock 1966, p.95) and for a given species rarely is the upper limit achieved in nature. It is an ecological truism that growth rate potential per se is a poor measure of 'success' (Slobodkin 1968; Sibly and Callow 1986, p.92) because fitness costs are involved in the trade-off. Nevertheless, bacteria come the closest of any taxon to organisms that multiply as fast as possible (Pardee 1961; Harder and Dijkhuizen 1982; see also Chapter 3), and selection for rapid growth rate seems to be the most important driving force.

5.6   Some Implications to Modular Organisms of Being Sessile

Organisms are either motile or sessile. As is the case for most classification schemes, the extreme examples—in this case, say a lion versus a lichen—are clear while the mid-range of the spectrum is not. For instance, sessile life forms have dispersal phases and often immature stages that are motile. 'Stationary' clonal organisms in effect feed at many locations by producing horizontal branching systems; nonmotile algae often move vertically in the water column by flotation mechanisms. Most people would consider the black bear to be motile, yet it spends about half of its life hibernating in a sleep-like state. Barnacles are unitary organisms, motile as microscopic larvae, but spend their adult lives sitting passively on rocks. Nevertheless, the demarcation is useful and valid as a first approximation. Many of the traits we intuitively accept as being fundamentally 'animal' or 'plant' are a consequence of motility, in particular the search for food and mates.

Sex and genetic recombination   Being motile or not carries interesting implications for genetic recombination. One option is to forego the search for a mate and reproduce (grow) asexually, relying primarily on mutation, including the various forms of mobile genes, to introduce variability among progeny (Chapter 2). There are some fungi for which a sexual state as conventionally defined is not (yet) known, and asexual reproduction is the dominant mode among bacteria, and some protists and lower invertebrates. Alternatively, but still acting alone, genes of the organism can mix in various combinations by some form of parasexual event as occurs in the fungi (Chapter 2) or engage in some form of 'self-compatibility'. This includes hermaphroditism, homothallism, or homostyly, common among some animals, fungi, and plants, respectively. Finally, sexual reproduction (including its analog in bacteria; see Chapter 2) does of course occur, but immobility requires that it be accomplished either by gene transport over more or less long distances in some sort of package (e.g., pollen; bacteriophage) or that the mating individuals come together by growth (e.g., proximal bacterial cells; hyphal fusions). In short, mobile organisms engage in activities that broadly constitute mate selection and courtship rituals; sessile organisms accomplish essentially the same thing by growth and passive dispersal mechanisms.

Competition and defense   Sessile organisms interact closely with only a subset of individuals in a population, namely those which happen to be in the immediate vicinity. This means that the effects of one neighbor on another can be specifically documented, whereas interactions among mobile organisms have to be assigned abstractly to 'density effects' (Harper 1981a).

The outcome of interactions between sessile organisms depends on the behavior of the species involved. Few generalizations can be made. Plants rarely actively defend territory, and their resource depletion zones (RDZs) tend to be invaded or overtopped (Harper 1985). The growth form of plants is to a large degree mutually accommodative: Genets of white clover (*Trifolium repens*) interweave in a meadow (Sackville Hamilton and Harper 1989; Schmid 1986), and Horn (1971, p.104) has said "the shape of a tree *in the forest* [emphasis added] is largely the shape of the space that it fills". Other sessile organisms, however, such as fungi (Cooke and Rayner 1984) or colonial metazoans (Buss 1987, pp.149-153) may actively defend against invasion. For these organisms, competitive ability is directly and primarily influenced by growth morphology (Buss and Grosberg 1990). The significance of microbial antibiosis in nature and the analogous phenomenon of allelopathy among plants remains controversial. However, it is at least logical that inhibitory chemicals could play a significant role in the defense and competition of sessile organisms (see the case study).

Response to environment   When exposed to 'hostile' environments, motile organisms may congregate for warmth (birds; social insects such as bees) or seek shelter or emigrate. Being stationary means having to adjust *in situ*. Therefore, while architecture is genetically set within broad limits, one would expect the morphological state of a sessile organism to reflect local environmental conditions (including nutritional aspects discussed below; for expansion of these ideas, see Chapter 7). Phenotypic plasticity is one such manifestation; it has already been noted how plants of the same age may vary by 50,000-fold in weight or reproductive potential (Harper 1977, p.25; see also Gatsuk et al. 1980; Tomlinson 1987). Consequently, the calendar age of a plant (and presumably all other sessile modular organisms) generally means little whereas it means a lot to a unitary organism. What is more important

A Case Study: On Being a Neighbor

One consequence of a sessile lifestyle is interactions among neighbors (e.g., Harper 1964). The organism cannot escape from predators by running away or from competitors by moving to a different pasture to graze. Thus, one would expect to find evidence of various defensive mechanisms (and resource exploitation strategies) in such life forms. For instance, competition may be mediated by chemicals, a phenomenon commonly referred to as allelopathy by botanists, and as antibiosis by microbiologists. Sage plants of the genus *Salvia* in California evidently inhibit the growth of other vegetation at least in part by an allelopathic mechanism (e.g., Muller 1966), although alternative explanations have been put forward (Bartholomew 1970). Likewise, the production of antibiotics is frequently alleged to be an artifact of laboratory culture conditions. Nevertheless, it is now clear that antibiotics can be produced in nature (Rothrock and Gottlieb 1984), and strong, multiple lines of evidence implicate their role in the antagonistic interaction among some microbes (reviewed by Handelsman and Parke 1989).

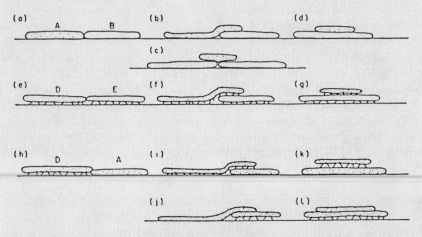

Figure 5.3 Schematic vertical sections through crustose (mat-forming; shown lying flat) and foliose (on stilts) lichens, showing the potential range of interactions. (a-d) Interactions between two crustose species, A and B. (e-g) Interactions between two foliose species, D and E. (h-l) Interactions between a foliose (D) and a crustose species (A). Foliose forms can slide over most crustose forms and thus stalemate conditions (h) or overtopping by a crustose species (j) are uncommon. (d,g,k,l) designate epiphytic growth which requires colonization of the underlying thallus by propagules containing both the algal and fungal components of the epiphytic species, a rather uncommon event. From Pentecost (1980).

Competitive (and other) interactions are also reflected by morphological behavior. In experiments on competition between the free-floating aquatic plants *Lemna* and *Salvinia*, the latter dominated because it overtopped *Lemna* by expanding its fronds above, and then down onto, the water surface (Clatworthy and Harper 1962). The outcome could be predicted neither by the carrying capacities nor by the intrinsic growth rates of the respective species in isolation. When lichens of various growth forms meet on a rock surface, different outcomes are possible, but most commonly the foliose (upright) types override the crustose (appressed) forms (Figure 5.3; Pentecost 1980). However, the foliose forms are more vulnerable to being removed by abrasion.

In the hydroid *Hydractinia* intraspecific competition between genetically unrelated colonies also takes on various phenotypic forms ranging from passive to aggressive (Figure 5.4; Buss and Grosberg 1990). The former response involves the secretion of a fibrous matrix

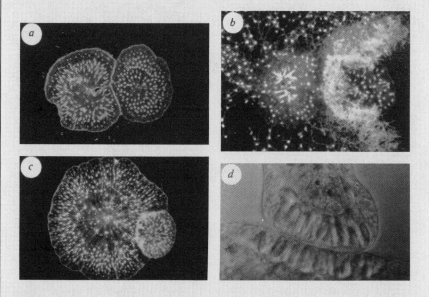

Figure 5.4 Intraspecific encounters between hydroid (*Hydractinia symbiolongicarpus*) colonies of unrelated genotypes. (a) Passive rejection between two colonies without stolons. (b) Aggressive rejection between two colonies with stolons. (c) An initially passive reaction that has escalated into an aggressive response. Hyperplastic stolons have developed at the margin of the larger colony where it contacts the smaller neighbor. (d) Nematocysts at the point of contact between stolons of neighbors. From Buss and Grosberg (1990). Reprinted by permission from Nature, vol. 243, p.63. © 1990, MacMillan Magazines, Ltd.

by both colonies and inhibition of growth at the interface. Active rejection is characterized by production of "a specialized organ of aggression", the hyperplastic stolon, at the tip of which nematocysts accumulate and discharge into the neighboring colony. So, as Buss and Grosberg commented, these simple invertebrates can display a range of aggressive reactions analogous to those of organisms which are much more developmentally and behaviorally complex.

In the fungi, intraspecific contact may be marked by tolerance, fusion (vegetative or somatic compatibility), and potentially the formation of a cooperative ecological unit (physiological individual). This phenomenon has become the standard test for 'self' versus 'nonself' in studies on the distribution of fungal genets (e.g., Dickman and Cook 1989; see also Chapter 2). If the interacting colonies differ genetically at more than a few loci, this is manifested by a darkened zone of interaction in which the hyphae are deformed (vegetative or somatic incompatibility). Interspecific responses may occur at a distance which reflects antibiosis or nutrient depletion (see **Resource acquisition**, below), or following contact. In the latter case, several outcomes are possible (see Chapter 2 in Cooke and Rayner 1984). Interestingly, these responses parallel the interactions among hydroids. The important practical significance of combative responses is that fungi and bacteria can be selected for their competitive, antagonistic ability and used as biological control agents to counteract plant pathogens. For further information see Cook and Baker (1983); Cullen and Andrews (1984); and Handelsman and Parke (1989). To cite but one example, the crown gall disease caused by *Agrobacterium tumefaciens* is controlled by a closely related organism, *A. radiobacter* strain K84 which carries a plasmid coding for the production of an antibiotic toxic specifically to virulent *A. tumefaciens* (e.g., Kerr and Htay, 1974).

is the stage of development. In other words, in response to an unpredicted (i.e., noncyclic) environmental challenge, the mobile organism typically shows behavioral changes, but essentially either lives or dies. In contrast, the entire ontogenetic and morphological program may be disrupted for the sessile organism. At the extreme, it may insulate itself from the environment in some form of structure (seed; resting spore) which can remain dormant for highly variable periods of time. The closest parallel that mobile organisms show to these flexible dormancy responses is a programmed quiescence of relatively fixed duration (hibernation; diapause).

Resource acquisition   The challenge for a sessile organism in acquiring resources is to strike a morphological compromise between optimal search and exploitation strategies. As noted in Chapter 3, how this is accomplished is as much a part of optimal foraging theory, broadly construed, as is the active pursuit of prey. Animals search their environment in organized fashion; modular organisms do so by by systematically locating their feeding sites. Clonal plants, for instance, accomplish this by altering feeding locations (e.g., leaves and roots at horizontal nodes) with spacers (rootless stems; nonphotosynthetic tissue) (A.D. Bell 1984). Bell visualizes four basic feeding behaviors: 1) a single (unitary) organism motile along some sort of foraging route (animal); 2) a single, immobile organism feeding at one site (plant); 3) a clonal organism feeding at many sites reached by growth of a branching system that potentially fragments; and 4) a 'community' of the same organism foraging as a unit along a defined branching structure, each feeder contributing to the organized whole (Bell uses as examples social insects such as the army ant, although slime molds and similar microbes would be the microbial analog).

At one extreme, if the foraging objective is to leave no resource site unexplored, a spiral growth pattern without branching is theoretically the most economical of effort (Chapters 2 and 4 in Stevens 1974). This is analogous to an efficient search strategy for locating an object such as a sunken ship. Fossil evidence shows that the primitive sea slug *Dictyodora* fed in a spiral pattern—a foraging behavior that allowed it to cover the whole of an area (Seilacher 1967). This is considered to be more efficient than a simple 'scribble' pattern, but it covers an area less efficiently than tight 'meandering' (Seilacher 1967) because the spaces between the spirals are not exploited.

At the other extreme, to exploit a pocket of an unlimited resource with minimal overlap, a linear growth pattern is best. The closest approximation to avoiding overlap while leaving no resource zone untapped is to branch and to change the branching pattern as growth proceeds (Harper 1985). Using a simple two-dimensional, hexagonal array of equi-spaced dots, Stevens (1974: Chapter 2) compared four basic patterns—the spiral, meander, explosion (center dot connected directly to each outlying dot), and various forms of branching—in terms of four geometric attributes: uniformity, space-filling, overall length, and directness. Each pattern has advantages. However, if the organism must search and exploit with limited biomass at its disposal, branching, which is both short and direct, is the best compromise. (Similar patterns result in a three-dimensional array but it is noteworthy that the spiral becomes a corkscrew or helix and can not completely fill space, unlike its two-dimensional counterpart.)

In interpreting theoretically optimal search strategies, the issue of compromises to multiple demands (see Chapters 1 and 3, and next section) needs to be kept in mind. Although at any one instant only one resource can be *the* limiting factor, in reality many resources will be limiting over time. Since resource depletion zones are dynamic and vary with the resource, a strategy that may be optimal for capturing a particular resource (e.g., readily diffusible nitrate ions in soil) will not be necessarily so for another (e.g., weakly diffusible phosphate ions in soil). There are other considerations. Search and exploit theory, as conventionally depicted, disregards the fact that the object sought may be providing clues such as nutrients that the searchers can home in on. The object may also resist being taken, perhaps by producing a barrage of antibiotics.

Extending the 'explore versus exploit' argument shows that an open architecture (longer spacers) would promote rapid colonization of new habitat, while a closed architecture (shorter spacers) would favor consolidation of acquired habitat. (This is conventionally depicted in terms of horizontal expansion but is equally true for vertical growth as represented, for example, by tightly or loosely packed canopies of corals and trees.) For plants, the ecological implications have been recognized at least since Salisbury (1942, pp.225-226) contrasted the aggressive vegetative spread of the creeping buttercup (*Ranunculus repens*) with the slow expansion of the figwort (*Scrophularia nodosa*). Lovett Doust (1981) coined the terms "guerrilla" and "phalanx" to describe these respective growth forms which were viewed as representing the opposite poles of a continuum of habits (Figure 5.5). The guerrilla architecture is characterized by long internodes, infrequent branching, spaced modules, and minimum overlap of RDZs. Other examples of guerrilla-type plants include white clover (*Trifolium repens*) and strawberry (*Fragaria vesca*). In the phalanx situation, internodes are short, branching is frequent, and modules are closely packed. Consequently space is tightly occupied and RDZs overlap. Other plant examples include various tussock grasses such as *Deschampsia caespitosa* (Harper 1985) and the common goldenrod (*Solidago canadensis*) (Smith and Palmer 1976).

Clearly the descriptors phalanx and guerrilla also fit the growth forms of other sessile organisms. Indeed, the specific epithets of some animals (e.g., the corals *Pectinia paeonia*, *Pavona cactus*) are derived from architecturally similar plants (Harper et al. 1986). Crustose (mat-forming) lichens grow as phalanxes as do some corals, bryozoans, and clonal ascidians; other corals and the foliose (aerially branched) bryozoans and lichens have, relatively speaking, a guerrilla-type habit (Harper 1981a; 1985). Of the six basic forms of sessile marine animals (Jackson 1979), the formation of 'runners' and 'vines' is essentially a guerrilla habit, while the 'trees', 'plates', 'mounds', and 'sheets' are to varying degrees phalanx in form.

With little imagination one can extend the terminology to certain microbial patterns. Because of the plasticity of microbial growth, the terms can be applied at two levels to growth form. First, as with other modular organisms, the descriptors can be used in a relative sense to compare different taxa. For example, certain fungi, particularly the Zygomycetes and Oomycetes, extend rapidly in guerrilla-like fashion. This is facilitated by their coenocytic nature which permits unimpeded cytoplasmic flow and rapid communication (at the

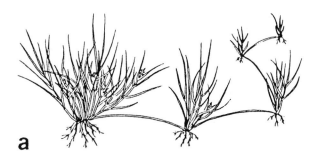

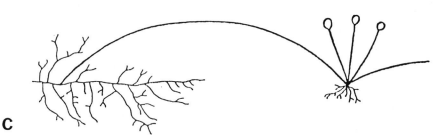

Figure 5.5   Clonal morphology of sessile, branched, modular organisms as represented by plants and fungi. (a) 'Guerrilla', or the extensive type, shown by buffalo grass (*Buchloe dactyloides*). (b) 'Phalanx', or the intensive type, shown by false buffalo grass (*Munroa squarrosa*). From Silander (1985, p.117, based on terminology of Lovett Doust 1981), *in* Population Biology and Evolution of Clonal Organisms, © Yale University Press, 1985. (c) An aerial stolon of *Rhizopus stolonifer* has arisen from behind the margin of a colony growing on nutrient agar, 'rooted' in an unexploited zone of the medium, and produced a new ramet. From Ingold (1965, p.32).

cost to the thallus of vulnerability to damage). *Rhizopus stolonifer*, as its name implies, colonizes substratum rapidly by stolon-like runners which develop rhizoids and tufts of sporangiophores at 'nodes' (Figure 5.5). It is the creeping buttercup of the microbial world. At the other extreme, some fungi such as the apple scab pathogen have a relatively dense growth form and expand slowly and in dense fashion across a standard laboratory medium or their natural substrata.

Second, the terms guerrilla and phalanx can also describe different growth phases of a particular organism (Andrews 1991). Basidiomycetes colonizing woody substrata in soil or on the forest floor, commonly do so in exploratory, guerrilla-like fashion by producing elongated mycelial cord or branching rhizomorph systems (Thompson and Rayner 1982; Thompson 1984; Rayner et al. 1985a). Once the fungus is within a nutrient cache a diffuse mycelial network tends to become established. This is strikingly similar to tropical lianas that initially produce rapidly growing 'searcher' shoots to find a suitable environment which is then colonized by densely branching shoots (Strong and Ray 1975). The extreme in plasticity is exhibited by the cellular slime molds which alternate between an assimilative phase of diffuse, free-living amoebae or plasmodial networks, and an aggregative, migratory (slug or grex) phase before reproduction.

Although plants and fungi share many attributes as sessile, branching, modular organisms, differences exist in their processes of resource capture. Mainly this is because, for the former, much of the resource moves *to the organism*. Sunlight impinges on leaves (although there is a race to avoid becoming shaded by one's neighbors), carbon dioxide and oxygen move by convective flow and diffusion, and inorganic nutrients move quickly by bulk flow driven by transpiration. Contrary to this, fungi typically move *to their resource*. Clonal dispersal by (asexual) spores to new environments is similar to local exploration of a fresh resource by hyphal tips (Jennings 1982). The arrangement of leaves, stems, and roots is well organized and structured, relative to hyphal ramification (as noted above; also see Chapter 7) is much more plastic and involves mutual cooperation. It is in this sense of communal effort that the fungi are like the social insects in resource capture (Figure 5.6), division of labor, and genetic interactions (Rayner and Franks 1987).

Horn (1971) interpreted the geometry of trees and the distribution and shape of leaves largely in terms of resource acquisition. He was not the first to attempt an analysis of growth form, although his work seems to overshadow (as well as disregard) what went before (e.g., Donald 1963; Moss 1964). Horn's synthesis provides an explanation for why succession starts with trees intolerant of shade and proceeds to more tolerant species, rather than simply starting with the latter. The concept is based on calculations showing that net photosynthesis of leaves increases with light intensity up to about 20% of full sunlight. Beyond this, photosynthesis is not enhanced by further increases in light. Hence, strategies for light interception will be important

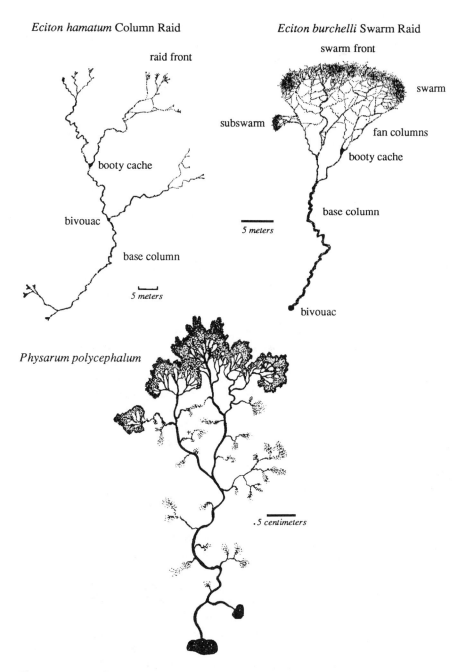

Eciton hamatum Column Raid

raid front

booty cache

bivouac

base column

5 meters

Eciton burchelli Swarm Raid

swarm front

swarm

subswarm

fan columns

booty cache

base column

5 meters

bivouac

Physarum polycephalum

.5 centimeters

Figure 5.6   Similarity in the architecture of growth. (Top) Raiding patterns of two species of army ants. From Rettenmeyer (1963, p.325), reproduced by permission of The Kansas Science Bulletin. (Bottom) Foraging pattern of the plasmodium of the slime mold *Physarum polycephalum*. Drawn from a photograph, courtesy of Tom Volk, of a plate culture.

whenever there is sufficient shade (foliage) to reduce light to less than this 20% threshold value. Leaf placement in crowns of plants was viewed as either monolayered—leaves being concentrated at the periphery with few gaps and little overlap, or multilayered—leaves being randomly scattered vertically and horizontally throughout the canopy. Where light is abundant as in open fields typical of early successional environments, the multilayer was considered to be adaptive. In shaded, late-successional environments, the monolayer form can theoretically grow faster. This is because the multilayer can produce several tiers of leaves in the open habitat without reducing intensity on the lowermost to less than the critical 20% level. In the shade, however, photosynthetic gains of the lowermost tier may not compensate for respiratory losses, so the advantage would swing to the monolayer design.

In rather sharp contrast, the growth form of sessile marine animals (Jackson 1979; McKinney and Jackson 1989) and seaweeds (Neushul 1972; Koehl 1986) has been interpreted relatively less with regard to nutrient acquisition and more in terms of resistance to water turbulence, bottom instability, boring organisms and predators, and in promotion of larval recruitment.

Compromises to multiple demands   It should be reemphasized that growth form, like other attributes of an organism, must be a compromise response to conflicting demands. Too often this seems to be overlooked in the excitment of projecting a favorite hypothesis for the evolution of form. Compelling reasons can be advanced intuitively for particular shapes. The evidence for the role of a specific force is almost always circumstantial and, although subject to confirmation by simulation and laboratory experimentation, does not preclude the possibility that other processes could have produced the same result (Jackson 1979). Hence, there must be continual interplay between observation and experimentation, coupled with cautious interpretation. Is the positioning of sea fans (a type of coral), which are generally oriented perpendicularly to the ocean currents, adaptive in that this arrangement maximizes food interception (Leversee 1976) or is it in fact the *only* physically stable orientation for stiffly pliable fans in currents (Wainwright and Dillon 1969)? Many influences, such as water movements and substrate stability in an aquatic environment, are covariant. Trees, because of their size, have to withstand strong winds, shed snow, translocate fluids in a long plumbing system, and disperse seeds, as well as display leaves for light interception and gaseous exchange. There must be direct trade-offs, for instance between the need to fix carbon and the need for nonphotosynthetic tissue to ensure support, and to provide height and a transport system. What is the role of each (and other factors) in the geometry of plants? Roots serve not only in acquisition of mineral resources but also in water absorption, anchorage, and translocation. The same argument of course applies to fungal hyphae, with the additional requirement that they obtain organic carbon. The shape of plant roots and fungal hyphae must reflect *all* these interacting and interdependent needs. For trees, Horn (1971, p.120) concludes that species typifying any particular successional stage should show a constellation of adaptations: as long as a certain means of dispersal, or an efficient root system, or some other attribute does not *preclude* a particular leaf shape, the characteristics should evolve together.

These morphological trade-offs by sessile organisms to multiple demands nicely illustrate the point made in Chapter 1 (see also synopsis in Chapter 8) that every genotype is necessarily a compromise among different and often opposing selection pressures. Traits have no significance in isolation from the whole organism (Dobzhansky 1956). Natural selection acts on what is available; perfection in form or any other attribute is not necessary for reproductive success and indeed cannot be achieved. All that is required is that that some individuals leave more descendants than others.

5.7   Form in the Natural World

What factors define shape in clonal and modular organisms? An infinite variety of forms is not what is seen the natural world, but rather variations on basic themes (Table 5.2). The repetition of shape in conspicuously different systems (e.g., spiral galaxies versus spiral seashells) is not coincidental; it occurs by the operation through nature of the laws of physics, in particular conservation of energy. Architecture merely reflects the consequences of such phenomena as expansion, packing requirements, stress, surface tension, cracking, and wrinkling. Viewed in these terms, a particular form conveys benefits or efficiencies (Stevens 1974) regardless of the size of the system, or the forces or materials involved: the overall result is a minimization of work or energy.

Table 5.2   Some similarities of form in the natural world

Shape	Represented by
Spiral	Whirlpools; galaxies; eddies; snail and seashells; tusks, horns, and antlers; DNA; phyllotaxis; certain plant viral inclusions; vine tendrils; floret packing in sunflowers
Branching system	Trees; streams; lightning; mycelium; alveoli and bronchioles; arteries; nerves; moose antlers; feathers
Hexagon (three-way joints)	Crystals; virus capsids; packed kernels; cells or bees' or wasps' nests; snowflakes; veinal networks in leaves; rhizome systems
Meander	Rivers; snake movement

Thus, different things, living and nonliving, can have the same spatial configuration. The superficially unrelated phenomena of fluid flow around a cylindrical obstacle, an electric dipole in a two-dimensional electric field, and stress trajectories around a circular hole can all be represented by the same picture (Figure 5.7) (Stevens 1974, p.65; see also Sugihara and May 1990). The underlying commonality is space. Similarly, the three-way joints in nature can result from surface tension (as among adpressed bubbles), from close packing (the cells in a wasp's nest), or from wrinkling (convolutions of the brain). The three mechanisms, though physically different, produce a similar form, the common element being that all result in minimum surfaces (Stevens 1974, Chapters 7 and 8).

Commonality in architecture among organisms can be illustrated by designs based on the sphere (and its variations) and among different kinds of branching systems. Many microorganisms are spheroidal, and the sphere, deformed to various degrees, forms the basic cell type of most multicellular organisms. This shape, as noted in Chapter 4, provides a simple yet robust, efficient structure and serves as a starting point for architectural innovations (e.g., rods, cylinders, branching systems). Unicellular organisms (other than the amoeboid or hard-walled forms such as the diatoms) that depart significantly from spherical have cilia or flagella which influence form as well as function (Thompson 1961, p.81). In the small phytoplankters, departure from a spherical design can help diminish the 'package effect' that reduces the effective specific absorption coefficient of pigment molecules (Kirk 1975a,b; 1976). Small size, nonisodiametric shape, and extensive vacuolation all increase the plasmalemma area per unit cytoplasm area, which promotes nutrient absorption and reduces sinking rate (reviewed by Raven 1986b; see also Chapter 4). A nonspherical shape also reduces sinking rate.

One way a spherical object could grow is by elongating and ultimately branching. Is the manner by which a mycelium branches similar to that of a tree canopy, an arterial or alveolar system or, for that matter, of river tributaries or lightning (Figure 5.8)? All are means of distributing something in space. Paths which might connect scattered points to a common origin obviously could take many forms. Comparing trees with streams and certain

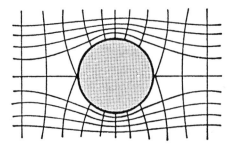

Figure 5.7   Three different physical phenomena—fluid flow about a cylinder; electric dipole in an electric field; and stress trajectories around a circular hole—all can be represented by the same diagram. From Stevens (1974, p.65).

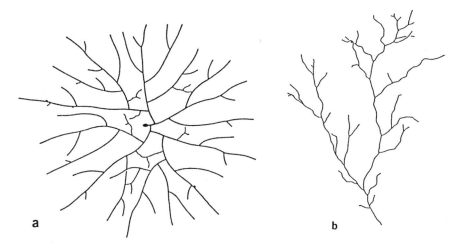

Figure 5.8 Branching patterns in two dimensions from (a) a representative animate object, a young colony of the fungus *Coprinus sterquilinus* (from Buller 1931, p.157); and (b) an inanimate object, a lightning stroke (from Stevens 1974, p.111; original photograph by E. Holbert).

random models, Leopold (1971) proposed that, subject to applicable constraints, the most probable form would minimize the total length of all paths and would be a branching system in which there was, at all potential points of branching, a specific probability that a branch would develop. Considerable data from diverse examples (see below) are consistent with this postulate. The real question is what sets the potential points of branching.

Attempts to analyze biological branching patterns and to define universal properties of branching systems are longstanding (MacDonald 1983). All branching structures can be described in topological (branch order and counting) and geometrical (branch length, angle, diameter) terms. All form either *networks* (closed paths) or *trees* (no closed paths). Cross-linking or anastomosis of branching in some systems (e.g., fungi, hydroids) constrains modelling them as trees. Branching structures may develop within defined limits or be potentially unlimited. The former include river drainage systems, constrained by topographic features, or arterial systems, constrained by body dimensions. The latter include botanical trees or fungal colonies. Various ordering methods exist, the major distinctions among them being whether the numeration is centrifugal (e.g., commencing at the mouth of a river and progressing upstream through the tributaries), centripetal (commencing at the periphery); cumulative or noncumulative; topological only or partly geometrical (for details see Chapter 12 in MacDonald 1983).

Geographers, because of their interest in describing and comparing river systems, were among the first to analyze branching structures. Horton (1945) noted numerous empirical regularities. His ordering method, or Strahler's (1952) variation of it (Figure 5.9), is still commonly used and illustrates one of several approaches. The basic rule for branch (= stream) analysis was that higher order streams can be fed only by lower order streams. So numbering begins at the periphery and proceeds centripetally. Streams of a first order have no tributaries; those of the second order have as tributaries only those of the first order, and so on (for useful comments and comparisons among systems, see Barker et al. 1973; McMahon and Kronauer 1976).

A logarithmic relationship results when either the number (Horton's First Law) or the length (Horton's Second Law) of branches of a given order are plotted against order number. The antilog of the slope in a plot of number of branches versus order number gives the *bifurcation or branching ratio*. In river systems (see Leopold 1971) this averages 3.5 which means that any stream of a given order has three to four branches of the next lower order, when numbered by a centripetal method. Similarly, the antilog of the slope in a plot of stream length gives the *length ratio*. In rivers this is about 2.3, meaning that any branch is typically 2.3-times longer than its average tributary in the next lower order. Leopold also showed that for rivers there can be comparable regularity in drainage basin size and stream order.

The various ratios resulting from these logarithmic relationships enable

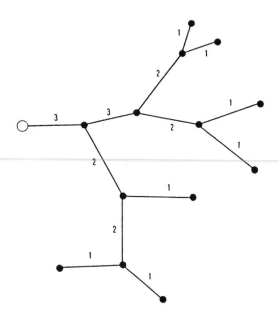

Figure 5.9 The Strahler method for ordering branching systems. From MacDonald (1983, p.112). Reprinted by permission from "Trees and Networks in Biological Models". © 1983, John Wiley and Sons, Ltd.

branching systems to be compared, irrespective of size or origin. The real question is how meaningful are such comparisons. Bifurcation ratios ranging from fairly high (6.5 for a *Fraxinus* sp. tree; Leopold 1971) to low (2.04 for certain bryozoans; Cheetham and Thomsen 1981) have been obtained (Table 5.3). Among microorganisms examined is the fungus *Thamnidium elegans* (Gull 1975), and Prosser (1983) cites in his review unpublished data by Gow and Gooday on mycelia of the yeast *Candida albicans* and by Allan and Prosser on the actinomycete *Streptomyces viridochromogenes* (Table 5.3). According to Prosser (1983) these microbes support the empirical log relationships. However, Horton's method was developed for rivers (confined systems; see above) and in a critical analysis Park (1985a,b) has questioned whether it can be applied appropriately to 'open' (unlimited) branching systems such as those shown by fungi and trees. (Note however that the degree to which one group is confined and the other is not is conjectural; for example, in nature neither trees nor fungi will be strictly unlimited, except possibly early in their growth phase, if then.) The branching patterns he analyzed for 14 fungal species agreed well with those for two tree species. Most fit Horton's law of branch number, but did not correspond with the law for branch length: Mean lengths of branches per branch order tended to show an arithmetical, not a logarith-

Table 5.3 Bifurcation ratios[1] for some branching systems

System	Ratio	Reference	Ordering method
Ash tree (*Fraxinus* sp.)	6.5	Leopold 1971	Horton
Fir tree (*Abies concolor*)	4.8	Leopold 1971	Horton
Apple tree (*Malus* sp.)	4.35	Barker et al. 1973	Strahler
Candida albicans (yeast)	3.86	Prosser 1983	?
Thamnidium elegans (third-order system; filamentous fungus)	3.8	Gull 1975	Horton
River drainage basin	3.5	Leopold 1971	Horton
Streptomyces viridochromyces (actinomycete)	3.1	Prosser 1983	Horton
Bronchial tree	2.81	Barker et al. 1973	?
Thamnidium elegans (fourth-order system)	2.6	Gull 1975	Horton
Cheilostome bryozoans	2.04-2.51	Cheetham et al. 1980	Strahler

[1]Average number of branches of a given order per branch of the next higher order (centripetal numbering system).

mic, relationship. The log function should be found only in situations where growth rate declines progressively in geometric fashion (Park 1985a,b). The arithmetical relationship would be expected when fungal colonies are young and growth conditions are unlimited. When crowding occurs in older cultures and growth rate declines, the length relationship may then change to logarithmic. Thus the pattern one sees depends, among other things, on age of the fungal colonies.

The general importance of specific values, such as branch ratios, remains to be seen. They may prove to be of utility only in a crudely descriptive, comparative context, and when used with caution. For neither trees (see review by Waller and Steingraeber 1985) nor fungi (Park 1985a,b) is the bifurcation ratio a species-specific constant. Indeed for fungi, branch ratios vary within an isolate and single determinations are probably meaningless, as are length ratios (Park 1985b). Nevertheless, the values can reveal variation between trees (early versus late-successional species) and parts of trees (ratio of first- to second-order branches) that may be ecologically significant; at a gross level they show differences between say, trees and bryozoans (Waller and Steingraeber 1986; but see caveats below). Alternatively, the values may have mechanistic significance, as in interpretations of how certain biological systems exhibit minimal resistance to flow. Horsfield (1980) compared flow in variously designed models of bronchioles with that in lungs of dogs and humans. Thompson (1961, pp.125-131; see also Stevens 1974) has shown how the form and arrangement of blood vessels is such that circulation occurs with minimal effort. (For a good discussion on the design of the red blood cell see Lehmann and Huntsman 1961.) The vessels are neither too small—thereby avoiding a large driving force—nor too large, because blood cannot be wasted. Analogous hydraulic conductance considerations apply to all biological piping systems (e.g., for translocation in roots see Fiscus 1986). The relevance of the log relationships for fungi, given the limitations noted above, pertains to optimizing efficiency in accessing nutrients while minimizing the amount of mycelium that needs to be synthesized to do the job (Prosser 1983). Within the context of plant structure, models (Horn 1971; Waller and Steingraeber 1985) have helped clarify how physical development relates to function, and hence have provided a framework for ecological interpretations. For example, the branching patterns of clonal plants affect the degree to which ramets can be integrated (Bell and Tomlinson 1980; Schmid et al. 1988).

To summarize this section, unrelated systems extend themselves in organized fashion by adding progressively smaller, more numerous parts. Clearly this commonality cannot be because the materials accessed or transported are identical or even similar. There is a more fundamental reason which Stevens (1984: Chapters 5 and 6) attributes to how space is utilized: If a branching system is to be distributed uniformly in space it must expand at the margins; this is manifested by a branching pattern in which progressively smaller branches at the periphery outnumber the larger, more central branches. It is worth noting that the rules of branching are the same whether material is being transported centripetally to a locus (e.g., converging stream tributaries), centrifugally to multiple, diffuse termini (electrons in a lightning flash), or both to and from the periphery (air; blood; nutrients and water in trees and hyphae). The particular shape any given pattern assumes is evidently dictated by the law of conservation of energy.

Of what conceptual value in ecology is the fact that branching systems share a common order? Analogies across branching species in different phyla seem tenuous and potentially misleading for three reasons. First, the underlying commonality, utilization of space, is a universal property not unique to biological structures. It is so general and vague as to be of little use in making meaningful parallels and distinctions. Second, form in different biological systems has been shaped to varying degrees by different evolutionary forces. Determining the forces involved and their relative contribution across species is extremely difficult if not impossible. In other words, superficially similar shapes can and do arise for very different reasons. Third, the biological patterns are not fixed, but vary with stage of growth of the organism, and numerous other factors such as module size, age, and interaction with other modules (see above and Waller and Steingraeber 1985). These constraints do not mean that productive insights cannot be obtained from studies of particular branching organisms, only that comparisons are prone to being overdrawn. What can be said is that the diversity of forms seen represents a limited subset of arrays, not any conceivable pattern; that these forms must reflect some sort of balance between the functional needs of an organism and the engineering means available to the organism for meeting those needs; and that physical form profoundly affects the ecology of both unitary and modular creatures.

5.8   Summary

Growth form, that is shape and mode of construction, sets fundamental limits on the biology of organisms. All living things are either basically unitary or modular in design.

Unitary organisms, represented by most mobile animals, follow sequential life cycle phases predictably and their number of appendages is fixed early in ontogeny. Their growth is noniterative (not based on a repeated multicellular unit of construction), and determinate (of strictly limited duration). They display generalized (systemic) senescence, and are generally unbranched. Reproductive value increases with age to some peak and then declines. The genetic individual (genet) and the physiological or numerical individual are the same entity, and are repeated only at the start of each new life cycle.

Modular organisms, including plants, many invertebrates, bacteria, and fungi, grow indeterminately (no genetically fixed upper limit) and in a manner much more subject to environmental influences, by the iteration of a unit of construction (module). Modules may remain attached to the body derived from the zygote or become detached and function as separate physiological individuals (ramets), as among modular clonal organisms. Thus, the ratio of genets:modules is 1:many. Modular organisms are frequently branched and are for the most part sessile or passively mobile. They typically display an internal age structure (e.g., a cohort of leaves is 'born', expands, and dies over a discrete block of time), show absence of generalized senescence, a nonsegregated germplasm, and an increasing reproductive value with age.

The modular paradigm has been developed for plants and many clonal animals (primarily invertebrates); little attempt has been made previously to place bacteria and fungi within the unitary/modular context. Fungi are modular organisms architecturally by virtue of their extension of size by iteration of a hyphal growth unit. Furthermore, some fungi (e.g., those undergoing heterokaryosis) are unique as modular organisms because a given physiological individual may become a true genetic mosaic in a way that other modular creatures cannot, even by somatic mutation (see also Chapter 2).

Bacteria generally grow in nature as microcolonies of cells. Architecturally, bacteria are modular organisms because the individual bacterial cell (module) is iterated indefinitely. (Note, however, that in eukaryotes the term module is reserved for multicellular units.) Genetic recombination in bacteria (Chapter 2) is nonmeiotic but in effect sexual. However, because formation of the recombinant is not tied to a particular divisional event, or a particular morphological structure, or a characteristic life cycle stage, the genet concept is even more vague than it is for the fungi.

The key evolutionary implications of modular design stem directly or indirectly from growth by iteration and sessility. They include: 1) high phenotypic plasticity (shape; size; reproductive potential; growth as a population event); 2) exposure of the same genet to different environments and selection pressures; 3) iteration of germ plasm and the potentially important role for somatic mutation; and 4) potential immortality of the genet. On balance, the modular versus the unitary distinction is probably more biologically significant than the demarcation based on size (microorganisms versus macroorganisms). Unfortunately, its implications are difficult if not impossible to ascertain because of interacting correlates: modular organisms are frequently clonal (and clonal organisms in turn are colonial or solitary), sessile, and branched.

For sessile organisms habitable space is a critical resource which affects survival, and it can limit size and hence reproductive potential. Implications of sessility include: 1) evolution of mechanisms for the individual to reproduce without a mate or to reach a mate and disperse progeny by transport (pollen; bacteriophage) or growth; 2) potentially strong interactions with neighbors; 3) inability to escape from adverse environments, hence means to cope in situ; and 4) development of resource depletion zones and consequently re-source search/exploitation strategies (e.g., guerrilla versus phalanx) from rel-atively fixed positions. Growth form of modular organisms must therefore be a compromise response by the organism to multiple demands which include nutrient acquisition in the face of competition from neighbors; tolerance of abiotic conditions such as water turbulence, wind, or snow loading; dispersal of gametes and zygotes; and plumbing (translocation) constraints.

What factors determine shape in modular and unitary organisms? One sees among architectural form in the natural world variations on basic themes (e.g., the spiral, sphere, or the hexagon) rather than an infinite variety of designs. Branching systems include lightning, rivers, arteries, hyphae, trees, bryozoans, and alveoli. All represent the uniform distribution in space of a particular pattern. Branches can be ordered systematically by various methods for analysis, and because all branching systems are superficially similar, anal-ogies are tempting. Comparisons are tenuous, however, and prone to being overdrawn for three reasons: 1) the underlying commonality (space) is so general as to be of no real value; 2) apparently similar architecture has arisen for very different reasons; and 3) branching parameters are neither necessarily species-specific nor fixed for the lifetime of an individual.

5.9 Suggested Additional Reading

Harper, J.L., B.R. Rosen, J. White (eds). 1986. The growth and form of modular or-ganisms. Phil. Trans. Roy. Soc. B 313: 1-250. Also, Proc. Roy. Soc. B 228: 109-224. Perspectives of various authors on modular growth and its implications.

Hegstrom, R.A. and D.K. Kondepudi. 1990. The handedness of the universe. Sci. Amer. 262(1): 108-115. An interesting overview of asymmetry in nature, from the mo-lecular to the organismal level.

Jackson, J.B.C., L.W. Buss, and R.E. Cook (eds). 1985. Population biology and evolution of clonal organisms. Yale Univ. Press, New Haven, Conn. Proceedings of a sym-posium. Interesting perspectives; inconsistent terminology; absence of microbes from the synthesis.

Sournia, A. 1982. Form and function in marine phytoplankton. Biol. Rev. 57:347-394. The opportunities and constraints of being small and essentially spherical.

Stevens, P.S. 1974. Patterns in nature. Little, Brown and Co., Boston. Fascinating pictorial compendium and stimulating discussion.

Sugihara, G., and R.M. May. 1990. Applications of fractals in ecology. Trends. Ecol. Evol. 5: 79-86

Thompson, D.W. 1961. On growth and form. Abridged edition edited by J.T. Bonner. Cambridge Univ Press, Cambridge, U.K. The classic book on form in nature.

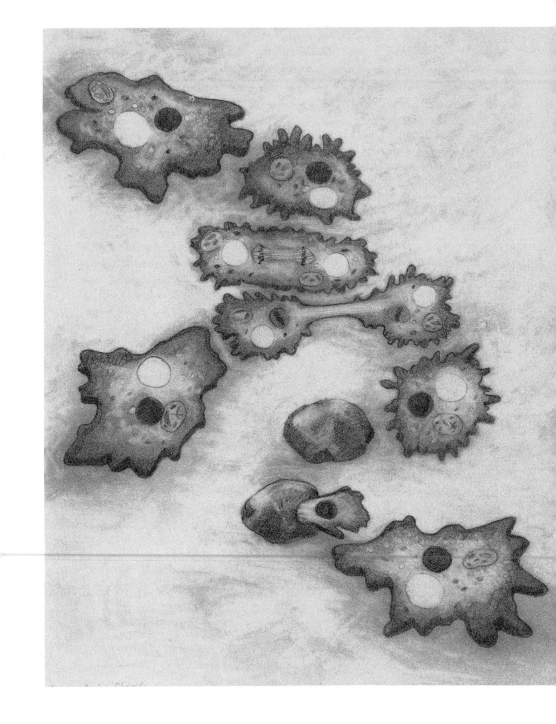

6

The Life Cycle

The "survival machines" of genes are the life cycles that link parental zygotes to progeny zygotes.

—HARPER AND BELL, 1979, p. 30

6.1 Introduction

The importance of the life cycle is captured succinctly by Bonner (1965, p.3) who calls it "the central unit in biology". He has gone on to examine many of the innumerable interesting questions that can be raised within the life cycle context. I shall consider only two issues. The first is the diversity in life cycles among both macro- and microorganisms, which prompts one to ask why such extreme variation has arisen—especially when this appears among fairly closely related taxa. The second concerns whether senescence is a property common to all life, as is generally presumed to be the case.

6.2 Simple Versus Complex Life Cycles

This discussion of life cycles is based on several premises. The first is that, like all other attributes, the life cycle is a compromise response by the organism. The ultimate 'goal' is to maximize the number of descendants. High numbers are facilitated by precious maturity, large litter size, and frequent litters. Thus, other things being equal, the inescapable consequence of some organisms leaving more descendants than others is rapid development and early reproductive maturity. This is seen most dramatically among the bacteria, noted previously (Chapter 5). The major constraints on molding the life cycle are allometric (e.g., macroorganisms simply cannot develop as rapidly as can microorganisms), phylogenetic (e.g., if gametes are nonflagellated they will have to be dispersed passively), and environmental (e.g., if free water is required for spore germination or fertilization, progress through the life cycle will vary accordingly). In practice, the role of the environment will vary from being highly influential (for organisms said to be 'plastic'; Via and Lande 1985) to negligible (organisms 'canalized'; Waddington 1942). Second, natural selection acts on the entire organism, that is, on the life cycle as a whole, so it is misleading to consider specific stages in isolation. One implication is that a deleterious feature can become fixed in one phase (see Section 6.3) if it is more than offset by a positive effect expressed at some other time, such that the number of descendants of the carrier is enhanced overall. Third, any stage has a dual role: to allow the organism to function in its immediate environment while being responsive to informative signals about future environments (see Chapter 7).

There is pronounced variation in life cycles among organisms and it is appropriate to start by asking how life cycles may have arisen. Some four billion years ago, primitive replicator molecules (gene precursors, probably in the form of catalytic RNA molecules) in the primeval soup probably gained a selective advantage by giving up their freedom and becoming protected within 'survival machines' of diverse forms: animals, plants, fungi, bacteria, and viruses (Chapter 2 in Dawkins 1989; however, see also Williams 1966, pp.134-138; for a detailed discussion, see Loomis 1988). Certain gene combinations were undoubtedly more beneficial in some environments than were others. Selection acted to make the combinations more permanent if successful and the unsuccessful simply disappeared. This provides a basis for the match between particular environments and life cycles of the organisms inhabiting them (see Chapter 7). Alternation of periods of gene stability with those of gene dissociation and recombination would have provided for the start of a vegetative/growth and sexual reproduction cycle (Williams 1966, p.137). Also, from the very outset, there would have been an association between environmental change and onset of the sexual phase. All that has happened since then is really just embellishment on this fundamental program.

We find similar improvements in the life cycle across widely divergent taxa. Very different organisms have found similar 'solutions' to common 'problems', perhaps because the ways the life cycle can be advanced are limited. For example, fungal spores and plant pollen, structures which play a major role in dissemination, are modified anatomically and physiologically. Fungi and many marine algae have conspicuous alternation of haploid and diploid generations. The angiosperms display a reduced gametophyte phase associated with internal fertilization analogous to vivipary in animals. One sort of embellishment is the complex life cycle. It appears in many cases (insects and amphibians being conspicuous exceptions) to have been the outcome of a sessile life style. The complex life cycle is a means of accommodating the conflicting needs of competitive growth at a fixed site or time, with genetic exchange and dispersal of offspring to new sites or to better conditions at a later time.

When the young of a species are born into essentially the same habitat as the adults, and do not undergo sudden ontogenetic change, the organism is said to have a *simple life cycle*. Examples include birds, humans, and other mammals. In the semantics of life cycles, complexity is generally considered by ecologists to be "an abrupt ontogenetic change in an individual's morphology, physiology, and behavior, usually associated with a change in habitat" (Wilbur 1980). An organism that passes through two or more such (irreversible) changes would thus have a *complex life cycle* (Istock 1967). Examples include the amphibians, most fish, and most marine organisms (Roughgarden et al. 1988), not to mention many microbes, the case for which is developed below (Figure 6.1). As it progresses through the life cycle, "a cohort of such organisms can be viewed as navigating a landscape of ecological niches" (Werner and Gilliam 1984, p.395), which is not to imply that organisms with a simple life cycle do not also display appreciable changes (e.g., in resource and habitat use).

The classic example of a complex life cycle is that of the frog, which occupies two distinct niches: that of an aquatic herbivore in the tadpole phase, followed by that of a terrestrial carnivore in the adult stage. As Wilbur (1980) notes, it is interesting that two species may be ecologically and morphologically more similar as tadpoles than either is to the adult frog it will become. To a degree this could be said also of species of fungi during the time that they occupy the same resource: In the 'juvenile' feeding phase, characterized by a diffuse hyphal network within a substratum, their shape and energy allocation patterns can be different from the 'mature' reproductive phase, which may involve massive aggregations of hyphal tissue thrust into a new (aerial) environment (see Chapters 3-5). While an analogy to the frog is dramatic, a closer comparison of the fungi is with ferns. In these plants, a prothallus (haploid gametophyte stage) produces many gametes, hence many zygotes are created, and each resulting sporophyte may produce many haploid spores. There is thus a multiplication phase in both stages of the life cycle. Tadpoles, in contrast, do not lay eggs.

The theory of complex life cycles was developed by zoologists. Unfortunately, with rare exception (Lubchenko and Cubit 1980), it has not been extended conceptually beyond the metazoans that undergo conspicuous metamorphosis—viz. amphibians, holometabolous insects, and marine invertebrates. However, many protists (e.g., the sporazoans; most algae) and many fungi fulfill Istock's and Wilbur's criteria outlined above. The heteromorphic algae, of which the red, brown, and green seaweeds are generally good examples, have two separate, ecologically distinct phases that may be so dissimilar morphologically that they have even been classified as separate species. For instance, in the red alga *Porphyra*, a haploid thallus of leafy sheets about 3-15 cm long growing on rocks in the intertidal zone alternates with

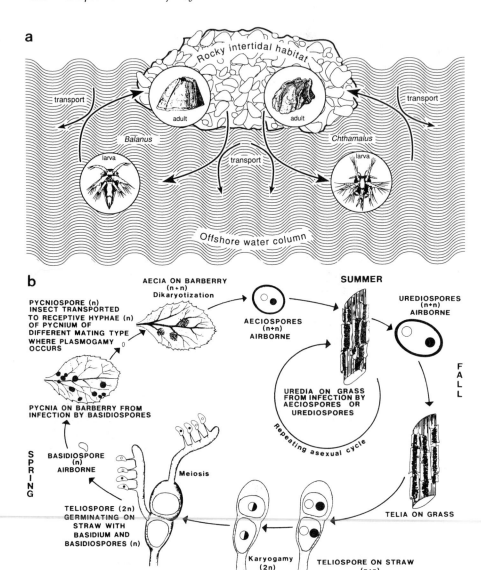

Figure 6.1   The complex life cycle as it pertains to macroorganisms and microorganisms. (a) The tiny, motile larval phases of barnacles (e.g., *Balanus* and *Chthamalus*) live in offshore waters; the sessile adults occupy the rocky intertidal zone. Based on Roughgarden et al. (1988). Barnacles reproduced from Darwin (1854; plates I and XIX) (b) The rust fungus *Puccinia graminis* alternates between the stems of grasses and the leaves of barberry, with multiplication phases on each. The five spore stages clockwise from the bottom right-hand corner are the teliospore, basidispore, pycniospore, aeciospore, and urediospore. Nuclear conditions are indicated as n, n+n (dikaryotic), and 2n. From Roelfs (1985, p.10).

an inconspicuous, shell-inhabiting sporophyte, the *Conchocelis* phase (Lubchenko and Cubit 1980). Frequently the morphological and ecological divergence within the algal life cycle parallels the alternation of haploid with diploid generations, although the phases are often self-perpetuating and neither morphological nor genetic alternation is obligatory (Chapter 11 in Dawson 1966; Lubchenko and Cubit 1980). A similar situation is true of many fungi. The ascomycetes typically have two distinct alternating phases, a perfect or teleomorph state and an imperfect (conidial) or anamorph state. In some instances the conidial stages have not been found and may not exist; conversely, many fungi are known only in an imperfect state (hence the name Fungi Imperfecti or Deuteromycetes) and are generally presumed to be conidial states of ascomycetes whose sexual phase remains undiscovered or has been lost during evolution.

Among the most complex life cycles are those of animal and plant parasites. These include certain rust fungi and many of the nematodes, trematodes, and protozoa that must use two or more hosts in sequence (Chapters 2 and 3 in Price 1980). Usually the sexual phase of the parasite life cycle is associated with a particular host while the asexual stage, functioning mainly to increase and disperse offspring, occurs with a different host. In insects, the functional significance of the sedentary larva or nymph is typically in terms of feeding (exemplified best in the sluggish, cylindrical caterpillar well suited for growth). This 'eating machine' is then transformed into what is by comparison a complex, active, nongrowing adult which functions in dispersal and mating. A similar situation pertains for the amphibians (Wilbur 1980). The converse is true for benthic invertebrates and algae, where the juvenile forms act in dispersal and the main growth stage (adult) is sessile (Figure 6.1a and Thorson 1950; Roughgarden et al. 1988). Thus for organisms with complex life cycles, important activities appear to be segregated by stage, the organism moving progressively from one functional compartment into another, broadly speaking by metamorphosis. It is amazing to reflect on the radical differences in size and shape (and on the underlying genetic program) that a single genet can display. This is nowhere more apparent than in the holometabolous insects, where the organism progresses from egg through larva to pupa and adult. Although other organisms such as the algae and fungi display remarkable differences in form through the life cycle, the genet is discontinuous because the organism moves from a haploid (and in some cases dikaryotic) to a diploid state.

Unlike the case for frogs, it is unclear for many taxa whether Istock's and Wilbur's criteria for a complex life cycle have been met. Water snakes, lizards, and numerous invertebrates all show size-related, ontogenetic shifts in resource use (Werner and Gilliam 1984), yet traditionally have been classified as having a simple life cycle. The essence of the complex life cycle would seem to be that an organism comes to occupy a new ecological niche in a new environment by undergoing pronounced ontogenetic changes in executing its life cycle. This is of course a matter of degree, and a continuum of life cycle types can be imagined. What is the interpretation when nymphs and adults live in the same microhabitat and eat the same foods? According to Broadhead (1958) there is no ontogenetic difference in niches. Higher plants which have a simple gametophyte stage (e.g., the Angiosperms and others without a conspicuous alteration of generations) represent a more complicated case but on balance one that seems closer to a complex life cycle. The seedling is only a small version of the adult and the habitats are similar (typically more so for annual plants and less so for large trees), a condition typical for a simple life cycle. However, other characteristics, including a transport phase, a possibly lengthy period as a quiescent embryo in the seed bank of the soil, and a population biology involving both adult-adult interactions as well as resupply (and outflow) of juveniles (seeds) are more likely to be found in organisms with a complex life cycle.

The case for parasites can also be equivocal. Here one must distinguish complexity in the epidemiological sense from that of the life cycle of the causal organism. Many parasites attack a single host species. Although the *epidemiology* of the resulting disease may be complex, it does not follow that the *incitant* has a complex life cycle. In other cases, several species may potentially be hosts, yet neither is their sequential involvement mandatory for completion of the parasite's life cycle, nor does the parasite have to undergo an irreversible ontogenetic change. The epidemiology of scrub typhus presents a fascinatingly intricate case that illustrates these points. The concurrence of events culminating in human infection depends on the presence of: 1) the bacterial pathogen, *Rickettsia tsutsugamushi*; 2) the vector, chiggers (trombiculid mites); 3) small mammals, particularly rats, on which the chiggers feed; and 4) transitional or secondary forms of vegetation ('scrub'), through which humans chance to pass. The intimate relationship of these four factors has led to the descriptor "zoonotic tetrad" (Traub and Wisseman 1974). However, chiggers are the primary reservoir of the causal organism, and the rickettsia need not pass through either rodents or humans to complete their life cycle. The cycle is, therefore, *not* complex in the ecological sense of Istock and Wilbur.

In contrast, the malarial parasite, *Plasmodium*, must pass through both the mosquito and human to complete its life cycle, during which it undergoes alternation of generations and several morphological states. Similarly, the liver fluke (class Trematoda) qualifies as having a complex life cycle by virtue of obligately passing through snails and humans (or usually another vertebrate such as dogs, cats, pigs, or sheep) to complete its life cycle. In the process the parasite passes sequentially from zygotes to ciliated larvae (miracidia), sporocysts, nonciliated larvae (rediae), tadpole-like larvae (cercariae), metacercariae, young and finally adult stages from which are produced the zygotes.

Many fungal plant pathogens (such as those belonging to the Ascomycetes, alluded to above) pass the growing season in an asexual phase which consists largely of repetitive conidiation on the leaves of the living host. These fungi then complete their life cycle with a morphologically distinct, sexual, over-wintering stage in leaf litter. Although there is no host alternation, the cycle does involve fungal morphogenesis and two distinct niches. Thus, according to Wilbur's and Istock's criteria, these organisms have a complex life cycle.

The plant parasitic rust fungi as organisms with a complex life cycle
The rust fungi illustrate nicely some general principles of the complex life cycle. As a group they have undergone some 250-300 million years of phylogenetic history since their ancestors became established as parasites of ferns in the Carboniferous Age (Savile 1971a; Anikster and Wahl 1979). Unlike most fungi, the rusts do not grow saprophytically in nature and are thus closely coevolved with the living host. During the course of their life cycle they are also distinctive in being able to exhibit, at least at some point in evolutionary time, numerous morphological states, including many spore forms (Figure 6.1b). Two significant features stand out in the life history of rusts as a group: Most species have numerous spore forms which appear in a definite order, and heteroecism (host alternation) is common (Savile 1976).

In the basic life cycle (Figure 6.1b) a complete sequence of spore types—which characterizes a so-called macrocyclic rust—consists of teliospores, sporidia (basidiospores), pycniospores (spermatia), aeciospores, urediospores, and teliospores again to complete the cycle. Thus, unlike the frog or the fern, multiplication occurs at *many* points in the life cycle. Apart from the basidiospores, which are borne naked on a modified basidium, all the spore stages are produced within specialized structures (sori) called, respectively, pycnia (spermogonia), aecia, uredia, and telia. Species lacking the uredial stage are termed demicyclic; where just the pycnial and telial stages or only the telial stage are present, the cycle is microcyclic. All the microcyclic rusts are of necessity also autoecious (basidiospores reinfect the same host species), although the converse is not true.

Puccinia graminis, the stem rust fungus, is distributed worldwide and has tremendous economic impact. Not surprisingly, it is not only the focal point of rust investigations, but is one of the most intensively studied of all organisms. The fungus is a macrocyclic, heteroecious species which forms uredia and telia on the Gramineae and alternate stages on various *Berberis* species or rarely on *Mahonia* (Figure 6.1b). This is in contrast to other life cycles represented in the *Puccinia* genus: for instance, *P. asparagi* is a macrocyclic species but autoecious, while *P. malvacearum* (hollyhock rust) is microcyclic. For *P. graminis* or any other heteroecious rust to complete its life cycle, geographic overlap of the alternate hosts (or effective dispersal of appropriate infective forms) is obviously a necessity. Details of the life cycles of various rusts can be found in Anikster and Wahl (1979), Bushnell and Roelfs (1984), and Roelfs and Bushnell (1985); here I focus instead on aspects that relate to the evolution of complexity or simplicity in the life cycle.

In numerous publications, Savile (1953, 1955, 1971a,b, 1976) reconstructs the evolution of the rusts based on evidence derived from comparative morphology, host relationships, and biogeography. The brief synopsis in the following paragraphs is based mainly on his work. Significantly, the rusts have apparently been derived invariably from parasitic ancestors (i.e., not secondarily from free-living forms as is believed to be the case for many parasitic animals). Probably the ancestral form arose as a parasite of ferns and moved with its host as drier subtropical regions were invaded from the tropical swamps (Savile 1955). The ancestral type was thus autoecious. Monokaryotic basidiospores produced from teliospores infected the fern fronds. It is the simple teliospores—originally merely rounded-up mycelial cells in or below the host epidermis—and the basidiospores, that were the first spore states. These indicate the origin of the rusts (Savile 1955, 1971a). The teliospore was and remains the site of nuclear fusion and survival (overwintering in temperate zones; oversummering in hot, dry Mediterranean climates). The dikaryotic stage was established by fusion of sexually compatible hyphae produced by the germinating basidiospores, and the cycle was completed by a new round of teliospore production. Specialized, dikaryotic (uredia) and monokaryotic (pycnia) states evolved later, the former facilitating long-range dispersal and the latter, nuclear transfer.

Having arisen fortuitously, heteroecism was maintained because susceptible fern tissue was scarce when basidiospores were being discharged. Whatever its origin, the advent of host alternation must have carried considerable benefit because the increased distance of travel between two hosts involves higher costs in population death than would travel between tissues of the same plant. (The practical implication of host alternation is that frequently an effective disease control strategy for heteroecious rusts is to eradicate the alternate host.) Heteroecism presumably fostered evolution of uredia and pycnia, but it was also the driving force behind formation of the aecium, the final stage to appear (Savile 1955, 1976). Gymnosperms were probably the ancestral alternate host and became the recipients of the pycnidial and primary uredial stages. These plants typically have a much tougher epidermis than do the ferns. This mechanical problem was met by various structural modifications of the primary uredinium, which became the aecium.

Just as environmental conditions promoted heteroecism and the associated life cycle stages, so did the environment set the stage for evolution of autoecism and the microcyclic rusts (Jackson 1931; Savile 1976; Anikster and Wahl 1979). While it was controversial for some time whether the microcyclic forms gave rise to the macrocyclic, or vice versa, it is now generally agreed that the body of evidence (primarily cytological and morphological) indicates that microcyclic forms descended from the corresponding macrocyclic species. There are two principal ways that this change may have occurred (Savile 1976). First, a macrocyclic, autoecious rust could develop on the aecial host of the parental heteroecious species. This phenomenon would give the new form greater ecological amplitude because it would be freed from regions of host overlap; the process would thus be a major means of speciation. Shortening of the life cycle could then follow gradually and be associated frequently with appreciable evolutionary change of the teliospore. Second, there may be an immediate conversion to the microcycle, again always on the aecial host.

The benefit of the microcycle is strikingly apparent in arctic, alpine, or desert areas where the growing season for the host and hence the parasite may be only a few weeks (Savile 1953, 1971a, 1976). (For an analogous discussion of environmental constraints as they influence amphibian life cycles, see Wilbur and Collins [1973].) Convergent similarities among rust species include not only autoecism and short-cycling, but also self-fertility and suppression of pycnia. Savile (1953, 1976) contrasts the life cycles of heteroecious rusts along a latitudinal gradient from temperate zones to the Arctic. In the former, most of the species have a macrocycle similar to that described above for *Puccinia*. Numerous uredial generations occur and the rusts can persist with the alternate hosts relatively far (many meters) apart. As one moves northward, time available for the repeating stage decreases progressively; near the tree line the identities of the plant species involved become obvious because the rust cannot survive unless the alternate hosts are within about 50 cm of each other. Just inside the tree line the plants must be contiguous. Beyond the tree line the only heteroecious rusts are those that can abbreviate the life cycle, for example by self-fertility or by producing dispersible (diasporic) rather than sessile teliospores, or by 'double-tracking' (Savile 1953, 1976) the life cycle: *Chrysomyxa*, for example, overwinters as a dikaryotic perennating mycelium rather than as a telium in the leaves of its evergreen host, *Ledum*. In the spring this mycelium produces uredia and telia nearly simultaneously. These sori mature to produce their respective spore forms. The urediospores infect *Ledum*; the teliospores germinate in situ to produce airborne basidiospores which infect the alternate host, *Picea*. The life cycle is thus abbreviated because gene recombination (teliospore/basidiospore phase) and dispersal/multiplication (urediospore phase) proceed concurrently.

So, what does the diversity of life cycles among the rusts tell us about the complex life cycle? First, by their heteroecism-to-autoecism and macrocycle-to-microcycle transitions, these fungi exemplify the postulate that natural selection acts to adjust the length of time spent in a particular life stage to maximize lifetime reproductive success. 'Telescoping' of the rust life cycle by the dropping of one or more stages is in effect an example of progenesis, a term used in developmental biology to refer to abbreviated ontogeny by accelerated (precocious) sexual maturation. Gould (1977, Chapter 9) reviews some of the many examples among animals, including animal parasites (e.g., the cestodes and certain copepods). There are numerous instances where a parasitic species is progenotic with respect to free-living relatives or related parasites. This has been accomplished by adopting a simplified life cycle, from what was originally a complex life cycle, by deleting the terminal host. Progenesis may be a life history specialization for *r*-selecting environments where the premium is on early reproduction as well as high fecundity (see the concept of reproductive value in Section 6.3). A parasitic lifestyle (e.g., Esch et al. 1990) is often one associated with *r*-selection (Gould 1977, pp.328-330; Andrews and Rouse 1982).

The parallel for the amphibia (Wilbur 1980, 1984) is in the timing of metamorphosis from egg to juvenile to adult. What is the optimal time to spend in each stage? Small ponds fed by runoff from rainfall provide nutrient-rich conditions for rapid growth of juveniles. At the same time, however, death from predators is high, the ponds are ephemeral, and the juvenile phase is prone to be killed by desiccation. In becoming an adult the organism escapes from this environment. If a stage is particularly vulnerable, selection pressure should act to shorten or eliminate it (Chapter 3 in Williams 1966). There are several instances where direct development has evolved independently among amphibians in montane and rainforest habitats (Wilbur 1980). This may have arisen because the advantages of rapid juvenile growth are offset by high mortality (Wilbur 1980). Abbreviation of the rust life cycle from the heteroecious, macrocyclic form to the autoecious, microcyclic type is directly analogous to elimination of either the terrestrial or the aquatic phase by amphibians. A notable difference is that while rusts can multiply at every stage, the frog cannot.

Costs, benefits, and the origin of the complex life cycle   Complex life cycles may allow organisms to exploit transient opportunities (Wilbur 1980) and forego certain compromises so as to concentrate instead on particular activities such as dormancy, growth, reproduction, or dispersal. As we have seen in the rusts, the uredium and subsequently the aecium evolved as the long-range, repeating, dispersal stages; the sexual stage is the telium; the pycnial stage is specialized for nuclear transfer. This complex life cycle and the associated morphological elaboration is unique among the fungi. Transient opportunities—for example in the form of an accessible alternate host at precisely the appropriate stage of susceptibility and in an environment conducive for infection—are clearly just as apparent for parasites with complex life cycles as they are for say, anuran larvae in resource-rich but temporary ponds, or insect larvae that feed on carrion or seeds. Thus, the opportunity is also the gamble for organisms with complex life cycles. A major cost in complexity is vulnerability to failure: the system is only as robust as its weakest link. This is apparent for complexity in terms of number of cell types seen earlier (Chapter 4), and is evident again here in the life cycle context.

Because organisms with complex life cycles exhibit compartmentation of stages, the opportunity arises for independent evolution in the different phases. The extent to which this may occur is contested (Istock 1967; Strathmann 1974; Bonner 1982b, pp.226-229). Based on theoretical life tables (i.e., those that account for age-specific survivorship) and tables of reproduction, Istock (1967) proposed that the adaptiveness of the two stages of a hypothetical organism with a larval and an adult form is largely independent. He suggested further that it is this independence that usually makes complex life cycles unstable over evolutionary time because the flow rates of individuals (e.g., of larvae from the larval to the adult environment; of progeny from the adult to the larval environment; Figure 6.1a) between the phases are unbalanced. A nice example of the interaction between the two phases is described for marine intertidal invertebrates by Roughgarden et al. (1988). The ecological advantages of the complex life cycle are viewed as being fully exploited only where both the larval and adult environments are saturated (Istock 1967). Even if achieved, this condition is very difficult to maintain, leading eventually to reduction or loss of one of the stages. Strathmann (1974), on the other hand, using marine invertebrates as an example, emphasized that evolutionary feedback acts in regulatory fashion to integrate the various phases. Slade and Wassersug (1975) criticized Istock's model on mathematical grounds and, because the basis rests on density-dependent regulation, they argued that it would not apply to r-strategists with complex life cycles. Although the model is conceptually appealing, it would seem unlikely to describe microbes (and various other organisms) because it depends on the premise of density-dependent mortality, which disregards the role of environmental factors. A more general alternative theory has yet to be devised.

Each stage of the organism with a complex life cycle must follow on compatibly from its predecessor, respond to its immediate environment, and in essence prepare the organism for the next stage and future conditions (Chapter 3 in Williams 1966). What influence does having to cope with grossly different environments have on the informational machinery of the organism? Do rusts with five life cycle stages on two obligatory hosts have correspondingly more DNA, or merely different programming controls, than those with only two stages on a single host (Andrews and Harris 1986)? In a similar context, Williams contrasted the simple life cycle of a sheep with the complex life cycle of the liver fluke within it. He went on to ask whether the informational content of the human embryo would have to be increased to accommodate life like a tadpole on a pond bottom. Questions of the size and regulation of DNA as it may pertain to life cycle complexity are especially intriguing since there is no consistent relationship between haploid DNA content and structural or physiological intricacy. As shown earlier (see Chapters 2 and 4 and Figure 4.6), some amphibian and plant cells have 30 times as much DNA as do human cells (see also Watson et al. 1987, p.622). While this makes speculation tempting, the matter cannot be resolved for several reasons: 1) as noted, there is great variation in DNA content among taxa; 2) life cycle complexity cannot be isolated as a factor and compared strictly against genome size; and 3) only a small and perhaps variable fraction of DNA is believed to code for essential proteins.

Finally, what is the origin of complex life cycles and why they have been maintained? Istock (1967) suggests that during the evolution of life there were probably episodes, such as the colonization of new habitats, when adoption of such cycles was particularly advantageous. As organisms underwent adaptive radiation, these opportunities were initially high (as in colonization of the ancient seas or the origins of many symbiotic associations; Istock 1967). Over geological time these changes presumably declined, while those for parasitic forms may have increased with increasing speciation (Istock 1967). In the rusts, for example, evolution of the complex life cycle seems merely to have been one mechanism whereby certain fungi were able to advance from the primeval swamp into otherwise inhospitable environments. The unusual complexity of the rust cycle is unique within the fungal world and, if assessed on its prevalence among taxa, must therefore be judged a relatively unsuccessful evolutionary experiment. Nevertheless, it persists. The rust/plant interaction seems to be an excellent example of Harper's (1982) observation that evolution generally pushes organisms to become more specialized; as a particular case, partners in the coevolutionary spiral drive each other into "ever deepening ruts of specialization" (see also Chapter 3 and Huffaker 1964).

Istock's prediction that complex life cycles should be unstable is consistent with observational evidence from the rusts and other taxa that current evolutionary trends are towards simplification. Why some such cycles are stabilized is unclear, but it may be simply that they will persist until all heritable phenotypic variation for improvement is exhausted by natural selection (Istock 1970). This appears to be nothing more than a specific case of the phylogenetic principle (Savile 1955) that within any lineage a period of elaboration is followed by one of simplification. Alternatively, complex life cycles could be maintained by: 1) limiting mortality equally in the various phases of the life cycle (Strathmann 1974), or 2) environmental heterogeneity (Slade and Wassersug 1975). The latter is illustrated elegantly by Lubchenko and Cubit (1980) who propose that the heteromorphic life history of certain marine algae can be interpreted as evolved responses to seasonal herbivory. The upright life cycle stages facilitate competition, colonization, and high rates of reproduction when grazing pressure is low; the crustose life cycle forms enable survival when grazing pressure is high. Because these respective properties are mutually exclusive, a single morphotype would be a poor compromise. The authors predict isomorphic algal species would predominate if grazing intensity were constant. Analogously, in the relatively aseasonal environment of the tropics, frogs and their anuran relatives have reverted to direct development (Wassersug 1974; Slade and Wassersug 1975); heteroecism (or dormancy) among the rusts would serve little purpose under such conditions.

6.3 Senescence

At the level of the individual, senescence is "a general title for the group of effects that, in various phyla, lead to a decreasing expectation of life with increasing age" (Comfort 1979, p.7). Being an organismal phenomenon, this is distinct from the programmed 'senescence' of cells and tissues that occurs routinely as part of ontogeny and morphogenesis. However, some of the mechanisms may be similar or identical, especially if senescence of the individual is also a genetically programmed process (see below and Sugawara et al. 1990). What is termed senescence of individuals in botany is usually called aging in medicine and zoology. However, synonymous usage of the terms can be ambiguous because only some of the many physiological changes accompanying aging are truly senescent in the sense of being deteriorative and thereby reducing the probability of survival. For instance, loss of hair and change in hair color, in themselves, are innocuous correlates of aging, whereas increase in blood pressure and decrease in auditory or visual acuity, reaction time, and bone strength are senescent changes (Figure 6.2).

At the population level, senescence is quantifiable statistically within a cohort as an increasing mortality (decreasing survivorship) rate with age. This presupposes that sufficiently large numbers of the organism pass through early to mid-life for the effects to be evident *numerically*, a condition occasionally but not often realized in the wild. The resulting hypothetical plot (Figure 6.3), with more-or-less extreme variation, is descriptive of most animals (indeed of unitary organisms generally). For interpretations of senescence, such curves have many shortcomings (Chapter 1 in Comfort 1979). For example, mortality factors should be random, not age-selective, while in practice all natural populations will be subjected to death from both random and age-distributed causes.

If the mortality rate is constant with age, the number of survivors declines exponentially with time (Figure 6.4). Obviously this may occur rapidly or very slowly depending on the coefficient of the exponent. Because a constant *fraction* of the population dies per unit time, the curve never reaches zero and this has led many authors (e.g., Buss 1983; Jackson 1985; Watkinson and White 1985) to infer that some genetic individuals must be immortal. This is a probabilistic event, however, and as Comfort (1979, p.27) points out "is not more significant than the 'potential' meeting of any pair of railway metals at infinity". Thus, while clonal organisms such as strawberries or corals can spread "theoretically without limit", actually this will occur only as long as the peripheral ramets encounter favorable conditions; eventually the environment would become hostile or they will be out-competed (Williams 1975, pp.26-27).

The only advantage nonsenescent individuals may have over those that senesce is that their chances of dying at any *particular* age do not increase with time. The demise of glass tumblers in a cafeteria follows an approximate nonsenescent decline (Brown and Flood 1947) as do many species of animals in nature. For instance, the annual mortality rate for most wild birds is between 30-60% (Comfort 1979, p.141-142). In captivity, birds generally live more than twice as long as mammals of the same size (Schmidt-Nielsen 1984,

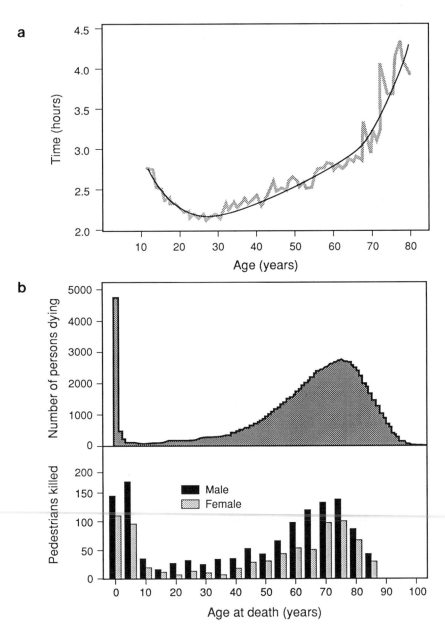

Figure 6.2 Manifestations of senescence. (a) Declining functional capacity with age indicated by the fastest times, by runner's age, in the New York Marathon. Gray line represents data; black line represents the trend. Redrawn from Walford (1983, p.79). (b) Distribution of deaths from all causes (top) is very similar to that of pedestrain deaths in road accidents (bottom), indicating progressive vulnerability with age. From Comfort (1979, p.25; original data from De Silva 1938, and Lauer 1952). Reprinted by permission of the publishers from ''The Biology of Senescence'', 3rd ed. © 1979 by Elsevier Science Publishing Co., Inc.

Figure 6.3 Survival curve of a cohort in which mortality increases with age, that is, for a population that senesces. From Comfort (1979, p.22). Reprinted by permission of the publisher from "The Biology of Senescence", 3rd ed. © 1979 by Elsevier Science Publishing Co., Inc.

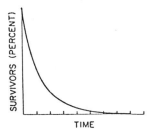

Figure 6.4 Survival curve for a cohort with a constant rate of mortality (50% per unit time) with age. From Comfort (1979, p.22). There is no evidence for senescence from such a plot. Reprinted by permission of the publisher from "The Biology of Senescence", 3rd ed. © 1979 by Elsevier Science Publishing Co., Inc.

p.147; longevity of captive mammals varies directly with body size), and senescence is detectable. Nonsenescent curves frequently imply simply that populations in the wild are pruned by high infant and subsequently high adult mortality before senescent decline can appear in actuarial terms. Therefore, such plots are not necessarily evidence against intrinsic senescence of the individual. An intriguing related point is that evidently there is no true menopause in nonhuman primates (Graham et al. 1979), presumably because death removes all females while still fertile, so that the trait has not had a chance to evolve as a secondary characteristic of long life.

Finally, for some organisms, there may be an increasing expectation of life over time (Pearl 1940; Haldane 1953; Chapter 1 in Comfort 1979) evidenced by declining mortality rates with age. Apparently this is the case for some fish and trees (Comfort, p.23). A constant or declining mortality rate would be anticipated in general for modular clonal organisms (see below). Note, however, that again this refers to death at any *specific* age. *Cumulative* probability for death must increase with length of life. Unless the mortality rate dropped to zero (and there is no evidence that it does or, given stochastic events, that it ever could), this would still mean that these organisms are at best nonsenescent and potentially very long-lived, *not immortal*. On theoretical grounds, senescence will "tend to creep in" (Hamilton 1966), but only if residual reproductive value (see below) does not go on increasing. Modular organisms will be selected for "infinite" (more realistically stated as indefinite) life if reproductive value exceeds present reproductive capacity—the case if individuals (e.g., the duckweeds) grow exponentially (Sackville Hamilton et al. 1987).

Variation among taxa in the occurrence of senescence is considerable. On balance, Comfort's (1979, p.109) conclusion for the invertebrates that senescence "probably occurs in every *group* [emphasis added] where the power of regeneration or fissile reproduction is less than total, or where body cells are not continuously and 'indeterminately' replaced" could be generalized to all biological taxa. However, the only complete actuarial data are for humans. Substantial information exists for certain other higher organisms such as zoo and domestic animals, annual and some perennial plants (such as trees), and for selected species used in research. Elsewhere, data are few and conclusions have been drawn largely by extrapolation and educated guesswork, occasionally based only on age of an organism at death. The latter conveys some information on longevity. As a single statistic, it is not a reliable measure of senescence, which is a dynamic, continuing process. Although some actuarial evidence can be compiled from existing records or by inference from mortality tallied across all age groups at one time (Comfort p.53), the best method is to follow a particular cohort under optimal conditions for its lifespan. This is usually technically difficult and sometimes operationally impossible to do.

As will become apparent in the following section, evolutionary interpretations of senescence require that a distinction can be made between soma and germ line (Williams 1957; Watkinson and White 1985) or, more broadly speaking, between parent and offspring, whether progeny are sexually or asexually produced (G. Bell 1984). Thus it is useful to consider organisms from the standpoint of their reproductive modes, which I discuss below.

Evidence for senescence among macroorganisms   Among animals there is either good actuarial evidence or a strong basis (Chapter 2 in Comfort 1979; Finch and Hayflick 1977; Cutler 1984) for inferring the occurrence of senescence in essentially all members of taxa reproducing exclusively or primarily sexually (ovigerous organisms; G. Bell 1984). This includes the vertebrates and many invertebrates such as the nematodes, rotifers, crustaceans, and insects. It is fascinating that nontumorigenic cells from these species (most evidence is from humans) grown in vitro have a finite divisional or doubling lifespan (the 'Hayflick limit') that correlates well in a relative sense with senescence in vivo (Goldstein 1974; Hayflick 1965,1977; Kirkwood 1984). Cells undergo fewer doublings when taken from older as opposed to younger organisms, as they do from organisms with shorter as opposed to longer life spans.

Finite divisional capacity may be associated with progressive loss of DNA from the specialized termini (teleomeres) of eukaryotic chromosomes. Teleomeres of cultured human fibroblasts shorten during aging (Harley et al. 1990), but whether this is a cause or consequence of senescence is unresolved. Human and to some extent other animal cells have thus become a routine in vitro experimental model for mechanistic studies of cellular senescence in vivo, the implicit assumption being that such extrapolations can be made (for discussion of this point, see e.g., Schneider and Mitsui 1976). The standard criterion of senescence ('aging') is lifespan of the experimental cell line as assessed by replicative capability of the cells. A further inference is of course that events at the *cellular* level can explain senescence manifested at the *organismal* level. G. Bell (1988, p.100) points out that cellular senescence is probably the consequence of senescence of the whole organism rather than the other way round.

Among invertebrates where clonal growth (asexual reproduction) is prominent in the life cycle, data are either strongly against or equivocal for senescence. The strength of the case against senescence appears to depend largely on two factors. The first is whether the products of clonal growth are relatively equal in size and require only limited development to reach reproductive maturity (paratomical fission), as opposed to being markedly smaller and much less differentiated than the parent (architomical fission) (G. Bell 1984). As discussed later, evolutionary theory predicts that strength of selection for senescence should decline in the order of ovigerous to architomical to paratomical organisms. G. Bell (1984) provided limited evidence from the culture of six freshwater invertebrates that survival decreased significantly with age in the ovigerous animals (two rotifers, an ostracod, and a cladoceran), but did not change for two paratomical oligochaetes. This is consistent with predictions. However, to differentiate clearly between paratomical and architomical categories seems ambiguous and difficult. A further complication is that the frequency, timing, and overall importance of asexual reproduction in the life cycle can be expected to vary within a particular taxon depending on the environment (e.g., for marine invertebrates see Hughes and Cancino 1985).

The second factor mitigating against senescence is that the clonal organism grows indeterminately, that is, without definite restrictions or limits because the ultimate size of the *clone* is in principle unlimited. It is the fate of the clone as the *genetic* individual that is of evolutionary significance. Because the extent of clones has rarely been mapped, not much can be said quantitatively about mortality rates. There is broad agreement that genets of clonal benthic invertebrates survive substantially longer than their constituent ramets (Cook 1985; Hughes and Cancino 1985; see also Section 5.5), and that fecundity and survivorship increase with size (of the physiological individual and, where known, of the genet; Jackson 1985). Sponges and corals, for instance, have been estimated at ages of from one to several centuries (Jackson 1985). With time, encrusting organisms may die locally, become subdivided, and thus exist as clonal fragments over an unknown area. Jackson's review shows that sponges, hydrozoans, bryozoans, ascidians, and corals periodically degenerate ('regress') locally, but regenerate from other areas and so do not senesce as a clone. This is in striking contrast to the rotifers, for example, where the zygote produces a species-specific number of cells after which subsequent cell division ceases (Buss 1985); the powers of regeneration are negligible and senescence is well documented (G. Bell 1984; Chapter 2 in Comfort 1979).

The senescence of plant organs such as leaves (Woolhouse 1984) and of annual plants after fruit maturation (Woolhouse 1967, 1984; Leopold 1980) has long been recognized, though poorly understood mechanistically (Sexton and Woolhouse 1984). After a variable period of vegetative growth these plants undergo one round of reproduction followed by death within a season, that is, they are monocarpic (but see Watkinson and White 1985) or semelparous in zoological semantics. Their fecundity rises steeply from zero to a peak and then drops precipitously (Figure 6.5a).

In contrast, the reproductive schedule of almost all perennials is polycarpic (iteroparous; i.e., they undergo multiple rounds of reproduction) because the genet can continue to form new meristems, which themselves reproduce, and so on. The fecundity schedule depends on whether the perennial has a single meristem or multiple apical meristems. In the former case, as represented by the coconut palm (*Cocos nucifera*), seed production can be expected to rise and then level off for an indefinite period (Figure 6.5b) (Watkinson and White 1985). Where there are multiple meristems, the outcome

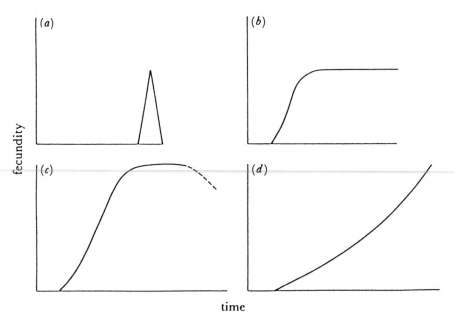

Figure 6.5   Reproductive schedules (relative fecundity over time) for (a) semelparous plants (i.e., those with one round of reproduction); (b) iteroparous plants (those with multiple rounds of reproduction) with a single shoot; and iteroparous plants with multiple shoots that are (c) aclonal, or (d) clonal. From Watkinson and White (1985).

depends on whether the plant functions as one physiological individual (aclonal; Figure 6.5c) or several (clonal; Figure 6.5d). In the former case fecundity increases more steeply than for plants with a single meristem; in some systems this increase may be curtailed (Watkinson and White 1985), but most commonly fecundity increases exponentially with size (possibly also with age) as is the case for the clonal plants. Clonal (and to a large degree aclonal perennial) plants are in this sense directly analogous to the clonal marine invertebrates discussed above, and in distinct contrast to the fecundity patterns of unitary organisms, which rise early to a maximum and then gradually decline (Cutler 1978; Charlesworth 1980). Similarly, unlike the Hayflick limit that determines the number of divisions of cultured animal cells (see above), cells taken from plants continue to grow indefinitely in vitro provided that they are subcultured at regular intervals and that growth conditions remain adequate (Murashige 1974; d'Amato 1985). Plant cells are also totipotent (provided that they retain a nucleus and living protoplast), that is, each retains the full genetic potential of the organism and differentiation is entirely reversible.

The implications are that survivorship rates of genets of semelparous plants are qualitatively similar to those of many unitary organisms (Watkinson and White 1985). In contrast, the exceptionally long-lived nature of iteroparous clonal and aclonal plants (for examples see Chapter 4 and Cook 1985), akin to that of clonal marine organisms, has prompted much speculation that senescence may not occur in this group. The shoot dynamics of many clonal perennials are well known (Harper 1977; Cook 1985; Sackville Hamilton and Harper 1989), but the clonal identities are often unknown. Localized death may not decrease survival probabilities of either aclonal or clonal iteroparous plants (Watkinson and White 1985). While rates of ramet mortality fluctuate over time for *Viola blanda*, rates of genet mortality *decline* continuously (Cook 1983). Antonovics (1972) found that the probability of genet mortality for established plants of the grass *Anthoxanthum odoratum* was independent of their age.

Evidence for senescence among microorganisms   Certain attributes of the population biology of microorganisms need to be kept in mind when one is attempting to come to grips with the issue of senescence. First, at least for bacteria, viability is defined operationally as the ability to grow to detectable levels in or on a recovery medium (Postgate and Calcott 1985). Survival is thus equated with the ability to divide repeatedly. By this criterion, all aclonal organisms would be dead! In fact, as judged by other criteria, nonculturable bacterial cells can be alive (e.g., Roszak et al. 1984). It is interesting that this is directly analogous to senescent human fibroblasts which remain viable in culture indefinitely, though they are incapable of division (Seshadri and Campisi 1990).

Second, it is the clonal lineage, not the individual cell, that is of interest.

This is comparable to the genet/ramet distinction in macroorganisms. (The term 'clone' has been used variously by bacteriologists and may or may not be equivalent to 'strain' [see e.g., Selander et al. 1987]. The latter term refers to descendants of a given isolation which frequently but not necessarily arise from a single colony.) For mortality statistics, the appropriate comparison would be *among* numerous clones, as opposed to the population dynamics *within* a single 'clone', which is what bacteriologists typically study, as in growth curve experiments. It is usual for microbe cultures to have an age structure, and consequently cells will not be physiologically uniform throughout the population (Yanagita 1977), but this is distinct from the issue of senescence. It is analogous to the age structure, from youthful shoot apices to decaying bases, among modules within a white clover genet.

Third, mapping the distribution of clones of microorganisms in nature is of necessity incomplete due to dissemination of clonal fragments (spores, bacterial cells). This is at least as difficult operationally as the tracking of plant and animal clones discussed above, but for different reasons. The problem in microbial ecology is the broad dispersal powers of bacteria and fungi (this also pertains to some of the protists), not the lack of sophisticated tracking methods (reviewed by Singer 1988; Trevors and van Elsas 1989).

Finally, the occurrence of dormancy and various physiological states of activity within a clonal population complicate the interpretation of senescence (e.g., Cooper et al. 1971). This is similar to modifications in the life cycle of reptiles and amphibians caused by diapause, diet, and temperature.

Microbes that divide by binary fission do not appear to senesce, but the evidence is mixed. For such organisms there is in effect no distinction between soma and germ cells: a mother cell either divides, in which case it disappears in the process and yields two identical fission products or, which amounts to the same thing, produces one daughter identical to itself. It has been debated (e.g., Hayflick 1977) whether the two products are indeed identical. There appear to be at least some differences in the case of bacteria (Dow et al. 1983) and in *Paramecium* (discussed later). Where there is inequivalence, one implication is that the product shortchanged may produce a defective line that senesces over time. However, the other fission product would continue the lineage. In culture and in nature microbes may starve or die from the accumulation of toxic products, but this is no more legitimately a case for senescence than would be the situation where 100 humans were locked in a food storehouse and mortality in the population followed after resources became exhausted. Where growth conditions are not limiting, as in turbidostat culture, bacterial clones persist indefinitely.

Diatom clones survive indefinitely and their population dynamics in culture are similar to those of bacteria. However, unlike bacteria, each cell is covered by a rigid silica wall formed in two components (valves) which overlap like the halves of a petri dish, the 'lid' being the epitheca and the 'bottom' the hypotheca. At division one progeny cell receives the parental epitheca and regenerates a new hypotheca. Hence this cell is the same size as the parent, as will be one of its descendants at each subsequent cell division. The other daughter cell receives the smaller hypotheca, which becomes its epitheca, and a new hypotheca is generated. This cell is thus smaller than the parent and will give rise to a lineage of progressively smaller cells (Yanagita 1977). Traditionally much has been made of this phenomenon, the implication being that at some point this series of cells will become critically small and prone to die. Even if this occurs, however, the lineage initiated at the first division and equal in size to the original parent could sustain the clone indefinitely. Furthermore, at least in some cases, the protoplast can escape from the shell and produce two large new valves.

Intraclonal 'aging' or localized senescence comparable to that in degenerating and regenerating benthic invertebrates unquestionably occurs in the fungi (Park and Robinson 1967; Trinci and Thurston 1976; for general remarks see Davies and Sigee 1984). On balance the evidence is against senescence of an entire clone, but there are a few instances in which this apparently occurs. It has been established from continuous growth experiments (in so-called racing tubes; Ryan et al. 1943; Gillie 1968) that clones can grow indefinitely (Fawcett 1925; Smith 1978; Perkins and Turner 1988), although not necessarily continuously (Bertrand et al. 1968). On the other hand, a few fungi, most notably *Aspergillus glaucus* (Jinks 1959) and *Podospora* spp. (Marcou 1961) seem incapable of repeated clonal propagation. (Senescence can be induced by genetic manipulation, as in various mutants or 'stopper phenotypes' of *Neurospora* [e.g., Munkres 1976], and undoubtedly several instances of reported senescence will prove to be caused by viral-like infections [Hollings 1978].) Holliday (1969; see also Lewis and Holliday 1970) attributed clonal senescence to irreversible, progressive infidelity in protein synthesis, in support of Orgel's (1963) "error catastrophe" hypothesis. Esser's group (e.g., Esser et al. 1984) has confirmed the existence of maternally (cytoplasmically) transmitted generalized senescence in all wild-type clones of *Podospora anserina* and is investigating the mechanism as are others (Wright and Cummings 1983). Mobile plasmids, alternately integrated within the mitochondrial chromosome, liberated and self-replicating, and incorporated within nuclear DNA (see 'Promiscuous DNA', in **Recombination**, Chapter 2) have been implicated for this fungus (the 'mobile intron model') and in certain strains of *Neurospora* (Perkins and Turner 1988).

The longevity of yeast cells reproducing by budding is fixed (Mortimer and Johnson 1959). Because bud scars accumulate, a mother cell can produce only a finite number of offspring after which it dies (Beran 1968). This has been frequently construed in the literature (e.g., Postgate 1976) as senescence, whereas in fact while the lifespan of the individual yeast cell is finite, that of the clone is not.

The evidence from protozoa is mixed. Some, perhaps most, species display senescence and limited lifespans. Others do not. Clones of the rhizopods (amoebas) grow indefinitely if cultural conditions are normal (Danielli and Muggleton 1959). Most clones of the ciliate *Tetrahymena* reproduce indefinitely (Smith-Sonneborn 1981). Lineages of *Paramecium* spp. evidently have species-specific lifespans fixed at several hundred fissions over up to several hundred days, accompanied by decline and eventual death (Sonneborn 1954; Smith-Sonneborn 1981; Takagi et al. 1987), or they may persist for thousands of generations without perceptible decline (Chapter 4 in Bell 1988). Bell, in the definitive review and analysis of this literature, concludes that isolated lines of protozoans do in general diminish in vitality (fission rate), but that such decline is not inevitable. Conjugation rejuvenates protozoa, as does sexual reproduction in all taxa, but of course then the clone ceases to exist as a genetic unit. G. Bell (1988, p. 134) concludes that senescence in protozoa is the consequence of accumulation of deleterious mutations, whereas in macroorganisms it results from the pleiotropic impact of early acting, beneficial genes (see below).

Evolutionary and nonevolutionary interpretations   What causes senescence? Viewed in a proximal (functional or mechanistic) sense, several explanations have been put forward by gerontologists over the past century. These fall into two general theories. First, the error catastrophe notion (Orgel 1963; see also Wattiaux 1968; Sokal 1970; Burnet 1974), mentioned above, proposes that accumulating somatic mutations together with faulty DNA repair, crosslinking of macromolecules, and free radical damage, lead ultimately to cessation of cell division (though such 'senescent' cells may remain viable indefinitely). Second, evidence is accumulating that senescence may be a genetically programmed event and, at least at the cellular level, a process of terminal differentiation (Seshadri and Campisi 1990; Goldstein 1990; Johnson 1990). For instance, when immortal and finite-lived human cell lines are fused, the majority of hybrids senesce (Pereira-Smith and Smith 1988). (Immortal lines originate from tumor cells which have escaped senescence and multiply indefinitely; see Chapter 25 in Watson et al. 1987.) Other studies have examined fusions between human fibroblasts and immortal hamster cells (Sugawara et al. 1990). All the hybrids that did not senesce had lost both copies of human chromosome 1. Introduction of a single copy of this chromosome to the immortal hamster cells caused them to cease dividing. Thus, genes on chromosome number 1 may encode products involved in senescence. However, no single mechanism appears capable of explaining the entire phenomenon, which is not surprising because probably several processes are involved. The question of whether senescence is programmed or results from accumulation of damage, or both events to varying degrees, remains open.

Mechanisms aside, evolutionary theories of senescence postulate either that the phenomenon is adaptive (beneficial) or nonadaptive. Since senescence is obviously deleterious for the individual, arguments are based at the level of the species (adaptive) or gene (nonadaptive). The former assert that senescence clears the stage of the old actors, leaving room for the rising young stars. This interpretation is not compelling for two reasons: first, as documented above, senescence is rare in nature; second, group selection pressure is weak at best (Maynard Smith 1976). Nonadaptive theories, which I review here, propose that senescence is the unavoidable consequence of the timing of gene action (Medawar 1946, 1952; Williams 1957; Hamilton 1966; Kirkwood and Holliday 1979). Wilson (1975, p.3) updated Samuel Butler's adage—that "a hen is but an egg's way of making another egg"—to the organism is only DNA's way of making more DNA. The soma is thus quite expendable. The interests of the gene are best served if it spreads within a population: There is no benefit in being locked inside a soma, even if it were immortal! Dawkins (1989, p.34) has said the same thing imaginatively as follows: ". . . the gene . . . does not grow senile; it is no more likely to die when it is a million years old than when it is only a hundred. It leaps from body to body down the generations, manipulating body after body in its own way and for its own ends, abandoning a succession of mortal bodies before they sink in senility and death."

Hence the concept of reproductive value underlies the nonadaptive theories. Reproductive value is measured as current reproductive output plus prospective reproductive output (Begon et al. 1986, pp.506-508). The latter is determined by the probability of survival to any particular age (survivorship) and the schedule of offspring production (fecundity) for each age group. If we ignore for a moment the numbers of offspring and look just at the likelihood that an organism such as a human will reproduce at any given age, the probability distribution is given by the solid line in Figure 6.6. This rises sharply from zero prior to sexual maturity to a peak followed by steep decline as potentially reproductive individuals are removed by death from the population. If we consider next the fecundity component of reproductive value, this will decline with adult age largely if not entirely *because* of senescent changes. So it cannot be invoked without circular reasoning to explain why the phenomenon occurs. However, the key is that over time, even *without* senescence, there will be a *cumulative* probability of death from random events. This dictates the decline in reproductive probability shown in Figure 6.6 because the likelihood of reproducing at an age clearly depends on the chances of surviving to that age (Williams 1957). In turn this means that a mutation affecting survival or fecundity early in life will have a greater effect than if expressed later when more of the potential carriers will be dead.

The concept of reproductive value can be expanded by using Fisher's (1958 pp.27-30) example for Australian women in 1911. The peak reproductive value was about age 18.5 years. The value for girls, though positive, is small prior to puberty because the prospect for children is remote and the female may die before reproducing. (Without social constraints and the custom of marriage, the peak should fall close to puberty.) With age past the maximum the probability of reproduction declines sharply (above paragraph and Figure

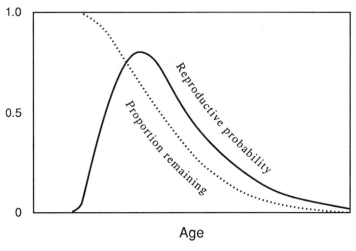

Figure 6.6   Relationship between age and probability of reproduction. Redrawn from Williams (1957, p.400). The solid line is the reproduction probability distribution curve, which measures the expectation that an organism will reproduce at any given given age. Area under this curve thus reflects the total expectation for reproduction. The dotted curve is the proportion of the total probability for reproduction remaining after any given age. It is also a measure of natural selection in force at any age because selection acts through reproduction and will be highest when the potential for reproduction is highest. This graph is similar but not identical to Fisher's (1930) concept of reproductive value, and to a plot of *rate* of reproduction versus age (cf. Figure 3 of Kirwood and Holliday 1979).

6.6). Thus, reproductive value also declines precipitously, but should not reach zero because even post-menopausal women contribute to survival of their offspring.

   It follows from this that individuals who reproduce early are more likely to have grandchildren by any specific future date than are those who postpone reproduction. Now, *if fecundity remains constant with age*, progeny produced later must contribute less to population growth (Lewontin 1965). Hence, selection will act, other things being equal, to favor those individuals that reproduce early, and to disfavor late reproduction. Senescence is related directly to *r*- and *K*-selection (Chapter 4), because in any *r*-selecting environment the first progeny (and the first ramets in a genet; Sackville Hamilton et al. 1987) are of greater reproductive value than those produced later. *Selection for precocity is really the same thing as selection for senescence*. Dawkins (1989: Chapter 3) has suggested facetiously that one way to extend the human life span would be to prohibit reproduction until middle age. Note that reproductive output is bought at the cost of resources otherwise channeled to upkeep of the soma (again we have the issue of trade-offs; see Gustafsson and Part 1990). To return to the point stated at the beginning: In teleological terms there is no gain in maintaining a soma whose proportionate contribution of genes to future generations is declining. Natural selection should favor a balance between resources shunted to maintain the soma and those used to produce progeny such that the soma remains functional only through its normal lifespan in the wild. Where organisms are protected (as in zoos and human society!) from environmental mortality, physical decline manifested as senescence will occur (Kirkwood and Holliday 1979, 1986).

The declining impact of genic effects with age can be expressed mathematically as follows. When a gene has mixed effects on fitness, the strength or coefficient, W, of selection for it will depend on the 1) magnitude, m, of its positive and negative effects; and 2) onset and duration of its effects, particularly with respect to the proportion, p, of the reproductive probability affected (Wright 1956; Williams 1957). This relationship can be described as:

$$W = (1+m_1p_1)(1+m_2p_2) \ldots (1+m_np_n)$$

where the subscripts designate the separate effects of the gene. A gene with positive effects early in the reproductive period will have a high p value and will tend to be selected even though it may later have deleterious effects, because these would have a lower p value. Although this example pertains specifically to side-effects (pleiotropy), the idea that the strength of genic effects is linked unavoidably to how it affects overall reproductive probability is not. In theory *this makes the onset of senescence inevitable except where reproductive value increases with age, as for modular organisms.*

There are three major nonadaptive theories for senescence. First, as Medawar (1952) argued, pressure to remove deleterious genes with late ages of action would be lower than for those acting early (mutation accumulation theory). The effects would become manifest as increasing mortality with age and there are numerous examples just in the human population: Huntington's chorea, multiple polyposis of the colon, and high blood pressure (Chapter 3 in Charlesworth 1980; see also Burnet 1974). In fact genes (e.g., any causing cancer in old persons) whose action is restricted to the phase of the life cycle beyond the reproductive period are effectively out of reach of natural selection. Second, the converse follows, that there should be a higher rate of incorporation of favorable mutations increasing survival at earlier ages (when reproductive value is high) than those genes acting later (Hamilton 1966). This may explain the origin and maintenance of tumor suppressor genes which participate in an array of tumor-related and normal functions (Sager 1989), including cell cycle control and signal transduction. Third, genes promoting survival early in life and associated with later costs should still be favored by natural selection because of their early beneficial effect (pleiotropy concept of Williams 1957; disposable soma concept of Kirkwood and Holliday 1979). These theories are not mutually exclusive. The key assumption underpinning them all is that genetic variability affects life history traits such as survivorship and fecundity. The key test is thus whether senescence changes as predicted when survivorship and reproductivity are manipulated experimentally (Rose 1985). It appears to do so (Rose and Charlesworth 1980, 1981). Rose (1985) reviews the experimental evidence, most of which supports the third model.

The three evolutionary models make similar predictions about the occurrence of senescence. These corollaries have been summarized most clearly by Williams (1957) whose main points follow. First, senescence is not expected where no distinction can be made between soma and germ plasm (recall G. Bell's [1984] experiment described in the preceding section). This would be the case for all unicellular microorganisms and for many multicellular microbes such as the fungi. A distinction cannot be made realistically in the fungi (Buss 1983, 1985) because nuclei and organelles move more or less freely through the organism. Buss also includes plants because the totipotency of their meristematic cells means that a new organism can potentially be regenerated from certain tissues at any life stage (somatic embryogenesis). Though true in theory this is limited in actuality because while all plants can be cloned, not all clone naturally. Wherever totipotency and cloning occur, the germ/soma demarcation will be blurred because of the reproductive potential of 'somatic' cells and heritability of somatic variability. The phylogenetic comparisons presented above for senescence appear to be fairly closely correlated with occurrence of somatic embryogenesis and clonality (e.g., see Tables 13.1 and 13.3 in Buss 1985). Williams (1957) observes that asexual *clones* should not senesce although the physiological individuals (or parts of physiological individuals, such as leaves on a tree) comprising them will do so; see also Chapter 4 and Jackson and Coates 1986). Buss's phylogenetic survey (1983) shows that preformistic development (cell lineages committed irreversibly to either germ or soma early in ontogeny) is absent in groups producing ramets.

Second, it follows that senescence should occur very slowly if at all in organisms whose fecundity increases with age (size). Conversely, senescence should start immediately after the age at first reproduction (as in annual plants) and proceed rapidly where growth is determinate and organisms do not increase in fecundity after maturity (Bidder 1932; Williams 1957; see also Caswell 1985). In age-structured populations fecundity usually increases with time for the cold-blooded invertebrates, reptiles, fish, amphibians, molluscs, perennial plants, and clonal organisms in general. As reviewed above, it is in these phyla where there is for the most part good evidence against senescence.

What, then, of apparent exceptions to the theoretical predictions? Partial senescence (at the level of the ramet or individual leaves or roots) is expected in clonal organisms; evidence of such decline in aquatic animals (Palumbi and Jackson 1983; Jackson 1985) and terrestrial plants (Shields and Bockheim 1981) is not counter to theory. Comfort (1979, p.151) suggests that the decline in some lines of protozoa is fundamentally unlike, and should not be compared with, senescence in metazoans. Where an entire fungal strain 'senesces', this appears to be due either to the cumulative effects of viral infection or to genetic incompatibility. Neither the situation for protozoa nor for fungi appears to be closely analogous to senescence in metazoans. A complication is that expression of senescence is not phenotypically unique. At least for microbes, it is impossible to prove operationally that it is not caused by exhaustion of some unidentified essential growth substance. It is also noteworthy that according to the model by Sackville Hamilton et al. (1987; see their Table 2), conditions exist under which senescence of the modular organism *is* predicted. One of these is where there are lethal factors that kill entire genets. Here there is no advantage to 'immortality' and resources are shunted from upkeep of the soma to reproduction.

Implications   A closer working relationship needs to be forged between evolutionists and gerontologists or physiologists in the study of senescence. Evolutionary theory to be robust must be founded on realistic mechanisms and, conversely, evolutionary insight would suggest to researchers that no single cause is likely to exist (because selection will always oppose the most senescent-prone mechanism; Williams 1957). Recognizing this would discourage myopic, single hypothesis, all-encompassing explanations. Not only could the causes be diverse, but we could expect them to be different in different phlya. For this reason, even the few microbes in which true senescence appears to occur (e.g., the nematodes) may be poor working models of senescence in higher organisms.

In the phenomenon of senescence we see once again the role of trade-offs in the ecology of organisms, and one more expression of the striking biological differences between modular (clonal) and unitary organisms. The modular/unitary dichotomy seems more fundamental than those between animal and plant, or micro- and macroorganism. Most of the traditional theories of population dynamics and evolution, including the evolution of senescence, have been developed for unitary organisms. For instance, where life forms are nonsenescent and long-lived, somatic mutation and other forms of somatic variation (e.g., gene rearrangements, amplification, regulatory changes; see Silander 1985) may be a significant evolutionary force (Chapter 2). Longevity provides both more opportunity for somatic changes to occur (more cells, more mitotic divisions, as in a tree or a bracken fern) and more time for them to be expressed in the lifespan of the product of the original zygote. This means that the original genet (if it can still be referred to as such) over time could become increasingly heterogeneous which might (Whitham and Slobodchikoff 1981), or might not (Klekowski and Godfrey 1989), be adaptive. It would be interesting to know the degree of genetic variation in young versus old clones of such organisms as sponges, corals, huckleberry, and creosote bush. The extent to which such polymorphisms occur and whether somatic changes can enter the germ lineage to a degree that is significant in *evolutionary* terms are important, fascinating issues which need to be resolved. We are on the threshold of an era of exciting discoveries that will unravel the distinctive ecologies of these two major classes of life forms.

6.4   Summary

The life cycle has been called the central unit in biology. In this chapter two of the many questions pertaining to life cycles are examined: why complex cycles as opposed to simple cycles have evolved, and whether all organisms are doomed to undergo senescence.

When the young are born into essentially the same habitat as the adults, and do not undergo sudden ontogenetic change, the organism is said to have a simple life cycle. Where organisms have two or more ecologically distinct phases separated by an abrupt ontogenetic change the life cycle is said to be complex. A classic example of an organism with a complex life cycle is the frog, which metamorphoses from a tadpole living in a pond to an adult living on land. Other examples include toads, the holometabolous insects (i.e., those having a complete metamorphosis), and various animal parasites and algae. The complex life cycle appears to be (or at least to have been at the many points in geological time when it evolved) a specialization that allows the organism to exploit certain opportunities, and forego compromises. Organisms with such cycles can concentrate instead on particular activities such as feeding or reproduction in any given stage. Like all specializations, the complex life cycle imposes constraints as well as opportunities on the organism (see also Chapter 3).

The rust fungi, some of whose life cycles consist of a sequence of five morphological states on two distinct hosts, illustrate many principles of the complex life cycle. The rust life cycle appears also to be an evolutionary dead-end and a prime example of how evolution tends to drive organisms to greater levels of specialization.

Complex life cycles present a paradox because in theory they should be unstable over evolutionary time but have remained fixed in major taxa even over geological time. Their maintenance may be merely a phylogenetic remnant, fostered by environmental heterogeneity. In the rusts and other organisms the complex life cycle is frequently abbreviated by the loss of one or more stages. This is analogous to progenesis (precocious sexual maturity in animals) and is evidently a modification favored by natural selection, particularly in *r*-selecting environments where the premium is on early reproduction as well as on high fecundity.

Senescence is the manifestation of various deteriorative effects that decrease the probability of survival with increasing age. It is depicted most clearly at the population level by actuarial statistics showing an increasing mortality rate over time. If mortality is constant with age, the number of survivors declines exponentially and there is no basis for inferring from the data that senescence occurs. Nonsenescent curves mean either that the organism intrinsically does not senesce or, more commonly, that insufficient numbers of the population in nature pass through early to mid-life for senescent effects to be detectable numerically.

Among macroorganisms that reproduce largely or exclusively sexually, a good case can be made for the occurrence of senescence. Organisms that senesce include the vertebrates, annual plants, and many invertebrates such as the nematodes, crustaceans, and insects. Modular organisms, whether they belong to invertebrate, perennial plant, or microbial taxa, do not appear to senesce at the clonal level, although evidence is equivocal in some cases.

Mechanistically, senescence probably results from multiple causal factors, most notably disfunction of the immune system leading to disease and death, faulty DNA repair, crosslinking of macromolecules, free radical damage (in aggregate the error catastrophe hypothesis); or from the products of senescence genes (the programmed event hypothesis). Evolutionarily, senescence should occur wherever the reproductive value of the individual diminishes with increasing age. Thus, a gene with positive effects early in the reproductive period, that is, when reproductive value is high, will tend to be selected even if it may later have deleterious effects. Likewise, because of the declining strength of selection with age, pressure to remove harmful genes which are expressed late in life would be lower than for those acting early.

The occurrence of senescence among unitary organisms and its general absence at the level of the genet among modular organisms is yet one more expression of the marked ecological difference between these two major groups of life forms. Most of the population dynamics and evolutionary theory have been developed for unitary organisms. Much remains to be done in sorting out the distinctive behaviors of these two classes, and the biological research needed transcends the domains of microbiologists, botanists, and zoologists.

6.5 Suggested Additional Reading

Bonner, J.T. 1965. Size and cycle: An essay on the structure of biology. Princeton Univ. Press, Princeton, N.J. The case for considering the whole life cycle, not just the adult, as the organism.

Charlesworth, B. 1980. Evolution in age-structured populations. Cambridge Univ. Press, Cambridge, U.K. Mathematical treatment of the evolution of life histories and senescence (see esp. chapter 5).

Comfort, A. 1979. The biology of senescence, 3rd ed. Elsevier, N.Y. The standard treatise on senescence.

Rose, M.R. 1985. The evolution of senescence. *In* Evolution: Essays in honour of John Maynard Smith. P.J. Greenwood, P.H. Harvey, and M. Slatkin (eds). Cambridge Univ. Press, Cambridge, U.K., pp.117-128. Brief review of the nonadaptive theories.

7

The Environment

The known facts of development and of natural history make it patently clear that genes do not determine individuals nor do environments determine species.

—R. Lewontin, 1983, p. 276

7.1 Introduction

This book started with a review of how organisms change genetically, because genetic variability provides the raw material for evolution. Since natural selection always occurs in a setting, I now conclude with the other half of the story, namely, how an organism's surroundings influence the evolutionary process. First, the issue is framed in terms of what is meant by an organism's environment. The remainder of the chapter then contrasts how different kinds of organisms experience their environments.

Environments are often loosely construed to mean only the purely physical or abiotic elements in which the organism is immersed. It is important to remember that both physical and biological elements are involved (although organisms other than humans do not make a distinction between the two!). The latter include surrounding individuals of the same or different species. Some organisms (most obviously symbionts or predators and their prey) have a direct influence on one another. The association between others may be only indirect or sporadic. The point is, first, that the environment drives natural selection by influencing which organisms survive and reproduce, how many offspring are produced, and of these, how many in turn survive and reproduce; second, and reciprocally, that natural selection is itself altered by the changing composition of the survivors.

The environment must always be considered from the *organism's* viewpoint (to the extent possible!), because what is biologically pertinent is determined by the organism. Such inferences, however, entail considerable interpretative risk. But it is clear that at some level what is environmentally relevant to a human will differ qualitatively and quantitatively from what is relevant to a butterfly or an elm tree.

7.2   The Environment and Organism Are Coupled

Continuous, reciprocal involvement between organism and environment means that the environment does more than merely set the evolutionary stage for the organism. In this important sense the conventional paradigm that genetically predestined 'normal' ontogenetic development simply unfolds in robot-like fashion against a conducive environmental backdrop is wrong, as Lewontin (1983) notes. Likewise, the phylogenetic paradigm of the environment posing 'problems' which the organism 'solves' is wrong (Lewontin 1983). Rather, the organism is both cause and consequence of a particular ontogenetic sequence; likewise, in phylogenetic change, the organism is both a driving force and is itself affected. In Lewontin's words (p.276), ". . . genes do not determine individuals nor do environments determine species". It is this emphasis on the dynamic, reciprocal, ongoing association between organism and environment that brings new life to the old elementary dogma that environment and genetics together determine the individual.

Consider the giant redwood tree. Its growth potential is determined over some range by its genes. The appearance of that tree will change through time from a microscopic zygote until its death, say 3,000 years later, as a 100-meter colossus. Its phenotype at a given time will reflect the state of the present and past genetic program, and the current as well as the past environment. It will be more or less phenotypically plastic at different stages and will be influenced to varying degrees by different environmental components such as pathogens or lightning. The phenotype would be different in a different environment. The tree will create its own environment by affecting, and in turn being affected by, animals, other plants, microorganisms, light, wind, soil nutrients, and so forth. What happens in the life history of unitary organisms is perfectly analogous to this giant modular organism, if less visibly dramatic! The differences between different groups of organisms are only in the specific ways they cope with a constantly changing environment and the rates at which they are able to change.

7.3   How Organisms Experience Environments

Life span, size, and growth form set the aggregate experience with the environment   There are no constant environments either in nature or in the laboratory. Differences occur in space (μm to km), and cyclic and noncyclic differences through time (seconds to centuries) always exist. For instance, geographic fluctuations range from the familiar regional and global phenomena (El Nino, greenhouse effect, ozone layer disruptions, acid rain) over hundreds or thousands of kilometers to microhabitat changes over micrometers, such as those surrounding particles of marine detritus (Alldredge and Cohen 1987; see also Poindexter 1981a,b). Conducive environments may also be fleeting, as is a shifting sand bar in the Mississippi River for plant colonists, or susceptible hosts for plant or animal pathogens. The real question is at what threshold level do global or local fluctuations become ecologically significant to an organism.

Each kind of organism experiences the environment differently. For a rotifer with a life of 10 days, weekly changes are comparable to yearly changes for a bird (Chapter 1 in MacArthur and Connell 1966). This returns us to the theme of physiological versus chronological time discussed in Chapter 4. The life cycles of animals in the Florida Everglades are coupled to the annual periodicity of the wet and dry seasons. The wood stork, for example, breeds when water levels are falling and it can easily obtain fish in the receding pools. The bird will not nest if this water cycle is disrupted (Kahl 1964). Analogously, in the annual cycle for many fungal pathogens of plants, the fungus overwinters as a saprophyte in plant debris where it undergoes sexual recombination; it then oversummers as a parasite, undergoing repeated rounds of asexual fragmentation in association with the living plant. Both phases are intimately tied to the seasonal activities of the host (Andrews 1991). In terms of a shorter time scale, epiphytic bacteria respond on the order of hours to changes in physical conditions (Hirano and Upper 1989). Growth rates over brief periods can be on the same order of magnitude (doubling times ca. 2-3 hours) as under optimal laboratory conditions.

Organisms adjust to short-term fluctuations relative to their life span by phenotypic plasticity (discussed in Section 7.4). Long-term oscillations influence gene frequencies because some individuals leave more descendants than others (Chapter 1 in MacArthur and Connell 1966). Genetic polymorphisms and ecotypes may occur (Section 7.4). Clonal organisms (see Section 7.6 and Chapter 5) persist and expand indefinitely, even though the life of a clonal unit (e.g., one bacterial cell) may be only on the order of minutes. Hence, as a genetic entity, each clonal organism experiences greater ranges in environmental variables than does the genet in most unitary organisms.

For comparison purposes, one can array a set of hypothetically important environmental variables against various life spans or generation times (Figure 7.1 and Istock 1984). Two aspects are evident. Every organism experiences many variables concurrently (Figure 7.1 top), and differences in life span (Figure 7.1 bottom) mean differences in aggregate experience with the environment. These relationships are preordained by the size and growth form of the organism. Organisms such as annual plants with a short life span relative, say, to a yearly pattern of seasonal change, are restricted to one

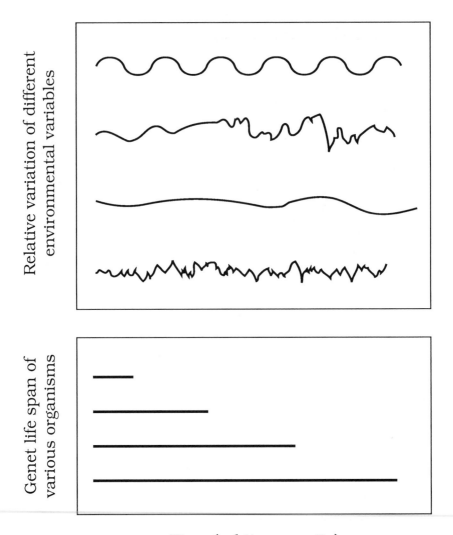

Figure 7.1   Because of different life spans or generation times (bottom), organisms have different aggregate experiences with the same set of hypothetical environmental variables (top). Modified after Istock (1984).

season and become specialized to it such that they could not develop at other seasons. Species that live over many seasons could be generalists—doing more or less well throughout the year—or specialists—flourishing in one season while remaining dormant or migrating in others (Section 7.5).

Organisms typically dampen environmental fluctuations. MacArthur and

Connell (1966, Chapter 1) cited the universal occurrence of homeostatic mechanisms as implying that metabolism is best conducted under relatively constant conditions. This is not surprising given that all living processes are driven by sets of chemical reactions. Macroorgansims, in particular the homeothermic metazoans, have more control (by virtue of their size and complexity; Chapter 4 and Bonner 1988) over their environments than do microorganisms. Homeostasis is most apparent in the exquisite mechanisms that mammals have for balancing temperature and blood and tissue chemistry. Bacterial parallels include the ability to activate or deactivate metabolic pathways by enzyme repression and derepression mechanisms (Section 7.4).

The idea of environmental grain   The most prominent effort to portray surroundings from the organism's standpoint is the notion of 'grain'. This attempts to relate environmental variation to an organism's size, longevity, and activity. Notice, then, that this idea involves more than simply a matter of relative scale of the organism. There are two components to consider: the environment per se and how the organism samples or experiences it. The matter of sampling will be influenced in turn by the various attributes of the organism, most obviously by size and growth form (which includes mobility; (Chapter 5).

Environments have been conceived, primarily with respect to how resources are presented, as being either fine- or coarse-grained patches in space or time as summarized below:

Fine-grained environment

Space: Motile individual moves among many patches, consumes resources in proportion in which they occur.*

Time: Since the environment undergoes change often in small increments, the effective environment is the average of the units.

Examples: Differences among fruits on forest floor for *Drosophila* adults (motile); motile larvae of barnacles; a bird foraging among many tree types; a bacterial or fungal clone feeding as a generalist.

Coarse-grained environment

Space: Nonmotile organism spends most or all of life in a patch.

Time: Environment varies over long periods relative to life span of organism; seasonal food.

---

*Grain refers to size and/or duration of patch relative to size and activity of the organism. Extended mainly from Levins (1968). With increasing size and motility of organism, coarse-grained environments may become fine-grained, and some fine-grained environments may become inconsequential (limiting case: all individuals experience same variation, no uncertainty; Levins 1969, p.3). See also Levins (1969), MacArthur and Connell (1966), MacArthur and Pianka (1966).

Examples: Differences among fruits on forest floor for *Drosophila* larvae (relatively immotile); individual bacterial or yeast cell on a leaf; sessile
adults of barnacles.

Fine-grained patches in space are small, and the organism does not really
distinguish among them. The patches are used in the proportion in which
they occur. The environment is fine-grained in time if it is experienced in
many small doses, or if larger fluctuations are encountered by a long-lived
organism over many years. Conversely, coarse-grained environments are sufficiently large so that the organism 'chooses' among them (space) or spends
its life in a single environment (time). An oak-hickory forest appears fine-
grained to a scarlet tanager, which forages in both oak and hickory trees, but
coarse-grained to a defoliating insect, which attacks only the oaks (Chapter
1 in MacArthur and Connell 1966). MacArthur and Pianka's (1966) theoretical
analysis predicts that specialist feeders are favored over generalists in fine-
grained systems and the converse in coarse-grained systems, where the generalist can compensate for its less efficient feeding with lower hunting time
(see also Chapter 3).

Mathematical models based on the grain concept have been developed
to describe selection in heterogeneous environments (see especially Levins
1968). There are at least two problems with such analyses in general and
Levins' in particular. The first is that it remains unclear what is being maximized by natural selection (Hamilton 1970). For example, at the population
level this could be the *mean* of individual fitness over all environments. Selection in this case eliminates extreme individuals, preserving those near the
population mean, and is said to be stabilizing. Alternatively, it could be acting
directionally to elevate the *minimum* of individual fitness in the population
over all environments. This is the so-called "maximin strategy" (Templeton
and Rothman 1974). To quote the authors (p.425), "the optimum population
is that population which maximizes its minimum fitness over all environments
instead of maximizing its average fitness. By adopting this strategy, the population further insures its survival by letting the worst conditions it experiences dominate in importance". Levins (1968) evidently assumes the former
mode of natural selection and his predictions have been criticized (Hamilton
1970; Strobeck 1975; Templeton and Rothman 1974).

Second, the grain concept is abstract and would seem useful only in a
very general way. It is quite difficult to apply to 'real' organisms in 'real'
environments. The level at which a resource or any other environmental
attribute appears fine- or coarse-grained is arbitrary. All facets of the environment interact with the organism in a multidimensional, mosaic-like fashion: Instead of a patchwork quilt, we have layers, each representing a different
parameter such as resource type, temperature gradient, predator size, and so
forth. Moreover, depending on the environmental characteristic, each layer
can be fine- or coarse-grained. How any particular organism views the sur-

roundings will vary with its growth stage. The sluggish caterpillar sees life in coarse-grained fashion, while the butterfly it will become flits about a fine-grained environment! Finally, different parts of the same modular organism (modules or in some cases ramets; Chapter 5) are exposed to potentially quite different environments. The mathematical models of Levins (1968) and others as applied to hypothetical situations overlook the fact that we do not know, and can never really know, very accurately how any other organism experiences the world.

7.4 How Organisms Respond to Environments

The consequences of size   Since much of what follows develops how size differences affect environmental relationships, it is worthwhile to start with some commonalities. Monod has said (personal communication reported by Koch 1976, p.47) that "what is true of *E. coli* is also true of the elephant, only more so", by which he evidently meant that they had biochemical reactions in common. Koch continues (p.47), "There never was a single *E. coli*, nor a single elephant, on which Darwin's law has not operated separately, and equally; it has done so on them and on all their ancestors. The law of survival of the fittest has been obeyed, and the little *E. coli* has survived". Both organisms have passed the screen of natural selection within the context of what is possible for each of them. For the elephant this has involved, "many more cells, much more DNA, more neurons, and the ability to walk and do other things that *E. coli* cannot do". For the bacterium, it has involved phenotypic and genotypic plasticity, that is, the ability to respond rapidly to environmental changes.

It should be added that what the elephant does as a genetic individual comprising one huge mass of coordinated, differentiated cells, *E. coli* does as a diffuse, essentially undifferentiated clone. Most of the elephant cells, being internal, are buffered from exterior fluctuations, but they do have to contend occasionally with such things as pathogens. The cells react in unison at the tissue or organ level under centralized control by neurons and hormones. As discussed earlier for unitary organisms (Chapter 5), the physiological and genetic individual are the same entity. How well the entire corpus responds to the range of environmental stimuli determines whether the elephant is sick or robust, whether it is short- or long-lived, and whether it will contribute significantly to the population gene pool. Being by far the largest land animal, the healthy adult elephant has no natural predators. Evidently the main environmental challenge it faces is to find food, a process which takes about three-quarters of the animal's time (Chapter 5 in Eltringham 1982). Each cell is a party to this venture and if the functional unit dies, all components die. The whole, multicellular elephant is affected.

In contrast, while cells of an *E. coli* clone may occur in aggregates, each is relatively more exposed than is an elephant cell to oscillations in the external environment. Each responds, and lives or dies, largely as a physiological individual. Serological and electrophoretic typing shows that resident strains may persist for weeks or months in a healthy human or other host (Hartley et al. 1977; Linton et al. 1978; Selander et al. 1987) despite loss of individual cells en masse by defecation. There is also turnover and sporadic reappearance of strains, the inoculum originating from a few cells which evidently persist in protected sites. Intra- as well as interspecific competition is presumably extreme. The environment is highly variable from the bacterial cell's perspective both in time and space. Koch (1971, 1987) postulates that there is a "feast and famine existence"—brief periods of glut alternating with chronic malnutrition. In spatial terms, the clone is subject to the environmental vagaries of, say, the colon as opposed to the ileum, the intestinal wall, or the lumen, the intestinal milieu of different hosts, and to sporadic doses of antibiotics (variation in time and space). Finally, there is life outside the host in soil, water, or feces. Savageau (1983) estimates crudely that an average *E. coli* cell spends about half its life in the intestine and about half on the surface of the earth.

Environmental signals of interest to *E. coli* are obviously different from those important to the elephant. To the bacterium, gravity is of no consequence, but Brownian motion, Reynolds number, and molecular diffusion coefficients are important. Natural selection will be primarily for growth. To grow in the highly competitive environment of the intestine means being able to remove nutrients, often at very low levels, before the host or competing microflora do; and efficiently to convert these metabolites into cell substances. Efficiency of nutrient removal is increased (within limits) by a decrease in size, asymmetry in shape, increase in the number of transport units per unit membrane, and increase in the capability of the transport mechanism (Koch 1971). Transport systems in *E. coli* work typically at about 1,000-times-lower concentrations than do those of yeast, algae, and the epithelium of macroorganisms (Koch 1976). Koch calculates (1971) that *E. coli* could theoretically clear 2,800 times its own volume per second in growth media at 37°C. He observes that one reason, if the bacterial cell were elephant-sized, it would starve to death in the midst of plenty is because of its inability to take up nutrients fast enough. The transport mechanisms have evidently evolved to a peak where further refinements would be useless, because the bacterium is constrained by viscosity of the fecal environment, which limits diffusion rate and, in turn, growth (Koch 1971, 1976; also, recall from Chapter 4 that viscosity is the denominator in the equation for Reynolds number. Therefore, an increase in viscosity will cause a decrease in Reynolds number).

Efficient conversion depends on the processes of protein synthesis, that is, on the efficiencies of transcription and translation. As an example consider protein synthesis. The rate of protein synthesis is directly proportional to the number of ribosomes, and each ribosome functions at a constant biosynthetic rate, regardless of the nutrient environment (Koch 1971; Koch and Schaechter 1984): Thus, regardless of the rate of cell division, each ribosome will wait the same length of time for a mRNA and take the same time to add an amino acid. While the cost to *E. coli* is that there is an excess of poorly utilized or nonfunctional ribosomes in very slowly growing cells, the benefit is that the bacterium is well equipped with ribosome machinery to get a head start for fast growth when a pulse of nutrients appears (Koch 1971; Koch and Schaechter 1984). Generation time is inversely proportional to size, and for *E. coli* in its intestinal environment is about 40 hours (Savageau 1983; Hartl and Dykhuizen 1984), while for the African elephant it is about one generation (birth to puberty) per 12 years (Chapter 4 in Eltringham 1982).

The preceding paragraphs consolidate to this: *E. coli* cannot control its environment, whereas an elephant, by virtue of its bulk and related complexity and homeostasis mechanisms, can modify its environment markedly (Smith 1954; Bonner 1965, pp.194-198). However, the bacterium can accommodate much more rapidly to changing environments (see e.g., Bennett et al. 1990). This tracking of environments entails many genotypic and phenotypic adjustments, including ultrastructural and morphological changes, enzyme inhibition or stimulation, and induction or repression of protein synthesis (Chapter 3 and Harder et al. 1984; Forage et al. 1985). So the issue of size as it relates to environment is in large part one of being either a well-buffered individual destined to change slowly in the face of adversity, or being vulnerable but capable of fast adjustment.

Phenotypic and genotypic variation   Sessile organisms cannot escape environmental vicissitudes. One would expect to see very good examples of adjustments to stimuli and extremes among these organisms. At the level of the individual, adaptations can be made in morphology, behavior, or physiology, as well as in the degree to which these attributes can be varied in response to the environment (phenotypic plasticity). At the population level, local, genetically specialized populations (ecological races or ecotypes) may occur.

Phenotypic plasticity (polyphenism or developmental polymorphism) refers to any kind of environmentally induced phenotypic variation (Stearns 1989b). It offers the major advantage of dampening the effects of selection by uncoupling the phenotype from the genotype (Stearns 1982). The potential range in phenotypic variation is set genetically and is trait-specific (Bradshaw 1965). Hence, plasticity in some attributes may enable stability in others. Within lineages the most common ultimate constraints on the extent of developmental polymorphisms are allometric (Chapter 4 and Stearns 1982). For example, upward plasticity in birth weight cannot exceed a limit without increase in size of the adult; if adult size increases, an increase in birth weight will necessarily follow, whether it is adaptive or not (Stearns 1982). Within a taxon, seed size in plants and spore size in fungi are relatively fixed: A large as opposed to a small plant or fungal fruiting body does not produce larger seeds or spores, but rather more of them (for caveats see Thompson and Rabinowitz 1989).

Where selection is directional (i.e., favoring phenotypes at one end of the range), plasticity can extend the response of a population with limited genetic variability. However, it seems to be more important where selection is disruptive (i.e., favoring individuals at both extremes of a range). Whenever there is noncyclic recurrent variation in time at intervals less than the generation time of the organism, adaptation can only be by plasticity (however, see later comments about mobile genetic elements). Environmental conditions may change rapidly for desert plants or for microbes exposed to sporadic pulses of nutrients. Disruptive selection can also occur over short distances, as for the clonal plant whose stolons encounter varying degrees of competition, or the aquatic plant which roots in the littoral zone, grows up through the water column, and finally produces leaves and flowers above the surface.

Plants, especially aquatic or desert species, are noted for their phenotypic plasticity (Bradshaw 1965; Jennings and Trewavas 1986), perhaps most strikingly shown by the environmentally determined leaf form called heterophylly (Figure 7.2 and Hutchinson 1975, pp.157-196). For instance, a single aquatic plant may bear two or more kinds of leaf, typically a dissected form in the water column and a laminar form above it. The finely dissected leaves increase the area-to-volume ratio. This means that tissues are more accessible to dissolved CO_2 for photosynthesis underwater (Hutchinson 1975, pp.146-148). Dimorphism among desert plants is related to water economy (Orshan 1963). Plasticity is also apparent in the resource allocation patterns common to all plants, either to vegetative biomass for foraging purposes, or to reproduction, depending on the prevailing environment (Abrahamson and Caswell 1982; Grime et al. 1986; Bazzaz et al. 1987). Other common manifestations of the enormous plasticity of plants are the differences in appearance of a plant grown in the sun versus the same genotype when grown in the shade (smaller, more closely spaced leaves in the sun; larger, more widely spaced and deeper green leaves in the shade), and the variation among flowers of the same individual (e.g., adaptations for open pollination in early season flowers versus adaptation by way of closed flowers produced later to ensure self-pollination).

A striking example of the sensory capability of plants is the report by Braam and Davis (1990) that *Arabidopsis* can sense numerous experimental stimuli such as sprayed water, touch, subirrigation, wind, darkness, and wounding. The expression of at least four touch-induced (TCH) genes is affected by these stimuli and additionally growth is inhibited by touch. The touched plants have shorter petioles, begin to bolt later, and develop shorter bolts. This implies that plants can detect, transduce, and respond to environmental signals such as wind and rain (see below, **Signals and genes**). The growth response is gradual and differs from the immediate morphological

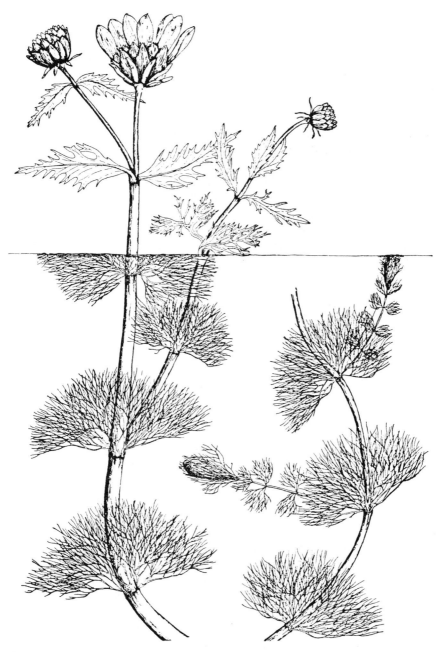

Figure 7.2   Heterophylly of submersed (below the line) and emersed leaves of water marigold (*Megalodonta beckii*). Dissection of the laminar form found in air into a capillary morphology enhances the rate of photosynthesis under water by increasing the area: volume ratio, which increases CO_2 assimilation from the surrounding water. Note that this rooted aquatic plant occupies three environments: sediment, water column, and aerial. From Fassett (1957).

reaction of such plants as the Venus fly trap (*Dionaea muscipula*) or the sensitive plant (*Mimosa pudica*). Nevertheless, *Arabidopsis* responds very quickly at the genetic level (within 10-30 minutes the amount of mRNA increases up to 100-fold). Three of the genes encode putative calmodulin and related proteins, which suggests that Ca^{+2} is involved in signal transduction and response. Thus, we now have the foundation for a mechanistic interpretation as to why plants may be able to respond rapidly and sensitively to environmental cues. The work begs the question whether other sessile, modular organisms (Chapter 5) respond similarly, and to what extent the environment may interact with (and alter genetically?) developing 'meristems' of such organisms before reproductive differentiation occurs.

What about microbial plasticity? Bacteria have extreme phenotypic plasticity but this is expressed mainly at the physiological level (Chapter 3). Being basically spheres or rods, they do not have the option of varying growth form appreciably and in this regard are like unitary animals. Fungi have shorter generation times than plants and are composed of fewer cell types (Chapter 4), all of which are modifications of the basic hyphal unit. Not surprisingly, then, the fungi are even more plastic than plants (see e.g., Smith and Berry 1978; Jennings and Rayner 1984; Moore et al. 1985; Rayner et al. 1985b). Some may generate complex organs such as rhizomorphs from diffuse or compact mycelium (Smith 1983; Jennings 1986), which is itself variable (Chapter 5) in many features such as internode length, branch angle, and growth pattern (Rayner and Coates 1987; Andrews 1991).

Some organisms can actually alternate their existence between a harmless commensal state and a virulent pathogenic condition. In certain fungi this occurs by a dimorphic transition, typically from a filamentous form in soil to a single-celled (yeast) state in blood or tissues of an infected host (Szaniszlo 1985; Shepherd 1988). The environmental trigger is temperature for some species (e.g., *Histoplasma capsulatum* grows in a mycelial form at 25°C, but as a yeast at 37°C); for others, nutritional factors and pH are also determinants of morphogenesis. Superimposed on this level of variability is the capacity of certain dimorphic species to switch colony morphology of the yeast phase in heritable and reversible fashion. *Candida albicans* is a diploid, pathogenic yeast, evidently lacking a sexual phase and meiosis. However, it has extensive phenotypic variability, switching in culture among at least eight morphotypes (Figure 7.3; Slutsky et al. 1985, 1987). When the fungus is plated onto a general growth medium, colonies of the original "smooth" type give rise spontaneously to variant phenotypes, which persist and interchange in successive clonal platings. Switching among the phenotypes at a fairly high frequency (10^{-1} to 10^{-4}; average 10^{-2}) and complete reversibility suggests changes in location of mobile genetic elements or alteration of gene expression within the cells (Slutsky et al. 1985, 1987). A genotype that can produce a relatively plastic phenotype should be favored over one that cannot (which does not mean that all plasticity is beneficial). In the case of *C. albicans* this

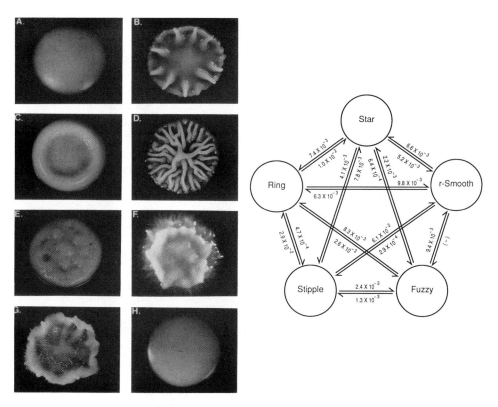

Figure 7.3 (Photos) The eight colony phenotypes of the pathogenic yeast *Candida albicans*: (A) "o-smooth"; (B) "star"; (C) "ring"; (D) "irregular-variable"; (E) "stipple"; (F) "hat"; (G) "fuzzy"; (H) "r-smooth". (Drawing) Interconversion (switching) occurs among many of the phenotypes; the frequencies are indicated. From Slutsky et al. (1985), from *Science 230*, 666-669. 1985. Copyright by the AAAS.

may relate to its ability to invade any host tissue yet also be able to live as a commensal in the digestive tract, although the implications of switching remain unclear. Similar switching behavior has been observed in the slime mold *Dictyostelium* (Soll and Kraft 1988), where it may be part of a survival strategy in the face of starvation. The prokaryotic animal pathogen *Mycoplasma hyorhinis* undergoes phenotypic switching in colony morphology and also in lipid-modified cell surface proteins. This mechanism confers antigenic diversity on the pathogen (Rosengarten and Wise 1990; see also Simon et al. 1980) and thus may be of survival value.

Nonsessile organisms can occasionally show extreme phenotypic plasticity. The freshwater snail (*Physella virgata virgata*) changes certain life history characteristics (grows to a larger size; delays reproduction) in response to a waterborne signal from its crayfish (*Orconectes virilis*) predator (Crowl and Covich 1990). This has the effect of decreasing size-specific predation. Other good candidates could be expected to be the holometabolous insects, because of their poorly mobile larval stage. Greene (1989) studied caterpillars on oak and found that the spring brood developed into mimics of the catkins on which they feed. The summer brood, which emerges after the catkins have fallen, developed into mimics of oak twigs. When caterpillars were raised on catkins, which are low in tannin, they developed into catkin morphs; when raised on leaves, which are high in tannin, they resembled twigs. Artificial diets of catkins supplemented with tannins in general caused the caterpillars to develop into twig morphs. Thus, the environmental signal for this developmental polymorphism seems to be related to the amount of tannin in the diet. Unlike switching in *C. albicans*, the advantages of this plasticity seem obvious and are well documented. The catkins, though short-lived, are nutritionally superior to leaves; the catkin morphs pupate more quickly, are larger at pupation, and have somewhat higher survival rates to pupation. The twig morphs are maintained in the population because they contribute to overall fecundity of the carrier. The two morphs impart survival value by making the insect extremely difficult for predators to distinguish from the respective plant parts.

Ecotypes, local matches between organism and environment, have been well documented in plants (Bradshaw 1972). The traditional approach to separating this form of variation from phenotypic plasticity has been to interchange representatives of local populations or to transplant them to a common environment (Clausen et al. 1948; Bradshaw 1959). If the major attributes of such transplants are maintained when they are grown in a new situation, then the evidence is for the genotypic component of variation. The most common observation is that phenotypic and genotypic variation, acting together, produce some intermediate result. Nevertheless, there is good evidence for localized, genetically determined variation in plants, for example with respect to heavy metal tolerance, flowering time, nutritional responses, and growth form along altitudinal gradients (summarized in Begon et al. 1986, pp.30-35). Selection can be strong for such local specialization, and hence margins of the populations can be sharply drawn despite gene flow across them.

Limits on variation in space are determined by dispersal characteristics. For plants, dispersal is approximately random and for all practical purposes exceedingly localized (Bradshaw 1972; Chapter 2 in Harper 1977). This implies that a given species is segregated into localized breeding populations, typically with a diameter on the order of a few meters, and that offspring inhabit sites similar to their parents. On balance the interesting consequence resembles habitat selection by unitary animals (Bradshaw 1972), where family units may live in the same vicinity. Interestingly, short-distance dispersal, both of sexual and asexual propagules, is also characteristic of modular animals such as the sponges, corals, bryozoans, and ascidians (Jackson and Coates 1986; see also Section 7.6).

The idea that environmental variation could directly impose heritable genetic change sounds heretical, but there is good evidence that this can happen in some organisms (Durrant 1962; Cullis 1983, 1987; see also McDonald's [1990] discussion of 'stress-induced mutations'). One example is flax, where exposure to suboptimal nutrient conditions can result in progeny with different growth properties from their parents. The differences include height and biomass, total nuclear DNA, and the number of genes coding for various RNAs (Walbott and Cullis 1985). Frequently, seeds from such plants breed true, producing stable lines (genotrophs), regardless of their nutrient regimen. Such transitions are associated with changes in DNA evidently resulting from activation of transposable elements. Hence, in these cases, changes in external environment cause an altered cellular environment which in turn somehow activates mobile genes. How widespread this phenomenon is phylogenetically remains unclear, as are the evolutionary implications. Plants do seem to be genetically much more plastic than animals and could potentially change in response to environmental stimuli in ways that animals cannot (Walbot and Cullis 1983). As noted in Chapter 2, there is also evidence that bacterial cells, when confronted by a particular environment, may be able to 'choose' which mutations occur (Cairns et al. 1988; Hall 1988; but see Davis 1989).

On a finer level than ecotypes are genetic polymorphisms. The most obvious and nearly universal polymorphism (in animals) is sex. Others include color, presence or absence of hairs or thorns, and numerous biochemical (e.g., blood group) distinctions (Chapters 6 and 7 in Wright 1978). Biochemical polymorphisms were not really appreciated until the 1950s, when electrophoretic studies of proteins revealed allozymic variants within local populations.

Why does such genetic variability persist? There is much circumstantial evidence but little experimental documentation that environmental heterogenity is a major factor in its being maintained (Hedrick et al. 1976; Hedrick 1986; Via and Lande 1985). Probably the best example is the resistance of particular cultivars of agronomic plants to genotypes of certain pathogens. This is the so-called gene-for-gene interaction (Chapter 3), a phenomenon that has been well characterized genetically (Person 1966; Flor 1971; Gilchrist and Yoder 1984; Vanderplank 1986; for general comments see Wolfe and Caten 1987). The other classic cases of single gene polymorphisms—'industrial melanism' among moths in Britain, and shell color and banding pattern in the land snail—show some anomalies with predictions and are less well understood (Hedrick et al. 1976).

In attempts to interpret the basis for polymorphisms, we must recognize that traits do not evolve in isolation (see Dobzhansky 1956 and Section 7.5).

What is produced will ultimately reflect direct selection on particular characters *as well as* selection on genetically correlated characters. For instance, Brodie (1989) reports that morphology is associated with anti-predator activity in garter snakes. Longitudinally striped forms usually flee when threatened, whereas the unstriped and spotted snakes rely on crypsis. This seems logical because vertebrate predators should more easily judge the motion of a heterogeneous pattern than no pattern or movement of a stripe. These genetic correlations (covariances) could be attributed to either pleiotropy (genes affecting multiple traits; for advantages, see Section 7.5) or linkage disequilibrium (nonrandom association of alleles at different loci) (Brodie 1989).

That polymorphisms defy simple explanation suggests that they are probably not maintained for any single reason, but rather reflect complex interactions of environment with multiple genes and the whole organism. (Here we enter the realm of epigenesis or how genotype and environment together translate into phenotype; see Section 7.5.) Lewontin (1983, p.277) comments that the straightforward examples of introductory genetics (Mendel's peas; phage mutants), where inheritance of a particular trait is clearcut, leave us with the misconception that genotype determines phenotype. The truth is that *"the vast majority of morphological, behavioral, and physiological differences among individuals do not 'Mendelize' "* (p.277; see also Plomin 1990). Ultimately, fitness itself is a phenotype, determined in complex fashion by interaction of many genes at many loci, all subject to environmental modification. In recognizing this, one cannot but wonder whether many biotechnology experiments with genetically engineered organisms are naive and misguided.

7.5   Traffic Lights Regulate Progress Through the Life Cycle

Pauses: dormancy   In the life cycle of most organisms there are one or more stages characterized by reduced metabolic activity and associated changes. These periods are variously referred to as dormancy (plants), 'shut down' (microbes), diapause (insects), hibernation (certain mammals), or estivation (certain birds, mammals, and lungfishes). While obviously each is unique in detail, all share the attribute of quiescence. Entry into such a phase is most apparent when the organism undergoes a morphological transition such as dropping its leaves, or forming spores or cysts.

Some form of torpor is phylogenetically widespread and continues to be maintained by natural selection in organisms which physiologically could have shorter generation times than they in fact have. This suggests that it is beneficial (Levins 1969). Dormancy is evidently a response to adverse or highly unpredictable conditions (Levins 1969; Chapter 3 in Harper 1977). If environments were always conducive and predictable, there would be no role for quiescence. Occasionally we see this: Seed germination is continuous in tropical mangroves (Harper 1977, p.74); bacteria grow indefinitely in log phase in turbidostat culture.

At the other extreme, a given phenotype cannot be highly fit in each one of many different environments. The options are either migration or conversion to a tolerant life stage. Clonal organisms have a third alternative which is basically to put a fraction of their ramets on a 'stand-by' condition, ready for the worst, while their counterparts continue active growth. For instance, among gram-negative bacteria (unlike gram-positive bacteria, this group does not form endospores), a proportion of the cells evidently can be in a resistant, nongrowing state (Koch 1987). Most higher animals are sufficiently motile that they can move to other surroundings, or modify their behavior and local environment by building shelter. For nonmotile organisms the choice is essentially to become resistant or to throw resistant progeny into the environment. These offspring may either outlast the local adversity or be transported by winds or water currents to a more favorable habitat. An intriguing idea is that mobility and the better physiological adaptation of animals to changing environments have replaced the need for dormancy (Bonner 1958). Bonner also observes that, among organisms producing spores or cysts, the dormant stage is closely associated with recombination. A major consequence is that one or more of the emerging products of meiosis (new genets) is likely to be suited to the new environment.

Shifting gears: developmental changes   Organisms respond to changing environments by decoding the relevant cue into some form of biological response. This is presumably what Lewontin (1983) means by saying that organisms transduce physical signals of the external world and, in so doing, create their own "statistical pattern" of the environment that differs from the background. Imagine how its surroundings as perceived through sonar might appear to a bat, compared with those of the same scene displayed by electromagnetic radiation in the visible range to a human, or in the 'photosynthetically active range' to a plant. The pictures transduced will be entirely different and what is of importance within them will be specific to the particular organism.

At the extremes there are two options in terms of timing the response. The first is to act as the environment changes. The disadvantage is that some lag phase in development is necessarily entailed. The organism is thus somewhat late for the event. The triggering of germination of soilborne plant pathogenic fungi by host exudates, or of seeds of certain desert plants by rainfall (Cohen 1967; Chapter 3 in Harper 1977) are examples. Note that both are sporadic events and that the environmental signal is itself the stimulus. Reasonable likelihood of continued survival is provided by the nature of the stimulus. For the fungus it is an appropriate kind and level of host nutrient; for the desert shrub it is a sufficiently heavy rain to leach chemical inhibitors from the seed ('rain gauge effect').

Alternatively, where environmental changes are predictably cyclic *relative to lifespan,* organisms will benefit by advance preparation (Levins 1968, 1969; Chapters 2 and 13 in Harper 1977). In this instance there is not necessarily a relationship between the physical form of a signal and the evoked response (Levins 1968, p.13, 35). Rather, the correlation between certain environmental parameters allows one to act as a predictor of another. Perhaps the classic example is regulation of life cycle activity in microtine rodents (voles) and their relatives (lemmings). Photoperiod, plant chemistry, and food quality appear to be the key environmental signals (Negus and Berger 1988). Presence of the compound 6-methoxybenzoxazoline (6-MBOA) signals a good food source in young grass shoots, and causes the vole (*Microtus montanus*) to initiate reproduction. Phenolic compounds, signalling a poor food source, tend to accumulate later in the season, and cause the animal to respond negatively.

Likewise, moulting in birds, and in dogs and many other mammals is not influenced by temperature per se, but by photoperiod (Lofts 1970; Follett and Follett 1981). Rainfall evidently is not the seasonal cue for tropical cicadas; Wolda (1989) speculates that the signal may be the photoperiod, which acts on the host plants of the insects. Photoperiod affects reproductive processes in numerous organisms, including birds, mammals, and many invertebrates. In birds it also triggers migratory behavior and associated physiological changes such as the laying down of fat. Hormones are the chemical messengers through which the birds transduce this physical signal. The life cycle phases (e.g., dormancy, diapause, estrus) of herbivores are similarly coordinated, which means that food is normally available when needed (Harper 1977, p.405). The disadvantage of regulation by a clock mechanism is of course an inability to respond to unusual opportunities, although this is counterbalanced by protection from 'false starts'.

Signals and genes   Regardless of the type of environmental trigger or the organism receiving it, the effect at the *cellular* level is similar in kind—namely, a reversible activation/deactivation of genes. What nutrients do as cell signals in bacteria or fungi (Chapter 3), hormones do in humans and in other macroorganisms. A fascinating recent example (Bartlett et al. 1989) is that at least some genes in deep-sea bacteria are regulated by hydrostatic pressure. A gene has been cloned from a bacterium isolated from a depth of 2.5 km (optimal pressure for growth of this barophilic organism is 280 atm). The gene encodes a major pressure-inducible protein, possibly functional in the outer membrane of the cell. The sensing device for this pressure signal is as yet unidentified, but the response system is an elegant example of functional modification for life at high pressure.

For proper gene expression, conducive physiological conditions (e.g., temperature, pH, ions) are essential. Thus, the cellular environment plays a major role in expression of mutations (Stebbins 1968), perhaps even more so than the external environment which is the usual focus of attention. Gene encoded products interact through pleiotropy and linkage disequilibrium in "informational relay" (Stebbins 1968) or "gene net" (Bonner 1988 p.144) fashion. Furthermore, often a particular pattern ("epigenetic sequence"; Stebbins 1968) of gene action is required during development in which otherwise unrelated genes coordinate their activities. This integrated process can be disrupted by an adverse cellular environment or by mutation at any stage in the hierarchy from transcription to assembly of membranes and organelles (Figure 7.4). Demands for coordinated gene activation/deactivation mean that the number of successful mutations which are possible is greatly reduced.

A consequence of the fact that developmental events are interconnected is that if a change in one character occurs it will be associated with changes in functionally related characters through a cascade effect (Cheverud 1984). In other words, because of this epigenetic sequence pattern, individual gene loci have pleiotropic effects (this is a good example of the developmental category of constraints discussed in Chapter 4). A bacterium can be assembled faster and more easily than can an elephant. Therefore, the constraints of informational relay and sequential gene action are presumably much less for bacteria than for elephants. Pleiotropic and related effects, however, *can be beneficial* in facilitating coordinated rather than individualized change of multiple phenotypic traits. Thus, despite its potentially adverse consequences, pleiotropy may be favored overall by natural selection (Riedl 1978; Cheverud 1984).

7.6   Habitable Sites and the Evolution of Gene Flow

Most of this chapter can be consolidated into a model which interprets the distribution and abundance of organisms in terms of habitable sites. Aspects of the habitable site idea were developed mathematically by Gadgil (1971) and extended conceptually by Harper (1981b). What follows here builds on this notion and depicts it descriptively and graphically, rather than mathematically.

Habitable space can be defined as a zone in which biotic and abiotic conditions allow an organism to become established and survive competitively. Obviously the clearest example of a habitable site is the place where an organism can complete its life cycle (i.e., including a reproductive phase). A site is not habitable to a species simply because it contains viable propagules, such as seeds or spores, of that organism. Habitable sites are thus distinct from 'sink' sites which may contain individuals—perhaps even consistently and in relatively high numbers—but only because the organisms are supplied as immigrants from a 'source' site nearby (Pulliam 1988). The colon but not a soil particle is a habitable site for *E. coli*. Thus, habitable space

Figure 7.4 Interaction of gene products in a hierarchical relay system for a prokaryote (sequence depicted by solid lines) or a multicellular eukaryote (solid plus dotted lines. A mutation could disrupt the sequence at any stage. Modified after Stebbins (1968).

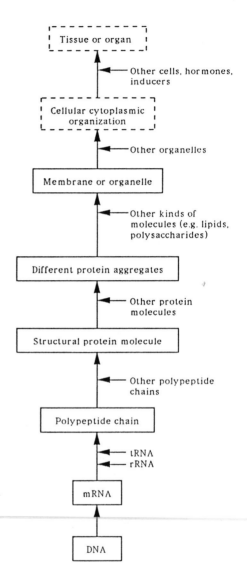

consists of localized, favorable conditions that occur as islands or patches within a biologically and physically hostile sea. Distribution and abundance will then be determined in large part by the match between characteristics of the organism (e.g., dispersal ability; intrinsic population growth rate) and environmental pattern (e.g., size, number, distance between, duration and carrying capacity of, habitable sites). As seen in the preceding sections, chances for a match are improved by phenotypic plasticity of individual colonizers, and by genotypic variation within the colonizing population to cope with the variability of sites. Occasionally, sites can have very different characteristics (see the case study).

Sites also have variable temporal and spatial components. This is perhaps best illustrated by considering how environments could appear in time and space to an organism (Southwood 1977). With respect to time they can be: 1) constant (favorable or unfavorable indefinitely); 2) predictably seasonal; 3) unpredictable; or 4) ephemeral (predictably short, favorable conditions followed by unfavorable conditions for an indefinite time). In space, habitats can be: 1) continuous (favorable area exceeds area that organism can reach regardless of dispersal mechanism); 2) patchy (unfavorable areas surrounding favorable islands which can be reached by dispersal); or 3) isolated (favorable but unreachable). Consider a migratory bird which breeds over the summer in northern Canada and passes the winter in the southern United States. The two sites are predictably seasonal and the bird's life cycle is adjusted to exploit both. On a local scale, the relative area occupied at each site will fluctuate annually depending on factors such as pressure from competitors and predators which also occupy the site. The life cycles of plant and animal pathogens which alternate habitats is directly analogous. The remarkable intersection in time and space between habitable sites and appropriate growth stage to colonize them is best seen among species with complex life cycles (Chapter 6; see also Chapter 3 in Price 1980). So, sites are highly plastic in size, may be discontinuous through time, and are not species specific. They can be depicted in three-dimensional form as meandering tubes of irregular diameter (Figure 7.5).

The dynamic aspect of habitable space results not only from changes in the organism or the site, but also from interactions between organism and site. It was noted previously (Section 7.2) that, strictly speaking, separation between organism and environment is impossible. As the organism moves through its life cycle it selects those parts of the site that are *relevant* to itself

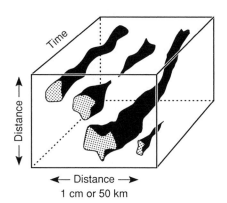

Figure 7.5   The habitable space concept projected through time. Species cohabiting within a site may or may not directly interact. In response to biotic or abiotic factors, any given species may increase (biomass or numbers) or decline possibly to temporary local extinction pending recolonization.

A Case Study: The American Legion, Cooling Towers, Ponds, and a Bacterium

There are probably few organisms with a more interesting ecology (not to mention notoriety) than *Legionella pneumophila*, the bacterium that causes Legionnaire's disease. This inconspicuous organism achieved international fame when 149 of some 4,400 persons attending a convention of the American Legion in Philadelphia in the summer of 1976 became ill. All developed to various degrees similar symptoms including fever, cough, muscle aches, chills, diarrhea, and pneumonia. More cases were discovered within the local populace. Eventually 221 people were affected and 34 died.

The mystery was unravelled through intensive, persistent, epidemiological and etiological studies (Fraser and McDade 1979; Bartlett et al. 1986). An early, significant finding was that the victims had only one thing in common: they lived in, or spent time in or near, the hotel where the convention was held. Eventually, the causal organism was cultured by an unorthodox procedure for bacterial studies, and identified, again by unusual methods for bacteria, and shown to be a new species. In particular, the culturing procedure was a pivotal development, not only in demonstrating the cause of the disease, but in establishing one aspect of the complex environmental relationships of the causal organism. *L. pneumophila* has fastidious nutritional requirements and grew in the laboratory only if culture media contained unusually high amounts of cysteine and iron.

Epidemiological evidence pointed to an airborne pathogen transmitted from a point source and suggested strongly that the singular feature of that particular hotel was its air conditioning system, which was probably the reservoir for the bacterium. In subsequent outbreaks (as well as in past episodes now known from serological evidence to have been caused by *L. pneumophila*), water in cooling towers or in evaporative condensers of air conditioners has been implicated unequivocally (as have other water systems, including domestic water supplies; Stout et al. 1985; Bartlett et al. 1986).

So, in terms of the environmental biology of this organism, we are faced initially with an intriguing paradox: How does a nutritionally demanding pathogen, which can be cultured in the laboratory only under the most exacting conditions, and in its human host multiplies primarily in lung tissue and blood, maintain substantial populations in water? A key discovery was that *Legionella* could grow in association with cyanobacteria, from the extracellular metabolites of which it evidently met all of its nutritional requirements (Tison et al. 1980). Subsequently, green algae, ciliated protozoa, and amoebae were shown to provide benefits as hosts in what appears to be a commensal or parasitic relationship (reviewed in Thornsberry et al. 1984, pp.323-332). This aspect of the life history of the pathogen is unclear and evidently complex. *Legionella* not only obtains nutrients from its microbial associates, but seemingly is protected from desiccation while in transit as an aerosol (Berendt 1981). Fliermans et al. (1981) isolated the bacterium from diverse aquatic habitats (e.g., temperature range 5.7-63°C; pH 5.0-8.5). *L. pneumophila* is now considered to be a thermotolerant component of freshwater microbial communities. Probably humans are merely incidental, 'mistaken' habitats in the life history of the microbe. It grows well in hot water tanks and distribution systems, particularly in areas where sediment (scale and organic floc) accumulate. Interestingly, the sediment evidently acts by promoting growth of other bacteria, which in turn enhance growth of *Legionella* (Stout et al. 1985).

Tolerance of warm environments and demanding nutritional relationships of the organism explain why air conditioning systems are ideal reservoirs. They provide warm, wet, lighted habitats, and often receive additional water from sources containing algae and other microorganisms. The cooling towers or evaporative condensers act as efficient scrubbers of microbes (including algae) from the air. A typical tower of the size for a large hotel would handle water at the rate of 1,000 gallons per minute. This hot water from the compressor unit is sprayed over splash bars and cooled by evaporation which is enhanced by air pulled through the spray with a fan. Airborne microbes are entrained in the spray and are expelled as drift, which is the source of airborne infection (Fraser and McDade 1979).

As is true of the case studies in previous chapters, this example presents an interesting problem in basic ecology with obvious practical implications. The organism grows in 'habitable sites' as diverse as portions of the human body and eutrophic lakes. It clearly has the ability to dramatically change its environment in the human host; in turn, survival of the bacterium in nature seems to depend on a complex commensal or parasitic relationship with other microorganisms which provide the appropriate nutritional and protective milieu. Knowledge of the environmental biology of *Legionella* provides a basis for the intelligent application of controls, such as elevating the temperature of domestic hot water supplies, or chlorination, which can be directed at the localized inoculum reservoirs involved in amplification and dispersal of the pathogen.

(Lewontin 1983) and *reorganizes* them: Twigs become nesting material for birds; trees are felled to become food and shelter and dams for beavers; lignin in wood is degraded by fungi and the monomers incorporated into fungal protoplasm (e.g., Kirk and Fenn 1982; fungal metabolites are also extruded and further modify the habitat, e.g., Harley 1971); the DNA of a plant host is reprogrammed by the crown gall bacterium to encode opalines and nopalines as nutrients for the pathogen. By such activities the organism modifies the site further, in the process making it more or less habitable by certain other species. In the extreme case, a successional sequence develops, as consortia of organisms are displaced by successors better adjusted to the space.

Distance between habitable sites is arbitrary and is related to distribution of favorable conditions relative to the size and growth form of the organism. For instance, colonizable patches may be separated by distance on the order of micrometers for bacteria or yeasts growing on a leaf surface, by meters for earthworms, or by kilometers for lions hunting on the Serengeti Plain. Clonal, modular organisms will explore their environment by growth, with or without fragmentation; unitary organisms will reach new sites by active, frequently directed, movement. That habitable space may exist unfilled is clear from the frequent success of intentional species introductions or inadvertent invasions (Mooney and Drake 1986).

Distance between islands will be a factor driving dispersal and molding life cycle features. The probability of colonizing any site is a function of the probability of propagule survival and the number of arriving propagules. Gadgil's (1971) models show that, as a result of dispersal, isolated, poorly accessible sites should be less crowded than an 'average' site. Episodic dispersal would produce unequal crowding at various sites. This, in turn, would lower the total population size of the species over its entire range.

Species have different tendencies to disperse and different dispersal characteristics. This will affect gene flow (used here to mean the exchange of genes among different populations of the same species). For instance, dispersal is generally strong and rigidly programmed in insects, evidently as an adaptive response to their short life cycles and ephemeral breeding sites (Wilson 1975, p.104). Migration is less programmed in vertebrates, but still predictably based on certain factors such as sex and age (e.g., young males are the most prone to disperse in baboon troops and lion prides; Wilson p.104). Microbes seem to be the ultimate nomads. Historical, geographical, or geological constraints are less important considerations in assessments of distributional patterns than they are for macroorganisms (Chapter 4 in Brock 1966). Small size facilitates long-distance movement by wind, water, or vectors across barriers that would be insurmountable for macroorganisms. A short generation time and high fecundity, which imply potentially high population densities, favor colonization. The fungi particularly seem to be the closest counterparts to the insects in programmed dispersal characteristics (timing in life cycle; specialized migratory forms), although there are obvious differences (e.g., fungal dispersal is largely passive; insect dispersal is frequently directed). Dispersal characteristics and colonization dynamics have been examined at length elsewhere (MacArthur and Wilson 1967; MacArthur 1972). Where dispersal is passive, as is typical in plants and microbes, the number of individuals declines exponentially (e^{-x}) with distance x from the source. Where dispersal is directed, as in the case of animal search patterns, the decline follows a normal distribution pattern, e^{-x-2} (Wilson 1975, pp.104-105). The two types will thus have very different effects on the rates of gene flow.

How has dispersal evolved? Since travel is risky, migration will only be favored by natural selection if the chances for finding a better site exceed those of colonizing a worse site and of death en route (Gadgil 1971). Overall, dispersal must be advantageous because it is exhibited in some form in all phyla. It can also entail considerable cost to the organism in expended time and resources. Birds migrate only after physiological preparation and training flights. In the slime mold *Dictyostelium* up to 90% of the formerly free-living cells are assigned to act as stem cells to elevate the fruiting body. They die as they are incorporated into the stalks (Whittingham and Raper 1960), so there must be considerable benefit in getting a few spores to a favorable habitat patch (Bonner 1982a). The benefit of dispersal to the *individual* (as opposed to the weaker argument for group selection at the species level) is the chance to colonize an empty habitat, and the initial mating advantage over the locals upon entering the new population (part of the process of 'migrant selection'; Wilson 1975, pp.103-105).

In unicellular organisms, dispersal may be incidental and the cells involved essentially no different from those remaining at the site of production. It is easy to see how chances for successful migration could be improved by modest physiological changes, such as evolution of a 'shut-down' phase (Dow et al. 1983), or by cells which became less sticky than their counterparts. The population would thus become partitioned into members suited for active growth or for survival and transport. A later development, perfected by the fungi, would involve cells (spores) packaged for transport with a nutrient reserve and thick walls and, in many cases, propelled on their way by elaborate release mechanisms (Ingold 1971). This is not unlike the segregation of insect populations into sedentary and mobile forms (nonwinged versus winged aphids; *solitaria* and *gregaria* phases of the African locust).

In other phyla, dispersal may have arisen coincidently with sex. The container enclosing the products of meiosis (Bonner 1958, discussed above) might have been an easily transported unit. Sessile, clonal organisms in effect sample new environments by growth. It was noted previously (Chapters 5 and 6) that while such organisms as strawberries and corals can in theory grow in unlimited fashion (Chapter 3 in Williams 1975), in practice it is only a matter of time before clonal individuals reach the boundaries of their habitable sites. Dispersal by clonal fragmentation is merely an extension of the growth process whereby the genet can move farther afield, free from the handicaps (e.g., transport and allocation problems; systemic toxins and infections) of operating as a physiological entity. Motile, unitary organisms typically select sites in an active, controlled event, whether this be accomplished by relatively primitive chemotaxis mechanisms (nematodes and their plant hosts), or by sophisticated olfactory, visual, and acoustic cues (vertebrates and higher invertebrates). Hence, the fundamental distinction in dispersal seems to be drawn along lines of growth form (Chapter 5), rather than kingdom.

Whatever its origin, dispersal can be viewed most productively as an opportunity met *gradually* by different organisms *interacting* with their environments. In evolutionary terms it seems no more likely that movement by one means or another to habitable sites suddenly became a 'problem' to be 'solved' by a species than did swimming in water become a 'problem' that seals 'solved' by losing their legs (Lewontin 1983). Seals developed flippers and in so doing likely incorporated water as a progressively greater component of their environment (Lewontin 1983). Analogously, a migratory phase joined a sedentary phase as an increasingly important component of the life cycle.

7.7   Summary

The 'environment' includes the physical and biological setting of an organism with which it is coupled and reciprocally interacting. This is superimposed on the internal cellular environment in which the genes must function. Environments both drive natural selection by their differential impact on survivorship and are in turn altered by the changing composition of the survivors. Organisms neither develop in mechanical fashion against a placid environmental backdrop nor do they create 'solutions' to 'problems' posed by the environment.

The organism determines what aspects of the environment are relevant and the thresholds at which a factor is significant. Organisms experience the same absolute fluctuations very differently, depending on the duration of their lifespan (see concept of physiological time, in section 4.5). For a bacterium with a generation time of hours to days, absence of nutrients or a 5°C degree drop in temperature for several hours would have major physiological implications. Either change over the same absolute time frame for an elephant with a generation time of 12 years would be negligible. For any given species, short-term fluctuations relative to life span typically induce phenotypic change in morphology, behavior, or physiology; long-term changes influence gene frequencies which reflect the fact that individuals leave different numbers of descendants.

Environments have been classified from the organism's perspective as fine-grained (experienced in many small doses and not actively sought out or avoided) or coarse-grained (sufficiently large so that the individual chooses among them or spends its whole life in one patch). Though conceptually appealing, the theory of environmental grain is abstract and of questionable pragmatic value. Mathematical models attempting to describe natural selection in patchy environments may be ill-founded because it is not clear what is being maximized by natural selection.

How organisms experience and respond to environmental fluctuations is affected by their size and growth form. Increasing size (of the physiological individual) is related directly to increasing complexity manifested in various ways: increase in number of cell and tissue types, and hence in the interactions among them; division of labor among cell types; formation of support structures; insulation and homeothermic mechanisms; centralized hormonal and neural control. Size is inversely proportional to generation time. While a bacterial cell is not as well insulated from its environment as an elephant cell (or, correspondingly, at the genet level, the bacterial clone or the whole elephant), the bacterium can track environmental fluctuations much more rapidly by phenotypic and genotypic changes. Modular organisms, composed of iterated parts and for the most part sessile (Chapter 5), respond differently to the environment than do unitary organisms: Resistant propagules (seeds, spores) may outlast local adversity or be transported to new sites, while the soma changes by addition or subtraction of modules. Unitary organisms, which are typically mobile, adjust by migration, and by physiological and behavioral mechanisms.

Organisms receive environmental cues and transduce these from a physical or chemical mode into an appropriate biological response. For example, the message that "day length is changing" is transduced via hormones to activate flowering in some plants, or reproductive activity in certain animals. Where environmental fluctuations are irregular, the response must occur directly as the environment changes. Where fluctuations are predictably cyclic, it is advantageous for organisms to 'anticipate' them by recognizing some form of early signal. Correlations between environmental parameters allow one form (e.g., photoperiod) to act as a predictor of another (e.g., temperature). As a result of natural selection, particular phases of life cycles tend to match those environmental conditions for which they are suited.

The distribution and abundance of species can be interpreted in large part by the match between organism (e.g., dispersal ability; intrinsic rate of population growth) and environmental pattern (size, number, spatial characteristics, and carrying capacity of habitable sites). Habitable sites are zones where the organism can develop competitively. Sites are dynamic in time and space due to changes in the organism, the environment, or the organism-environment interaction. Species invasions and successful introductions show that habitable space may exist unfilled. Distance between habitable sites is a factor driving dispersal and molding life cycle features. Gene flow patterns (exchange of genes between different populations of the same species) will be quite different for populations of organisms that disperse passively and decline in numbers logarithmically from a source pool (such as plants, microbes, and clonal benthic invertebrates) and those that search in directed fashion (most unitary animals), represented by a normal distribution pattern of decline. Despite the high mortality that occurs during a migratory phase, dispersal is presumably favored by natural selection because in some form it occurs in all phyla. Dispersal appears to be an ancient evolutionary development that has been progressively improved upon by various mechanisms.

7.8   Suggested Additional Reading

Bonner, J.T. 1988. The evolution of complexity by means of natural selection. Princeton Univ. Press, Princeton, N.J. How increase in size and complexity of organisms has evolved, and implications for environment-organism interaction.

Harder, W., L. Dijkhuizen and H. Veldkamp. 1984. Environmental regulation of microbial metabolism. *In* The microbe 1984. Part II, Prokaryotes and eukaryotes. (D.P. Kelly and N.G. Karr, eds.). Cambridge Univ. Press, Cambridge, U.K., pp.51-95. How microbes react physiologically to changing environments.

Lewontin, R.L. 1983. Gene, organism and environment. *In* Evolution from molecules to men (D.S. Bendall, ed.). Cambridge Univ. Press, Cambridge, U.K., pp.273-285. A reminder that organisms both make and are made by their environments.

8

Conclusion: Commonalities and Differences in Life Histories

There is something fascinating about science. One gets such wholesale returns of conjecture out of such a trifling investment of fact.

—Mark Twain, 1874, p. 156

8.1 Levels of Comparison

Because the life history of every species is at some level distinctive, the difficult task in attempting to forge a synthesis is finding meaningful commonalities. On one hand, if important differences are overlooked the analogies are false and misleading. Indeed, this must happen frequently because what is truly ecologically important may become known only in hindsight. Francis Crick has said, "There isn't such a thing as a hard fact when you're trying to discover something. It's only afterwards that the facts become hard" (Judson 1979, p.114). A good case in point is Hutchinson's (1959) famous observation that average individuals from geographically overlapping (sympatric), congeneric species showed regular differences in various morphological features such as length of proboscis in bumblebees or length of skull in squirrels (the so-called Hutchinson 'ratio' or 'rule'). The ratio was about 2.0 for weight differences; the corresponding ratio for length, which varies as the cube root of weight, is about 1.3. In an era when the role of competition as a force shaping communities was in its ascendancy, these ratios were frequently seized on "as prima facie evidence that communities were organized according to the principles of limiting similarity" (Eadie et al. 1987). Horn and May (1977) showed subsequently, however, that the 'rule' also applied to such inanimate objects as an ensemble of musical instruments and the wheel size of children's bicycles or tricycles. The ratios are now generally regarded as being artifactual (Roth 1981; Eadie et al. 1987).

239

On the other hand, sweeping generalities set at a level to avoid peculiarities of specific systems are usually so bland as to be trivial and uninformative. This dilemma is shared by model builders who always must compromise between accuracy and generality. As seen above, the problem is confounded by the intermix of fact and dogma in ecology. A distinction between fact and dogma is rarely made. Conclusions are commonly based on correlation rather than causation, and attractive 'explanations', if repeated often enough, become 'factual'.

It is well known that organisms are quite similar in terms of their cellular function and that similarities across taxa diminish as one moves through higher levels of organization (Table 8.1 and Slobodkin 1988). This principle is but one of many examples of one of the general laws of integrative levels, namely that the higher the level the smaller the population of instances (Feibleman 1954): Each succeeding level embodies the attributes of its predecessors as well as imparting some of its own (emergent) properties. Hence, organisms can be expected to be more distinctive as entities than the molecules of which they are mutually composed, and communities yet more different than are organisms.

Table 8.1 Some levels at which all biota can be compared, the bases for comparison, and the anticipated similarities

Level	Comparative basis	Expected commonalities
Cell	Mechanics; physiology; function; biochemistry; genetics	Numerous: ATP; genetic code and its replication and translation; metabolic cycles; cell division
Organism	Mobility; homeostatic ability; size; shape; mechanical design; life history patterns, incl. resource allocation	Each organism unique in detail; many analogous* (e.g., foraging, reproductive, dispersal mechanisms); and some homologous* features (e.g., forelimbs of chickens, humans, dogs, whales; cellulose in cell walls of members of Kingdom Plantae)
Community	Major organizing forces (e.g., competition, predation; mutualism; abiotic factors); recuitment (open vs closed); equilibrium vs nonequilibrium	Few: patterns such as species-abundance distribution may be similar but for different reasons; organizing forces may be the same but relative roles differ; size-frequency distribution for unitary but not modular organisms

*Analogous implies functional and/or morphological similarity among taxa, but not a common descent; homologous is similarity resulting from common ancestry.

8.2   On Being a Macroorganism or a Microorganism

The major frame of reference for this book, as the title implies, has been size differences across major groupings of taxa, that is, micro- versus macroorganisms. Size is clearly an ecologically important attribute of an organism and influences profoundly its interactions with the environment. Microbes are affected primarily by intermolecular forces while large creatures live in a world governed by gravity. The direct consequences of this were discussed in Chapter 4 and were developed, along with many indirect consequences, in each of the other chapters. Most if not all attributes of an organism are influenced by its size, and much of the variation in life histories is correlated with size.

One cannot compare organisms strictly on the basis of size, however, because this factor does not occur as an independent variable. First, organisms are locked into developmental and ecological channels established by phylogenetic differences, the most obvious of which are design constraints (Chapters 1 and 4). A dog and a fish of equivalent weight have very different life histories *and evolutionary potentials*. Second, even among geometrically similar organisms of a common phylogeny (such as different species of lizards), shape varies with increasing volume leading to proportionate rather than progressively diminishing changes in surface area. Efficiency in biological (exchange) processes dictates an approximately constant ratio of surface to volume. The same relationship means that supporting elements (bones, stems) must be proportionately thicker as size increases if the organism is not to fall under its own weight.

Within the scale from micro- to macroorganism, increasing size also means increasing complexity (more cell types) and concomitantly increasing division of labor and centralized control; increasing independence of the external environment (homeostatic ability); and increasing chronological time to maturity and between generations (although ratios of life history features to *physiological* time seem to be constant at least in unitary organisms; Chapter 4). Population densities (and intrinsic growth rates) of microorganisms are consequently much higher than those of macroorganisms. Since for a given nucleotide sequence the rates of base substitution are approximately constant per unit time across lineages (molecular evolutionary clock; Chapter 2), favorable mutations will spread more rapidly in a bacterial than in an elephant population. Put differently, this means that microorganisms, though more exposed to environmental variation (Chapter 7), can evolve more rapidly in response to it.

8.3   Natural Selection As the Common Denominator

As noted in Chapter 1 the general property which unites organisms is that all are shaped by evolutionary pressures. The major evolutionary process underpinning comparisons in this book is natural selection. This is as opposed to other forces such as founder effects, archetype (phylogenetic) effects, genetic drift, or pleiotropic effects (for discussion of these and others see Harper 1982). Admittedly, it is difficult to know the degree to which any specific feature has been shaped by natural selection (i.e., adaptively), or is maintained because of developmental, phylogenetic, or allometric constraints (Section 4.2). Nevertheless, where similar, functionally important characteristics appear across phylogenetic lines (convergent evolution) it is difficult to see how these could arise other than primarily, overall, by selection acting on fitness differences (through adaptation or *ab*aptation, to use the terminology of Harper 1982). As a metaphor it might be said that phylogeny and pleiotropy set the length of the biology text and confine the plot, whereas the contents of each page are dictated largely by adaptation acting on species attributes. Hence, just as many genes or gene families (e.g., histone, cytochrome, actin, myosin, globin) have been conserved in widely divergent organisms through a long evolutionary history, analogous responses or 'strategies' to similar 'problems' are shared by organisms, regardless of their size (Table 8.1).

It is therefore an unfortunate mistake that demarcation based on size, implicitly or explicitly, has acted to channel so much of the thinking in ecology. This situation is similar to but more serious than the counterproductive polarization of ecology into functional camps and evolutionary camps.

The principle of trade-offs as the universal currency   The general model presented at the outset (Figure 1.2 and Chapter 1) as a common basis for comparisons was the organism as an input/output system. Each life history is a compromise to multiple demands on limited time, resources, and options. Time and energy spent foraging are lost to reproduction. Biomass must be allocated to growth or maintenance or reproduction. Hence, improvement in one trait, or allocation of some limiting resource such as energy to one function, is typically at the expense of some other. If this were not the case every organism would operate at its optimum phylogenetic limits, an observation inconsistent with a wealth of evidence from both microbial ecology and macroecology (see especially Chapters 1, 3, 5, and 7; also Stearns 1989a). This, the most fundamental ecological concept, might be called the Principle of Trade-Offs.

Natural selection should favor organisms that are cost-efficient *overall* in resource acquisition, allocation, and expenditure. As such, they should be competitively superior and leave more descendants. (Note that while the principle of allocation is intuitively acceptable, it is usually not clear to what extent any particular allocation pattern translates ultimately into numbers of descendants; Chapter 3.) The activities of an organism on which selection acts can be subdivided for convenience into five arbitrary categories (Southwood 1988): 1) tolerance of the physical environment; 2) defense; 3) foraging; 4) reproduction; and 5) escape in time or space. Although the input/output scheme, as a graphic model, does not explicitly recognize the functional categories 1, 2, and 5, they are implicitly a part of it. Costs assignable to these categories will have a negative impact on foraging and a positive or negative effect on reproduction. For instance, with respect to defense, time spent by a bird avoiding predators or, with respect to escape, time spent by a fungus ensconced as a sclerotium, in both instances is time lost to feeding. These diversions may or may not ultimately increase the number of descendants (Chapter 3 and Andrews 1991).

Various levels of ecological generality can be recognized for each of the above categories in comparisons among taxa. Take, for example, adjustment to the physical environment. The most general comparison entails only a slight restatement of the principle of trade-offs, viz., microorganisms as well as macroorganisms must compromise allocations to physiological adjustment with other demands on their time and resources. This is a valid generality, but being a universal statement conveys little useful information in any particular context. A second, more specific level of comparison could be that, in the short term, bacteria respond rapidly to changes in the environment by altering metabolic pathways, whereas elephants change their behavioral patterns. For both the microorganism and the macroorganism this implies costs as well as benefits. Being a more specific statement than the first, it carries useful predictions, but these predictions are mostly limited to bacteria and elephants. Finally, at the third and most specific level, one could postulate that bacterial species A adjusts to a temperature increase by producing certain heat shock proteins (at the expense of other metabolites) while elephant species B spends proportionately more time resting in the shade (at the expense of foraging). Note that at this level a given response does not necessarily apply even to closely related taxa, much less from microorganism to macroorganism. A different species of bacterium may switch on different metabolic pathways or enter a temporary 'shut-down' phase (Chapter 7); a different elephant species might spend more time lolling in rivers or emigrate from the region.

By similar reasoning, phenotypic and genotypic changes in response to high temperature over the long term could also be compared. For instance, the membrane structure of an aquatic bacterium in a hot spring might be altered structurally to make it more stable, while one or more of the heat dissipation mechanisms of the elephant might become more efficient. Again, both types of changes are trade-offs against alternative demands in the respective organisms. At this third level we have the most specific, testable predictions. Though such statements have little generality (and in this sense what is true for the elephant is obviously *not* true for the bacterium), they are directly relatable up through the higher tiers of generality to a universal principle. The same comparative argument could be developed as a conceptual hierarchy for each of the other four patterns noted above on which selection acts—defense, escape, foraging, and reproduction.

Growth form as the integrator of life histories across size categories   If size establishes fundamental differences at the level of the individual, then growth form is probably the best unifier. Instead of slicing the biological world into halves based on size into micro- and macroorganisms, it can be cut along the axis of growth form into modular or unitary organisms. Such a scheme constitutes a rational as opposed to an arbitrary demarcation and is more biologically justifiable than is one based on size.

The unifying feature of growth form is that microorganisms share with a whole set of macroorganisms a modular life style. Foremost among such characteristics are sessility, indefinite growth by iteration, and totipotency or somatic embryogenesis (Chapter 5). As developed earlier, the main ecological and evolutionary implications include: 1) replacement of active mobility with growth by extension, passive dispersal mechanisms, and dormancy; 2) exposure of the same genetic individual to different environments (hence to different selection pressures); 3) close interaction among subsets of a population (neighbor effects); 4) high phenotypic plasticity; the potentially important (but as yet largely undocumented) role of somatic mutation in evolution; and 5) usual absence of senescence at the level of the genet. Bacteria, overall, have limited morphological diversity as modular organisms and must respond principally at the subcellular (biochemical) level; fungi and planktonic algae use both physiology and geometry to advantage. In contrast, architecture plays a major role in the life history of the modular macroorganisms such as the corals and plants.

By varying their growth form directly in response to environmental variation, modular organisms make trade-offs visible in a way that their unitary counterparts do not. Consider a fungus which can either forage for nutrients by a diffuse mycelial network, *or* extend rapidly away from resource depletion zones into new terrain by aggregating its hyphae into telephone cable-like strands (rhizomorphs), *or* divert biomass into wind-disseminated spores and the specialized structures built to broadcast them. Dispersal in some form is evident in all organisms, but the costs are shown most dramatically within the modular category. In dictyostelid cellular slime molds, where the stalk of the fruiting body may consist of dead cells sacrificed for the cause, mortality may exceed 90 percent of the cell mass (Bonner 1982a). At the other extreme, among the higher fungi (e.g., Aphyllophorales of the Basidiomycetes), resources are diverted into massive bracket conks which exist purely for dispersal and are ultimately destined to die. Moreover, the wood-rotting fungus may produce conks prolifically, thus expending its finite resource quickly in brief but 'riotous living'. Alternatively, it may 'opt' to stagger reproduction over time, thereby prolonging the food base but risking loss to competitors or by mishap. Among plants, the compromise is between the competing needs of photosynthetic and nonphotosynthetic tissue. Among sessile marine invertebrates, it is between biomass allocated for support against turbulence, for food interception, or for production of progeny. While the modular life style makes the concept of trade-offs especially compelling, the driving forces and the relative roles of each in the evolution of form are by no means clear.

What limitations do growth form impose on the extent to which basic ecological concepts can be generalized? A good example is the logistic or sigmoid equation developed by Verhulst (1838) to describe growth of human populations. The sigmoid model and variations of it have been used almost universally as descriptors of population increase (Chapter 4; for microbial analogs see Andrews and Harris 1986). In terms of a unifying ecological principle, 'logistic theory' at the population level is as basic a concept as is that of trade-offs at the level of the individual. It is the foundation for fundamental sequels such as the Lotka-Volterra model of interspecific competition, and r- and K-selection (Chapter 4). Although density-dependent regulation of populations in some form is generalizable, as is r/K theory, the predicted outcome may vary depending on growth form. Sackville-Hamilton et al. (1987) propose that the effects of r- and K-selection may not be same on unitary as on clonal (modular) organisms. For instance, whereas genets with a phalangeal growth form (Chapter 5) are expected to behave according to the theory, those with the guerrilla habit are not. In the latter case, r-selection favors nonreproductive (nonsexual), so-called 'immortal' genets that spread clonally unless there is a lethal condition which kills all stages in the module-to-module cycle. K-selection acts similarly, provided the environment is reasonably stable over many generations (if it is not, then K-selection should favor reproduction). These predictions are of course opposed to convention which associates greater reproduction with r-selection and competive growth with K-selection (Chapter 5). The authors observe that the genets of some higher plants are 'immortal' and never reproduce. It is also interesting that numerous fungi appear to grow indefinitely as clones. For many members of one entire group (Fungi Imperfecti) a sexual stage has yet to be found, and others (Mycelia Sterilia) evidently lack conidia and other reproductive propagules (Andrews 1991). Regardless, the significance of the paper by Sackville Hamilton et al. (1987) in the context here is that growth habit, *not* size per se, may greatly influence response to selection pressure. Hence, evolutionary predictions need to be clustered with respect to growth form, rather than segregated for microorganism as opposed to macroorganism.

8.4 Microbial Ecology and Macroecology Are Complementary

To conclude, I return to my earlier contention (see Preface) that macroecology is essentially phenomena in search of mechanistic explanation, whereas microbial ecology is experiment in search of theory. Slobodkin (1988 p. 338) has commented that "ecology may be the most intractable legitimate science ever developed". Despite the rich theoretical base of plant and animal ecology, few clear generalizations emerge which are other than trivial, and most if not all the precepts are contentious. Macroecology remains a field dominated by correlation. This is not surprising, given the complexity of higher organisms and their habitats, and the interactions among phenotypic traits. The consequence is that controlled experiments are difficult, if not impossible, and conclusions are often not robust (for criticism of ecological experimentation see Hairston 1990). The long-running debates about such matters as: 1) the relationship between diversity and stability (e.g., Murdoch 1975); 2) the factors that set the number of trophic levels (Chapter 2 in Pimm 1982); 3) the explanation for increasing species richness with decreasing latitude (Stevens 1989); 4) the forces shaping community structure (e.g., Diamond and Case 1986); 5) the evolution of life histories (e.g., Boyce 1988), including the role of adaption versus constraint in shaping phenotypic features (e.g., Bonner 1982a); and 6) the putative role of allelopathy in competition between neighboring plants (Muller 1966; Bartholomew 1970) are but a few of the more conspicuous examples. One solution is to emphasize more manipulative field experiments (rather than observational studies), but even these are difficult to interpret mechanistically because of the complexity of macroorganisms.

Microbial ecologists, meanwhile, have been preoccupied primarily by autecological studies of 'pet' microbes, generally under laboratory conditions. They have shown little taste for familiarizing themselves with the literature in macroecology or of using it as a guiding force in experimentation. The ironical consequence is that, while microbial systems exist that could be used to answer rigorously many ecological questions which cannot be addressed mechanistically by plant and animal models, they remain unexploited (for examples see Chapter 4, and Andrews and Harris 1986). The common university practice of sequestering microbiologists in departments unto themselves (and of placing plant and animal ecologists in botany and zoology departments, respectively) has promoted isolationism. The upshot is that the artificial distinction between microorganism and macroorganism is made even more extreme, different concepts of what 'ecology' supposedly is have emerged, and different vocabularies exist to describe fundamentally the same phenomena. Finally, the gulf between microbial and macroecology has been widened by the fundamental difference in methodological approach alluded to in Chapter 1.

There are two ways to undertake biological experimentation. The first is to ask a question and then choose the best organism to answer it. Hence, fundamental advances in genetics have been made with *E. coli* and *Drosophila* as models, not with elephants or redwood trees, or even myxobacteria. Because life histories are unique whereas the basic cellular processes are not, microbes can never be as good ecological models as they have been physiological/genetics models. We might expect them to be useful in such studies as those on optimality theory, density-dependent selection, colonization processes, the role of timing (heterochrony) in the life cycle, and generalists versus specialists. Some examples are developed later. Obviously, microorganisms can be expected to be poor models for any activity or process which is distinctive of higher organisms, especially higher *unitary* organisms, viz., studies of life span, age-related phenomena (life history tables), gene flow, and behavioral ecology.

The second approach is to use an organism that is inherently interesting to the investigator to address meaningful general questions. Thus, E.O. Wilson has employed ants to great advantage in the study of sociobiology and optimal foraging theory. John Tyler Bonner has used cellular slime molds in elegant investigations of basic questions pertaining to size, morphogenesis, chemical signalling systems, heterochrony, convergent evolution, and phylogenetic constraint versus adaptation in shaping organism patterns. The resounding message here is that, given sufficient imagination and knowledge of the biological literature on the part of the investigator, virtually any experimental subject can be used to gain certain fundamental insights into general biological principles.

An example of how microbial ecology and macroecology are mutually complementary is developed below. First, I suggest how general models relating growth form to body size for modular organisms could be used conceptually in understanding the growth dynamics of filamentous microorganisms. Second, I examine the converse situation, how the precise manipulations possible with microbes can answer basic ecological questions.

Consider how a fungus grows. In general this is by extending a network of absorptive, filamentous strands over or through nutritive substrata (Chapter 5 and Andrews 1991). What determines the size of an individual colony? In large part, the answer probably depends on energetic limitations and on allometric scaling factors (Chapter 4) that relate increasing volume (demand on resources) to increasing area (supply capacity). In this context we might ask when is it better to fragment (which to a fungus means generally to divide by sporulation) into multiple smaller colonies rather than continue to grow as a single large one. While the answer could be sought and then interpreted solely within the realm of fungal physiology and morphology, there is a fascinating parallel among the marine invertebrates which may prove instructive. Corals, sea anemones, and echinoderms are among the modular marine organisms which are suspension feeders or sessile predators. They obtain food by exposing a prey-capturing surface to water currents (Sebens 1982). Rates of contact with prey will depend on surface area available for interception or capture of food. Growth forms vary and include species which grow as solitary organisms of indeterminate size (e.g., the sea anemone *Anthopleura xanthogrammica*) as well as those in which the size both of individuals and colonies (clones) is indeterminate (e.g., *Anthopleura elegantissima*). The latter case is obviously very similar in growth habit to most fungi. In a series of elegant papers, Sebens (1979, 1980, 1982) has investigated the relationships among body size, growth form (colonial or solitary), prey capture rates, and metabolic cost. He comments (1982, p.220): "an animal with potentially indeterminate growth will reach a size limit set by energetic considerations which are themselves a result of the organism's geometry and the particular habitat in which it is located. *Clonal and colonial growth represent an escape from isometric growth and thus an escape from such size limitations*" [emphasis added]. Building on Sebens' work, McFadden (1986) showed that the per polyp food capture rates *decreased* as colony size of a soft coral *increased*. So, a growth strategy that involves repeated rounds of fragmentation in modular organisms may maximize nutrient uptake per unit biomass of the overall genet. This provides a fascinating framework for tests with the fungi.

What does experimentation with microorganisms hold of conceptual value for the ecology of macroorganisms? One answer is that by virtue of being more amenable as research subjects (e.g., fast generation times; control over replication of conditions and variables), microbes provide excellent models to test ecological theory. Bacteria provide a good system to test optimality theory (e.g., the metabolic burden or maintenance costs associated with 'excess' gene functions can be tested with prototrophic and auxotrophic forms—see Chapter 3 and Zamenhof and Eichhorn 1967; Helling et al. 1981). As another instance, the favorable experimental features of the fungi together with availability of innumerable leaf 'islands' of different sizes and stages of colonization (Andrews and Kinkel 1986), facilitate tests of island biogeographic theory for microbial species dynamics on leaves (Andrews et al. 1987; Kinkel et al. 1987). As yet another example, there are obvious parallels in the recruitment dynamics of organisms such as barnacles living on rocks in the marine intertidal zone (Roughgarden et al. 1985, 1988) and the immigration dynamics of microorganisms such as fungi on leaf surfaces (Kinkel et al. 1989). Among such analogies are that: 1) both systems are open and gene flow patterns are likely to be similar; 2) although barnacles are unitary organisms and fungi are modular, both are essentially sessile as 'adults'; 3) immigrants (barnacle larvae from coastal waters; fungal spores from the atmosphere) can be deposited anywhere, rather than invading across boundaries; 4) similar population densities are attained on the respective substrata; 5) competition is probably for space, and growth of the individual (barnacle) or clone (fungus) can interfere with recruitment; and 6) density-independent mortality or erosion removes organisms, opening sites for further recruitment. The main question for microbial ecologists then is what factors determine population kinetics of fungi (or bacteria) on leaf surfaces, and whether models such as those by Roughgarden et al. (1985, 1988) for population dynamics of organisms colonizing substrata in the ocean ('fouling communities') can provide insight into similar processes on the surfaces of plants.

Another answer is that microbes can serve as paradigms for investigating life history features because they are unique in offering a relatively close, direct linkage between genotype and phenotype. Their developmental and structural simplicity means that the problems of pleiotropy and epigenetic effects ("gene net" of Bonner 1988, pp.144 and 174-175; "informational relay" of Stebbins 1968), though real, are comparatively small relative to those of macroorganisms. For instance, imagine that we are interested in whether a particular phenotypic trait is responsible for the competitive dominance of species A over species B. To take a practical example from plant pathology, suppose further that organism A is seen to inhibit B in an agar plate assay, evidently by production of an antibiotic. We notice also that under field conditions, application of populations of A also controls a disease incited by B. At this observational level there is merely a *correlation* between antibiotic production and competitive suppression, just as among macroorganisms there may be a *correlation* between, say, body size and spatial location of species along a resource gradient. In the latter case interspecific competition has often (and occasionally erroneously) been inferred as the responsible mechanism (see examples cited by Roth 1981; Connell 1983; Price 1984; Eadie et al. 1987).

The cause for the pattern can be determined directly and often at the level of the gene in microorganisms in a way that it cannot for the macroorganism analog. First, the antibiotic from A could be purified, tested for an effect on B, and demonstrated to be present at inhibitory levels at the field microsites where antagonist A and pathogen B interact. Second, two complementary mutational analyses could be performed to implicate antibiosis as the mechanism (Handelsman and Parke 1989). Mutants of A lacking inhibitory activity to B should not produce the antibiotic and should fail to prevent disease development. Complementation of the mutants to restore antibiotic production should also restore competitive dominance and disease suppressing ability. Second, mutants of the pathogen B could be made and tested in the same manner as those of antagonist A. B mutants insensitive to the antibiotic should cause disease in the presence of A; restoration of sensitivity should coincide with failure to cause disease. Finally, the remote possibility of a correlative rather than a causal relationship between antibiotic production and biocontrol activity can be effectively eliminated by examining many mutants (Handelsman and Parke 1989).

The degree to which studies in microbial ecology will prove to be of heuristic value in macroecology, and vice-versa, remains to be seen. The fundamental commonalities reviewed in this book provide many intriguing points of departure for research. At the same time, it is evident that there will always be examples (and always a level) of comparison which just 'don't fit'. Nature displays a bewildering richness of expression, but is surprisingly conservative in underlying pattern. On balance, more is to be gained than lost by adopting a broad conceptual perspective in our individual studies.

8.5 Summary

Although the life history of every species is ultimately unique, homologies and analogies exist at some levels of comparison. Microorganisms and macroorganisms interact with their environments differently as a result of size differences over orders of magnitude. Because of their small sizes, microorganisms are governed largely by intermolecular forces, whereas macroorganisms are affected mainly by gravity (Chapter 4). With increasing dimensions inevitably come increasing complexity (number of cell types; division of labor) and other correlates of size, among them increasing generation time, hence decreasing rates of genetic change at the population level (Chapter 2) but increasing homeostatic ability of the individual with respect to environmental change (Chapter 7). Nevertheless, use of size to conceptualize the division of the living world is artificial. A more biologically defendable basis is to integrate across size by delimiting with respect to growth form into modular and unitary organisms (Chapter 5).

A universal basis for comparison is the individual as an input/output system wherein trade-offs are necessary to accomplish conflicting allocation demands among growth, maintenance, and reproduction for finite time and resources (Chapters 1, 3, and 5). The life cycle itself, like all other attributes, is a compromise response by the organism (Chapter 6). Within phylogenetic constraints, natural selection should lead to increasing overall cost efficiency. Cost-efficient creatures should be competitively superior and should thereby leave more descendants. Within this general principle, progressively more specific statements or postulates can be made as bases for comparisons across taxa. Eventually a point is reached where contrasts are meaningful only at the intra-generic or intra-specific level. Thus while *r*- and *K*-selection, for instance, is a general ecological concept, the predictions may vary with respect to growth form of the individual, and within the grouping of growth form the mechanistic basis for an *r*- or *K*-'strategy' may well differ at the guild or at the species level.

Macroecology remains, in part of necessity, an endeavor dominated by correlation. Microbial ecology lacks to date the strong theoretical basis and conceptual integrity of plant and animal ecology. It does offer the potential for controlled experimentation and for mechanistic interpretation. Life history features can be dissected at a genetic level not possible with the more complex macroorganisms. A broader perspective applied to studies with *any* organism would promote insight acquired by seeing things from a novel vantage point. It would also foster an appreciation that beneath a bewildering array of expressions there is fundamental integrity in nature. This would advance the whole discipline of ecology.

Bibliography

Abrahamson, W. G., and H. Caswell. (1982). On the comparative allocation of biomass, energy and nutrients in plants. *Ecology, 63,* 982-991.

Adams, J., and P.W. Oeller. (1986). Structure of evolving populations of *Saccharomyces cerevisiae*: Adaptive changes are frequently associated with sequence alterations involving mobile elements belonging to the Ty family. *Proc. Natl. Acad. Sci. (USA), 83,* 7124-7127.

Adler, J. (1976). The sensing of chemicals by bacteria. *Sci. Amer., 234*(4), 40-47.

Ahmadjian, V., and M.E. Hale (eds.). (1973). *The Lichens.* NY: Academic Press.

Alberts, B., D. Bray, J. Lewis, M. Raff, K. Roberts, and J.D. Watson. (1989). *Molecular Biology of the Cell, 2nd ed.* NY: Garland.

Alldredge, A. L., and Y. Cohen. (1987). Can microscale chemical patches persist in the sea? Microelectrode study of marine snow, fecal pellets. *Science (USA), 235,* 689-691.

Allen, T. F. H. (1977). Scale in microscopic algal ecology: a neglected dimension. *Phycologia, 16,* 253-257.

Anderson, E. S. (1968). The ecology of transferable drug resistance in the enterobacteria. *Annu. Rev. Microbiol., 22,* 131-150.

Anderson, J. B., R.C. Ullrich, L.F. Roth, and G.M. Filip. (1979). Genetic identification of clones of *Armillaria mellea* in coniferous forests in Washington. *Phytopathology, 69,* 1109-1111.

Andrews, J. H. (1984). Life history strategies of plant parasites. D. S. Ingram, and P. H. Williams (eds.), *Advances in Plant Pathology, vol. 2.* (pp. 105-130). NY: Academic Press.

251

Andrews, J. H. (1991). Life-history patterns of the fungi. G. C. Carroll, and D.T. Wicklow (eds.), *The Fungal Community 2nd ed.* NY: Marcel Dekker (in press).

Andrews, J. H., and R.F. Harris. (1986). r- and K-selection and microbial ecology. *Adv. Microb. Ecol., 9,* 99-147.

Andrews, J. H., and D.I. Rouse. (1982). Plant pathogens and the theory of r- and K-selection. *Amer. Nat., 120,* 283-296.

Andrews, J. H., and L.L. Kinkel. (1986). Colonization dynamics: The island theory. N. J. Fokkema, and J. van den Heuvel (eds.), *Microbiology of the Phyllosphere.* (pp. 63-76). NY: Cambridge Univ. Press.

Andrews, J. H., L.L. Kinkel, F.M. Berbee, and E.V. Nordheim. (1987). Fungi, leaves, and the theory of island biogeography. *Microb. Ecol., 14,* 277-290.

Anikster, Y., and I. Wahl. (1979). Coevolution of the rust fungi on Gramineae and Liliaceae and their hosts. *Annu. Rev. Phytopathol., 17,* 367-403.

Antonovics, J. (1972). Population dynamics of the grass *Anthoxanthum odoratum* on a zinc mine. *J. Ecol., 60,* 351-165.

Arndt, K. T., C. Styles, and G.R. Fink. (1987). Multiple global regulators control HIS4 transcription in yeast. *Science (USA), 237,* 874-880.

Atwood, K. C., L.K. Schneider, and F.J. Ryan. (1951). Selective mechanisms in bacteria. *Cold Spring Harbor Sympos. Quant. Biol., 16,* 345-355.

Bainbridge, B. W. (1987). *Genetics of Microbes, 4th ed.* NY: Chapman and Hall.

Baker, H. G. (1972). Seed weight in relation to environmental conditions in California. *Ecology, 53,* 997-1010.

Barbault, R. (1988). Body size, ecological constraints, and the evolution of life-history strategies. M. K. Hecht, B. Wallace, and G.T. Prance (eds.), *Evolutionary Biology, vol. 22.* (pp. 261-286). NY: Plenum Press.

Barker, S. B., G. Cumming, and K. Horsfield. (1973). Quantitative morphometry of the branching structure of trees. *J. Theor. Biol., 40,* 33-43.

Barrett, S. C. H. (1989). Waterweed invasions. *Sci. Amer., 261*(4), 90-97.

Bartholomew, B. (1970). Bare zones between California shrub and grassland communities: the role of animals. *Science (USA), 170,* 1210-1212.

Bartlett, C. L. R., A.D. Macrae, and J.T. Macfarlane. (1986). *Legionella Infections.* London: Arnold.

Bartlett, D., M. Wright, A.A. Yayanos, and M. Silverman. (1989). Isolation of a gene regulated by hydrostatic pressure in a deep-sea bacterium. *Nature (London), 342,* 572-574.

Bateson, G. (1963). The role of somatic selection in evolution. *Evolution, 17,* 529-539.

Baumberg, S. (1981). The evolution of metabolic regulation. M.J. Carlile, J.F. Collins, and B.E.B. Moseley (eds.), *Molecular and Cellular Aspects of Microbial Evolution (32nd Sympos. Soc. Gen. Microbiol.).* (pp. 229-272). Cambridge, UK: Cambridge Univ. Press.

Bazzaz, F. A., N.R. Chiariello, P.D. Coley, and L.F. Pitelka. (1987). Allocating resources to reproduction and defence. *BioScience, 37,* 58-67.

Beckman, C. H. (1987). *The Nature of Wilt Diseases of Plants.* St. Paul, Minnesota: Amer. Phytopathol. Soc.

Begon, M., J.L. Harper, and C.R. Townsend. (1986). *Ecology: Individuals, Populations and Communities*. Sunderland, Mass.: Sinauer Assoc.

Bell, A. D. (1984). Dynamic morphology: A contribution to plant population ecology. R. Dirzo, and J. Sarukhan (eds.), *Perspectives on Plant Population Ecology*. (pp. 48-65). Sunderland, Mass.: Sinauer Assoc.

Bell, A. D., and P.B. Tomlinson. (1980). Adaptive architecture in rhizomatous plants. *Bot. J. Linn. Soc., 80*, 125-160.

Bell, G. (1984). Evolutionary and nonevolutionary theories of senescence. *Amer. Nat., 124*, 600-603.

Bell, G. (1982). The *Masterpiece of Nature. The Evolution and Genetics of Sexuality*. Berkeley: Univ. California Press.

Bell, G. (1988). *Sex and Death in Protozoa. The History of an Obsession*. NY: Cambridge Univ. Press.

Bennett, A. F., K. M. Dao, and R. E. Lenski. (1990). Rapid evolution in response to high-temperature selection. *Nature (London), 346*, 79-81.

Benveniste, R. E. (1985). The contributions of retroviruses to the study of mammalian evolution. R.J. MacIntyre (ed.), *Molecular Evolutionary Genetics*. (pp. 359-417). NY: Plenum Press.

Benveniste, R. E., and G.J. Todaro. (1974). Evolution of type-C viral genes. Inheritance of exogenously acquired viral genes. *Nature (London), 252*, 456-459.

Benveniste, R. E., and G.J. Todaro. (1975). Segregation of RD-114 and FeLV-related sequences in crosses between domestic cat and leopard cat. *Nature (London), 257*, 506-508.

Beran, K. (1968). Budding of yeast cells, their scars and ageing. *Adv. Microb. Physiol., 2*, 143-172.

Berendt, R. F. (1981). Influence of blue-green algae (cyanobacteria) on survival of *Legionella pneumophila* in aerosols. *Infect. Immunol., 32*, 690-692.

Berg, D. E., and M.M. Howe (eds.). (1989). *Mobile DNA*. Washington, D.C.: Amer. Soc. Microbiol.

Beringer, J. E., and P.R. Hirsch. (1984). The role of plasmids in microbial ecology. M.J. Klug, and C.A. Reddy (eds.), *Current Perspectives in Microbial Ecology*. (pp. 63-70). Washington, DC: Amer. Soc. Microbiol.

Bernstein, H., F.A. Hopf, and R.E. Michod. (1988). Is meiotic recombination an adaptation for repairing DNA, producing genetic variation, or both? R.E. Michod, and B.R. Levin (eds.), *The Evolution of Sex: An Examination of Current Ideas*. (pp. 139-160). Sunderland, Mass.: Sinauer Assoc.

Bernstein, H., H. Byerly, F. Hopf, and R.E. Michod. (1985). DNA repair and complementation: The major factors in the origin and maintenance of sex. H.O. Halvorson, and A. Monroy (eds.), *The Origin and Evolution of Sex*. (pp. 29-45). NY: Liss.

Bertrand, H., K.J. McDougall, and T.H. Pittenger. (1968). Somatic cell variation during uninterrupted growth of *Neurospora crassa* in continuous growth tubes. *J. Gen. Microbiol., 50*, 337-350.

Beschel, R. E. (1958). Lichenometrical studies in West Greenland. *Arctic, 11*, 254.

Bhat, K. K. S., and P.H. Nye. (1973). Diffusion of phosphate to plant roots in soil. I. Quantitative autoradiography of the depletion zone. *Plant and Soil, 38*, 161-175.

Bidder, G. P. (1932). Senescence. *Brit. Med. J.*, 2, 583-585.

Bodmer, W. F. (1970). The evolutionary significance of recombination in prokaryotes. H.P. Charles, and B.C.J.G. Knight (eds.), *Organization and Control in Prokaryotic and Eukaryotic Cells (20th Sympos. Soc. Gen. Microbiol.).* (pp. 279-294). Cambridge, UK: Cambridge Univ. Press.

Boer, P. H., C.H. Adra, Y.-F. Lau, and M.W. McBurney. (1987). The testis-specific phosphoglycerate kinase gene *pgk* -2 is a recruited retroposon. *Mol. Cell. Biol.*, 7, 3107-3112.

Bonner, J. T. (1958). The relation of spore formation to recombination. *Amer. Nat.*, 92, 193-200.

Bonner, J. T. (1963). *Morphogenesis: An Essay on Development*. NY: Atheneum.

Bonner, J. T. (1965). *Size and Cycle: An Essay on the Structure of Biology*. Princeton, NJ: Princeton Univ. Press.

Bonner, J. T. (1968). Size change in development and evolution. *J. Paleontol.*, 42 (Part III), 1-15.

Bonner, J. T. (1974). *On Development. The Biology of Form*. Cambridge, Mass: Harvard Univ. Press.

Bonner, J. T. (1982a). Evolutionary strategies and developmental constraints in the cellular slime molds. *Amer. Nat.*, 119, 530-552.

Bonner J. T. (ed.) (1982b). *Evolution and Development*. NY: Springer-Verlag.

Bonner, J. T. (1988). *The Evolution of Complexity by Means of Natural Selection*. Princeton, NJ: Princeton Univ. Press.

Bonner J. T., and H.S. Horn. (1982). Selection for size, shape, and developmental timing. J. T. Bonner (ed.), *Evolution and Development*. (pp. 259-276). NY: Springer-Verlag.

Borst, P., and G.A.M. Cross. (1982). Molecular basis for trypanosome antigenic variation. *Cell*, 29, 291-303.

Boyce, M. S. (1984). Restitution of *r*- and *K*-selection as a model of density-dependent selection. *Annu. Rev. Ecol. Syst.*, 15, 427-447.

Boyce, M. S. (ed.). (1988). *Evolution of Life Histories of Mammals*. New Haven, Conn.: Yale Univ. Press.

Boyden, A. A. (1953). Comparative evolution with specific reference to primitive mechanisms. *Evolution*, 7, 21-30.

Boyer, H. W. (1971). DNA restriction and modification mechanisms in bacteria. *Annu. Rev. Microbiol.*, 25, 153-176.

Braam, J., and R.W. Davis. (1990). Rain-, wind-, and touch-induced expression of calmodulin and calmodulin-related genes in *Arabidopsis. Cell*, 60, 357-364.

Bradshaw, A. D. (1959). Population differentiation in *Agrostis tenuis* Sibth. I. Morphological differentiation. *New Phytol.*, 58, 208-227.

Bradshaw, A. D. (1965). Evolutionary significance of phenotypic plasticity in plants. *Adv. Genet.*, 13, 115-155.

Bradshaw, A. D. (1972). Some of the evolutionary consequences of being a plant. *Evol. Biol.*, 5, 25-47.

Brandon, R. N. (1984). The levels of selection. R. N. Brandon, and R.M. Burian (eds.), *Genes, Organisms and Populations*. (pp. 133-141). Cambridge, Mass.: MIT Press.

Breese, E. L., M.D. Hayward, and A.C. Thomas. (1965). Somatic selection in perennial rye grass. *Heredity, 20,* 367-379.

Brenner, D. J., and S. Falkow. (1971). Molecular relationships among members of the Enterobacteriaceae. *Adv. Genet., 16,* 81-118.

Brent, L., L.S. Rayfield, P.Chandler, W. Fierz, P.B. Medawar, and E. Simpson. (1981). Supposed lamarckian inheritance of immunological tolerance. *Nature (London), 290,* 508-512.

Brian, A. D. (1952). Division of labour and foraging in *Bombus agrorum* Fabricus. *J. Anim. Ecol., 21,* 223-240.

Brisson-Noel, A., M. Arthur, and P. Courvalin. (1988). Evidence for natural gene transfer from gram-positive cocci to *Escherichia coli. J. Bacteriol., 170,* 1739-1745.

Broadhead, E. (1958). The psocid fauna of larch trees in northern England—an ecological study of mixed species populations exploiting a common resource. *J. Anim. Ecol., 27,* 217-263.

Brock, T. D. (1966). *Principles of Microbial Ecology.* Englewood Cliffs, NJ: Prentice-Hall.

Brock, T. D. (1971). Microbial growth rates in nature. *Bact. Rev., 35,* 39-58.

Brock, T. D. (1985). *A Eutrophic Lake. Lake Mendota, Wisconsin.* NY: Springer-Verlag.

Brock, T. D., K.M. Brock, R.T. Belly, and R.L. Weiss. (1972). *Sulfolobus*: a new genus of sulfur-oxidizing bacteria living at low pH and high temperature. *Arch. f. Mikrobiol., 84,* 54-68.

Brock, T. D., and M.T. Madigan (1988). *Biology of Microorganisms, 5th ed.* Englewood Cliffs, NJ: Prentice Hall.

Brodie, E. D., III. (1989). Genetic correlations between morphology and antipredator behaviour in natural populations of the garter snake *Thamnophis ordinoides. Nature (London), 342,* 542-543.

Brown, G. W., and M.M. Flood. (1947). Tumbler mortality. *J. Amer. Statis. Assoc., 42,* 562-574.

Buchanan-Wollaston, V., J.E. Passiatore, and F. Cannon. (1987). The *mob* and *ori*T mobilization functions of a bacterial plasmid promote its transfer to plants. *Nature (London), 328,* 172-175.

Buller, A. H. R. (1931). *Researches on Fungi. Vol. IV. Further Observations on the Coprini Together with Some Investigations on Social Organisms and Sex in the Hymenomycetes.* London: Longmans, Green and Co.

Burdon, J. J. (1987). Phenotypic and genetic patterns of resistance to the pathogen *Phakopsora pachyrhizi* in populations of *Glycine canescens. Oecologia, 73,* 257-267

Burdon, J. J., N.H. Luig, and D.R. Marshall. (1983). Isozyme uniformity and virulence variation in *Puccinia graminis* f.sp. *tritici* and *Puccinia recondita* f sp. *tritici* in Australia. *Aust. J. Biol. Sci., 36,* 403-410.

Burnet, S. M. (1974). *Intrinsic Mutagenesis: A Genetic Approach.* NY: Wiley.

Burnett, J. H. (1968). *Fundamentals of Mycology, 2nd ed.* London: Edward Arnold.

Burnett, J. H. (1975). *Mycogenetics: An Introduction to the General Genetics of Fungi.* NY: Wiley.

Burns, T. P. (1989). Lindeman's contradiction and the trophic structure of ecosystems. *Ecology, 70,* 1355-1362.

Bushnell, W. R., and A.P. Roelfs (eds.). (1984). *The Cereal Rusts, vol I. Origins, Specificity, Structure, and Physiology.* NY: Academic Press.

Buss, L. (1986). Competition and community organization on hard surfaces in the sea. J. Diamond, and T.J. Case (eds.), *Community Ecology.* (pp. 517-536). NY: Harper and Row.

Buss, L. W. (1983). Evolution, development, and the units of selection. *Proc. Natl. Acad. Sci. (USA), 80,* 1387-1391.

Buss, L. W. (1985). The uniqueness of the individual revisited. J.B.C. Jackson, L.W. Buss, and R.E. Cook (eds.), *Population Biology and Evolution of Clonal Organisms.* (pp. 467-505). New Haven, Conn.: Yale Univ. Press.

Buss, L. W. (1987). *The Evolution of Individuality.* Princeton, NJ: Princeton Univ. Press.

Buss, L. W., and R.K. Grosberg. (1990). Morphogenetic basis for phenotypic differences in hydroid competitive bahavior. *Nature (London), 343,* 63-66.

Butler, G. M. (1984). Colony ontogeny in basidiomycetes. D.H. Jennings, and A.D.M. Rayner (eds.), *The Ecology and Physiology of the Fungal Mycelium.* (pp. 53-71). Cambridge, UK: Cambridge Univ. Press.

Butler, G. M. (1966). Vegetative structures. G.C. Ainsworth, and G.C. Sussman (eds.), *The Fungi: An Advanced Treatise.* (pp. 82-112). London: Academic Press.

Cairns, J. (1963). The chromosome of *Escherichia coli. Cold Spring Harbor Sympos. Quant. Biol., 28,* 43-46.

Cairns, J., J. Overbaugh, and S. Miller. (1988). The origin of mutants. *Nature (London), 335,* 142-145.

Calder, W. A. III (1984). *Size, Function, and Life History.* Cambridge, Mass.: Harvard Univ. Press.

Calow, P., and C.R. Townsend. (1981). Energetics, ecology and evolution. C.R. Townsend, and P. Calow (eds.), *Physiological Ecology: An Evolutionary Approach to Resource Use.* (pp. 3-19). Sunderland, Mass.: Sinauer Assoc.

Camerow, J. R., E.Y. Loh, and R.W. Davis. (1979). Evidence for transposition of dispersed repetitive DNA families in yeast. *Cell, 16,* 739-751.

Campbell, A. (1981). Evolutionary significance of accessory DNA elements in bacteria. *Annu. Rev. Microbiol., 35,* 55-83.

Campbell, J. H. (1982). Autonomy in evolution. R. Milkman (ed.), *Perspectives on Evolution.* (pp. 190-201). Sunderland, Mass.: Sinauer Assoc.

Carlile, M. J. (1979). Bacterial, fungal and slime mould colonies. G. Larwood, and B.R. Rosen (eds.), *Biology and Systematics of Colonial Organisms.* (pp. 3-27). NY: Academic Press.

Carlile, M. J. (1980). From prokaryote to eukaryote: Gains and losses. G.W. Gooday, D. Lloyd, and A.P.J. Trinci (eds.), *The Eukaryotic Microbial Cell.* (13th Sympos. Soc. Gen. Microbiol.). (pp. 1-40). Cambridge, UK: Cambridge Univ. Press.

Carlile, M. J. (1980a). Positioning mechanisms—the role of motility, taxis and tropism in the life of microorganisms. D.C. Ellwood, J.N. Hedger, M.J. Latham, J.M. Lynch, and J.H. Slater (eds.), *Contemporary Microbial Ecology.* (pp. 55-74). NY: Academic Press.

Carlile, M. J. (1987). Genetic exchange and gene flow: Their promotion and prevention. A.D.M. Rayner, C.M. Brasier, and D. Moore (eds.), *Evolutionary Biology of the Fungi*. (pp. 203-214). Cambridge: Cambridge Univ. Press.

Carlos, M. P., and J.H. Miller. (1980). Transposable elements. *Cell, 20,* 579-595.

Caswell, H. (1985). The evolutionary demography of clonal reproduction. J.B.C. Jackson, L.W. Buss, and R.E. Cook (eds.), *Population Biology and Evolution of Clonal Organisms*. (pp. 187-224). New Haven, Conn.: Yale Univ. Press.

Caten, C. E. (1972). Vegetative incompatibility and cytoplasmic infection in fungi. *J. Gen. Microbiol., 72,* 221-229.

Caten, C. E. (1987). The genetic integration of fungal life styles. A.D.M. Rayner, C.M. Brasier, and D. Moore (eds.), *Evolutionary Biology of the Fungi*. (pp. 215-229). Cambridge, UK: Cambridge Univ. Press.

Charlesworth, B. (1980). *Evolution in Age-Structured Populations*. Cambridge, UK: Cambridge Univ. Press.

Charlesworth, B. (1985). The population genetics of transposable elements. T. Ohta, and K. Aoki (eds.), *Population Genetics and Molecular Evolution: Papers Marking the Sixthtieth Birthday of Motoo Kimura*. (pp. 213-232). Tokyo: Japan Sci. Soc. Press.

Charnov, E. L. (1976a). Optimal foraging: attack strategy of a mantid. *Amer. Nat., 110,* 141-151.

Charnov, E. L. (1976b). Optimal foraging: the marginal value theorem. *Theor. Pop. Biol., 9,* 129-136.

Cheetham, A. H., and E. Thomsen. (1981). Functional morphology of arborescent animals: strength and design of cheilostome bryozoan skeletons. *Paleobiology, 7,* 355-383.

Cheetham, A. H., L.-A.C. Hayek, and E. Thomsen. (1980). Branching structure in arborescent animals: models of relative growth. *J. Theor. Biol., 85,* 335-369.

Cheverud, J. M. (1984). Quantitative genetics and developmental constraints on evolution by selection. *J. Theor. Biol., 110,* 155-171.

Chew, R. M., and A.E. Chew. (1965). The primary productivity of a desert shrub (*Larrea tridentata*) community. *Ecol. Monogr., 35,* 355-375.

Clarke, P. H. (1982). The metabolic versatility of pseudomonads. *Antonie van Leeuwenhoek, 48,* 105-130.

Clatworthy, J. N., and J.L. Harper. (1962). The comparative biology of closely related species living in the same area. V. Inter- and intraspecific interference within cultures of *Lemna* spp. and *Salvinia natans*. *J. Exp. Bot., 13,* 307-324.

Clausen, J., D.D. Keck, and W.M. Hiesey (1948). *Experimental Studies on the Nature of Species. III. Environmental Responses of Climatic Races of Achillea*. Washington, DC: Carnegie Inst. (Publ. no. 581).

Clewell, D. B., and C. Gawron-Burke. (1986). Conjugative transposons and the dissemination of antibiotic resistance in streptococci. *Annu. Rev. Microbiol., 40,* 635-659.

Cody, M. L. (1986). Structural niches in plant communities. J. Diamond, and T.J. Case (eds.), *Community Ecology*. (pp. 381-405). NY: Harper and Row.

Cohen, D. (1967). Optimizing reproduction in a randomly varying environment when a correlation may exist between the conditions at the time a choice has to be made and the subsequent outcome. *J. Theor. Biol., 16,* 1-14.

Cohen, J. B., and A.D. Levinson. (1988). A point mutation in the last intron responsible for increased expression and transforming activity of the c-Ha- ras oncogene. *Nature (London), 334,* 119-124.

Comfort, A. (1979). *The Biology of Senescence, 3rd ed.* NY: Elsevier.

Condit, R., and B.R. Levin. (1990). The evolution of plasmids carrying multiple resistance genes: The role of segregation, transposition, and homologous recombination. *Amer. Nat., 135,* 573-596.

Condit, R., F.M. Stwart, and B.R. Levin. (1988). The population biology of bacterial transposons: a priori conditions for maintenance as parasitic DNA. *Amer. Nat., 132,* 129-147.

Connell, J. H. (1961). The influence of interspecific competition and other factors on the distribution of the barnacle *Chthamalus stellatus. Ecology, 42,* 710-723.

Connell, J. H. (1983). On the prevalence and relative importance of interspecific competition: evidence from field experiments. *Amer. Nat., 122,* 661-696.

Cook, R. E. (1983). Clonal plant populations. *Amer. Sci., 71,* 244-253.

Cook, R. E. (1985). Growth and development in clonal plant populations. J.B.C. Jackson, L.W. Buss, and R.E. Cook (eds.), *Population Biology and Evolution of Clonal Organisms.* (pp. 259-296). New Haven, Conn.: Yale Univ. Press.

Cook, R. J., and K.F. Baker. (1983). *The Nature and Practice of Biological Control of Plant Pathogens.* St. Paul, Minnesota: Amer. Phytopathol. Soc.

Cooke, R. C., and A.D.M. Rayner. (1984). *Ecology of Saprotrophic Fungi.* London: Longman.

Cooper, A. F. Jr, and S.D. van Gundy. (1971). Senescence, quiescence, and cryptobiosis. B.M. Zuckerman, W.F. Mai, and R.A. Rhode (eds.), *Plant Parasitic Nematodes. Vol. II. Cytogenetics, Host-Parasite Interactions, and Physiology.* (pp. 297-318). NY: Academic Press.

Cousins, S. H. (1980). A trophic continuum derived from plant structure, animal size, and a detritus cascade. *J. Theor. Biol., 82,* 607-618.

Cowan, S. T. (1962). The microbial species—a macromyth? G.C. Ainsworth, and P.H.A. Sneath (eds.), *Microbial Classification (12th Sympos. Soc. Gen. Microbiol.).* (pp. 433-455). Cambridge, UK: Cambridge Univ. Press.

Crow, J. F. (1985). The neutrality-selection controversy in the history of evolution and population genetics. T. Ohta, and K. Aoki (eds.), *Population Genetics and Molecular Evolution.* (pp. 1-18). Tokyo: Japan Scientific Societies Press.

Crow, J. F. (1988). The importance of recombination. R.E. Michod, and B.R. Levin (eds.), *The Evolution of Sex: An Examination of Current Ideas.* (pp. 56-73). Sunderland, Mass.: Sinauer Assoc.

Crowl, T. A., and A.P. Covich. (1990). Predator-induced life-history shifts in a freshwater snail. *Science (USA), 247,* 945-951.

Cullen, D. C., and J.H. Andrews. (1984). Epiphytic microbes as biological control agents. T. Kosuge, and E.W. Nester (eds.), *Plant-Microbe Interactions. Molecular and Genetic Perspectives, vol. 1.* (pp. 381-399). NY: Macmillan.

Cullis, C. A. (1983). Environmentally induced DNA changes in plants. *CRC Crit. Rev. Plant Sci., 1,* 117-131.

Cullis, C. A. (1987). The generation of somatic and heritable variation in response to stress. *Amer. Nat., 130 (supplement),* 562-573.

Cutler, R. G. (1978). Evolutionary biology of senescence. J.A. Behnke, C.E. Finch, and G.B. Moment (eds.), *The Biology of Aging*. (pp. 311-360). NY: Plenum Press.

Cutler, R. G. (1984). Evolutionary biology of aging and longevity in mammalian species. J.E. Johnson, Jr. (ed.), *Aging and Cell Function*. (pp. 1-147). NY: Plenum Press.

D'Amato, F. (1985). Cytogenetics of plant cell and tissue cultures and their regenerates. *CRC Crit. Rev. Plant Sci., 3*, 73-112.

Damuth, J. (1981). Population density and body size in mammals. *Nature (London), 290*, 699-700.

Danielli, J. F., and A.L. Muggleton. (1959). Some quantitative states of amoeba with special reference to life-span. *Gerontologia, 3*, 76-90.

Darwin, C. (1854). *A Monograph on The Sub-Class Cirripedia, with Figures on All The Species*. London: The Ray Society.

Darwin, C. (1859). *On the Origin of Species by Means of Natural Selection or the Preservation of Favoured Races in the Struggle for Life*. London: Murray.

Davies, I., and D.C. Sigee (eds.). (1984). *Cell Ageing and Cell Death*. Cambridge, UK: Cambridge Univ. Press.

Davis, B. D. (1989). Transcriptional bias: A non-Lamarckian mechanism for substrate-induced mutations. *Proc. Natl. Acad. Sci. (USA), 86*, 5005-5009.

Davis, R. H. (1966). Mechanisms of inheritance. 2. Heterokaryosis. G.C. Ainsworth, and A.S. Sussman (eds.), *The Fungi: An Advanced Treatise*. (pp. 567-588). NY: Academic Press.

Dawes, C. J. (1981). *Marine Botany*. NY: Wiley Interscience.

Dawkins, R. (1982). *The Extended Phenotype*. Oxford, UK: Oxford Univ. Press.

Dawkins, R. (1989). *The Selfish Gene, 2nd ed*. Oxford, UK: Oxford Univ. Press.

Dawson, E. Y. (1966). *Marine Botany: An Introduction*. NY: Holt, Rinehart and Winston.

De la Mare, W. (1920). Miss T. (*In The Collected Poems of Walter de la Mare 1901-1918. Vol. II. Originally published as part of "Peacock Pie", 1913*). London: Constable and Co.

De Silva, H. R. (1938). Age and highway accidents. *Sci. Month., 47*, 536-545.

Deininger, P. L. (1989). SINEs: Short interspersed repeated DNA elements in higher eukaryotes. D.E. Berg, and M.M. Howe (eds.), *Mobile DNA*. (pp. 619-636). Washington, DC: Amer. Soc. Microbiol.

Dial, K. P., and J.M. Marzluff. (1988). Are the smallest organisms the most diverse? *Ecology, 69*, 1620-1624.

Diamond, J. M. (1973). Distributional ecology of New Guinea birds. *Science (USA), 179*, 759-769.

Diamond, J., and T.J. Case (eds.). (1986). *Community Ecology* NY: Harper and Row.

Dickinson, C. H. (1971). Cultural studies of leaf saprophytes. T.F. Preece, and C.H. Dickinson (eds.), *Ecology of Leaf Surface Micro-organisms*. (pp. 129-137). NY: Academic Press.

Dickman, A., and S. Cook. (1989). Fire and fungus in a mountain hemlock forest. *Can. J. Bot., 67*, 2005-2016.

Dirzo, R., and J. Sarukhan (1984). *Perspectives on Plant Population Ecology*. Sunderland, Mass: Sinauer Assoc.

Dobzhansky, T. (1956). What is an adaptive trait? *Amer. Nat.*, *90*, 337-347.

Dobzhansky, T. (1973). Nothing in biology makes sense except in the light of evolution. *Amer. Biol. Teacher, March*, 125-129.

Doelle, H. W. (1975). *Bacterial Metabolism, 2nd ed.* NY: Academic Press.

Donald, C. M. (1963). Competition among crop and pasture plants. *Adv. Agron.*, *15*, 1-118.

Doolittle, R. F., D.-F. Feng, M.S. Johnson, and M.A. McClure. (1989). Origins and evolutionary relationships of retroviruses. *Quart. Rev. Biol.*, *64*, 1-30.

Doolittle, W. F., and C. Sapienza. (1980). Selfish genes, the phenotype paradigm and genome evolution. *Nature (London)*, *284* (601-603).

Dougherty, E. C. (1955). Comparative evolution and the origin of sexuality. *Systemic Zool.*, *4*, 145-169.

Dow, C. S., R. Whittenbury, and N.G. Carr. (1983). The 'shut down' or 'growth precursor' cell—an adaptation for survival in potentially hostile environments. J.H. Slater, R. Whittenbury, and J.W.T. Wimpenny (eds.), *Microbes in Their Natural Environments (34th Sympos. Soc. Gen. Microbiol.).* (pp. 187-247). Cambridge, UK: Cambridge Univ. Press.

Downs, C., and W.E. McQuilkin. (1944). Seed production of Southern Appalachian oaks. *J. Forestry*, *42*, 913-920.

Drake, J. W. (1974). The role of mutation in microbial evolution. M.J. Carlile, and J.J. Skehel (eds.), *Evolution in the Microbial World (24th Sympos. Soc. Gen. Microbiol.).* (pp. 41-58). Cambridge, UK: Cambridge Univ. Press.

Dryja, T. P., T.L. McGee, E.Reichel, L.B. Hahn, G.S. Cowley, D.W. Yandell, M.A. Sandberg, and E.L. Berson. (1990). A point mutation of the rhodopsin gene in one form of retinitis pigmentosa. *Nature (London)*, *343*, 364-366.

Duncan, K. E., C.A. Istock, J.B. Graham, and N. Ferguson. (1989). Genetic exchange between *Bacillus subtilis* and *Bacillus licheniformis*: variable hybrid stability and the nature of bacterial species. *Evolution*, *43*, 1585-1609.

Durrant, A. (1962). The environmental induction of heritable changes in *Linum*. *Heredity*, *27*, 277-298.

Durrant, A. (1971). The induction and growth of flax genotrophs. *Heredity*, *27*, 277-298.

Durrant, A. (1981). Unstable genotypes. *Phil. Trans. Roy. Soc. Lond. B*, *292*, 467-474.

Dykhuizen, D. E., and D.L. Hartl. (1983). Selection in chemostats. *Microbiol. Rev.*, *47*, 150-168.

Eadie, J. M., L. Broekhoven, and P. Colgan. (1987). Size ratios and artifacts: Hutchinson's rule revisited. *Amer. Nat.*, *129*, 1-17.

Eagle, H. (1955). Basal media for growth of the HeLa cell line in culture. *Science (USA)*, *122*, 501.

Edelstein, L. (1982). The propagation of fungal colonies: A model for tissue growth. *J. Theor. Biol.*, *98*, 679-701.

Ellingboe, A. H., and J.R. Raper. (1962). Somatic recombination in *Schizophyllum commune*. *Genetics*, *47*, 85-98.

Ellis, J. (1982). Promiscuous DNA—chloroplast genes inside plant mitochondria. *Nature (London)*, *299*, 678-679.

Eltringham, S. K. (1982). *Elephants*. Poole, Dorset, UK: Blanford Press.

Engels, W. R. (1983). The P family of transposable elements in *Drosophila. Annu. Rev. Genet., 17*, 315-344.

Esch, G.W, A.O. Bush, and J.M. Aho (eds.). (1990). *Parasite Communities: Patterns and Processes*. NY: Chapman and Hall.

Esser, K., U. Kuck, U. Stahl, and P. Tudzynski. (1984). Senescence in *Podospora anserina* and its implications for genetic engineering. D.H. Jennings, and A.D. M. Rayner (eds.), *The Ecology and Physiology of the Fungal Mycelium.* (pp. 343-352). Cambridge, UK: Cambridge Univ. Press.

Fassett, N. C. (1957). *A Manual of Aquatic Plants*. Madison, Wisc.: Univ. Wisconsin Press.

Fawcett, H. S. (1925). Maintained growth rates in fungus cultures of long duration. *Ann. Appl. Biol., 12*, 191-198.

Fedoroff, N. V. (1983). Controlling elements in maize. J.A. Shapiro (ed.), *Mobile Genetic Elements*. (pp. 1-63). NY: Academic Press.

Fedoroff, N. V. (1989). Maize transposable elements. D.E. Berg, and M.M. Howe (eds.), *Mobile DNA*. (pp. 375-411). Washington, DC: Amer. Soc. Microbiol.

Feibleman, J. K. (1954). Theory of integrative levels. *Brit. J. Philos. Sci., 5*, 59-66.

Fenchel, T. (1975). Character displacement and coexistence in mud snails (Hydrobiidae). *Oecologia, 20*, 19-32.

Fenchel, T., and L. Kofoed. (1976). Evidence for exploitative interspecific competition in mud snails (Hydrobiidae). *Oikos, 27*, 367-376.

Finch, C. E., and L. Hayflick (eds.). (1977). *Handbook of the Biology of Aging*. NY: van Nostrand Reinhold.

Fincham, J. R. S. (1983). *Genetics*. Boston, Mass.: Jones and Bartlett.

Fiscus, E. L. (1986). Belowground costs: hydraulic conductance. T. Givnish (ed.), *On the Economy of Plant Form and Function*. (pp. 275-297). NY: Cambridge Univ. Press.

Fisher, R. A. (1958). *The Genetical Theory of Natural Selection, 2nd ed*. NY: Dover.

Fliermans, C. B., W.B. Cherry, L.H. Orrison, S.J. Smith, D.L. Tison, and D.H. Pope. (1981). Ecological distribution of *Legionella pneumophila. Appl. Environ. Microbiol., 41*, 9-16.

Flor, H. H. (1956). The complementary genic systems in flax and flax rust. *Adv. Genet., 8*, 29-54.

Flor, H. H. (1971). Current status of the gene-for-gene concept. *Annu. Rev. Phytopathol., 9*, 275-296.

Fogg, G. E. (1986). Picoplankton. *Proc. Roy. Soc. Lond. B, 228*, 1-30.

Follett, B. K., and D.E. Follett (eds.). (1981). *Biological Clocks in Seasonal Reproductive Cycles (Proc. 32nd Sympos. Colston Res. Soc.)*. NY: Wiley.

Forage, R. G., D.E.F. Harrison, and D.E. Pitt. (1985). Effect of environment on microbial activity. A.T. Bull, and H. Dalton (eds.), *Comprehensive Biotechnology, vol. 1*. (pp. 251-280). NY: Pergamon Press.

Fraser, D. W., and J.E. McDade. (1979). Legionellosis. *Sci. Amer., 241(4)*, 82-99.

French, D. L., R. Laskov, and M.D. Scharff. (1989). The role of somatic hypermutation in the generation of antibody diversity. *Science (USA), 244*, 1152-1157.

Freter, R. (1984). Factors affecting conjugal plasmid transfer in natural bacterial communities. M.J. Klug, and C.A. Reddy (eds.), *Current Perspectives in Microbial Ecology*. (pp. 105-114). Washington, DC: Amer. Soc. Microbiol.

Gadgil, M. (1971). Dispersal: Population consequences and evolution. *Ecology, 52*, 253-261.

Gadgil, M., and W.H. Bossert. (1970). Life history consequences of natural selection. *Amer. Nat., 194*, 1-24.

Garrett, S. D. (1973). Deployment of reproductive resources by plant-pathogenic fungi: An application of E.J. Salisbury's generalization for flowering plants. *Acta. Bot. Indica, 1*, 1-9.

Gatsuk, L. E., O.V. Smirnova, L.I. Vorontzova, L.B. Zaugolnova, and L.A. Zhukova. (1980). Age state of plants of various growth forms: A review. *J. Ecol., 68*, 675-696.

Gause, G. F. (1932). Experimental studies on the struggle for existence. I. Mixed population of two species of yeast. *J. Exptl. Biol., 9*, 389-402.

Gibbons, R. J., and J. van Houte. (1975). Bacterial adherence in oral microbial ecology. *Annu. Rev. Microbiol., 29*, 19-44.

Gilchrist, D. G., and O.C. Yoder. (1984). Genetics of host-parasite systems: A prospectus for molecular biology. T. Kosuge, and E.W. Nester (eds.), *Plant-Microbe Interactions. Molecular and Genetic Perspectives, vol. 1*. (pp. 69-90). NY: Macmillan.

Gillie, O. J. (1968). Observations on the tube method of measuring growth rate in *Neurospora crassa*. *J. Gen. Microbiol., 51*, 185-194.

Gingerich, P. D. (1983). Rates of evolution: Effects of time and temporal scaling. *Science (USA), 222*, 159-161.

Givnish, T. (1986a). Biomechanical constraints on self-thinning in plant populations. *J. Theor. Biol., 119*, 139-146.

Givnish, T. J. (ed.) (1986b). *On the Economy of Plant Form and Function*. Cambridge, UK: Cambridge Univ. Press.

Gliddon, C. J., and P.H. Gouyon. (1989). The units of selection. *Trends Ecol. Evol. (TREE), 4*, 204-209.

Golden, J. W., S.J. Robinson, and R. Haselkorn. (1985). Rearrangement of nitrogen fixation genes during heterocyst differentiation in the cyanobacterium *Anabaena*. *Nature (London), 314*, 419-423.

Goldstein, S. (1974). Aging in vitro: growth of cultured cells from Galapagos tortoise. *Exp. Cell Res., 83*, 297-362.

Goldstein, S. (1990). Replicative senescence: The human fibroblast comes of age. *Science (USA), 249*, 1129-1133.

Gorczynski, R. M., and E.J.Steele. (1980). Inheritance of acquired immunological tolerance to foreign histocompatibility antigens in mice. *Proc. Natl. Acad. Sci. (USA), 77*, 2871-2875.

Gorczynski, R. M., and E.J. Steele. (1981). Simultaneous yet independent inheritance of somatically acquired tolerance to two distinct H-2 antigenic haplotype determinants in mice. *Nature (London), 289*, 678-681.

Gottschal, J. C., S.C. DeVries, and J.G. Kuenen. (1979). Competition between facultatively chemolithotrophic *Thiobacillus* A2, an obligately chemolithotrophic *Thiobacillus* and a heterotrophic spirillum for inorganic and organic substrates. *Arch. Microbiol., 121,* 241-249.

Gottschalk, G. (1986). *Bacterial Metabolism, 2nd ed.* NY: Springer-Verlag.

Gould, S. J. (1966). Allometry and size in ontogeny and phylogeny. *Biol. Reviews, 41,* 587-640.

Gould, S. J. (1977). *Ontogeny and Phylogeny.* Cambridge, Mass.: Belknap Press of Harvard Univ. Press.

Goulden, C. E., and L.L. Hornig. (1980). Population oscillations and energy reserves in planktonic Cladocera and their consequences to competition. *Proc. Natl. Acad. Sci. (USA), 77,* 1716-1720.

Graham, C. E., O.R. Kling, and R.A. Steiner. (1979). Reproductive senescence in female nonhuman primates. D.M. Bowden (ed.), *Aging in Non-Human Primates.* (pp. 183-202). NY: Van Nostrand Reinhold.

Grant, P. R. (1986). *Ecology and Evolution of Darwin's Finches.* Princeton, NJ: Princeton Univ. Press.

Gray, M. W. (1983). The bacterial ancestry of plastids and mitochondria. *BioScience, 33,* 693-699.

Greene, E. (1989). A diet-induced developmental polymorphism in a caterpillar. *Science (USA), 243,* 643-646.

Gregory, P. H. (1984). The fungal mycelium. D.H. Jennings, and A.D.M. Rayner (eds.), *The Ecology and Physiology of the Fungal Mycelium.* (pp. 1-22). Cambridge, UK: Cambridge Univ press.

Grilione, P. L., and J. Pangborn. (1975). Scanning electron microscopy of fruiting body formation by myxobacteria. *J. Bacteriol., 124,* 1558-1565.

Grime, J. P., J.C. Crick, and J.E. Rincon. (1986). The ecological significance of plasticity. D.H. Jennings, and A.J. Trewavas (eds.), *Plasticity in Plants (Proc. 40th Sympos. Soc. Exptl. Biol.).* (pp. 5-29). Cambridge, UK: Company of Biologists, Cambridge Univ.

Gull, K. (1975). Mycelium branch patterns of *Thamnidium elegans. Trans. Brit. Mycol. Soc., 64,* 321-324.

Gustafsson, L., and T. Part. (1990). Acceleration of senescence in the collared flycatcher *Ficedula albicollis* by reproductive costs. *Nature (London), 347,* 279-281.

Hairston, N. G., Sr. (1990). *Ecological Experiments: Purpose, Design and Execution.* NY: Cambridge Univ. Press.

Haldane, J. B. S. (1953). Some animal life tables. *J. Instit. Actuaries, 79,* 83-89.

Haldane, J. B. S. (1956). On being the right size. J.R. Newman (ed.), *The World of Mathematics, vol. II.* (pp. 952-957). NY: Simon and Schuster.

Hale, M. E. Jr. (1983). *The Biology of Lichens, 3rd ed.* London: Arnold.

Hall, B. G. (1988). Adaptive evolution that requires multiple spontaneous mutations. I. Mutations involving an insertion sequence. *Genetics, 120,* 887-897.

Hamilton, W. D. (1966). The moulding of senescence by natural selection. *J. Theor. Biol., 12,* 12-45.

Hamilton, W. D. (1970). Review of the book "Evolution in Changing Environments". *Science (USA), 167,* 1478-1480.

Handelsman, J., and J.L. Parke. (1989). Mechanisms in biocontrol of soilborne plant pathogens. T. Kosuge, and E.W. Nester (eds.), *Plant-Microbe Interactions. Molecular and Genetic Perspectives, vol 3.* (pp. 27-61). NY: McGraw-Hill.

Harberd, D. J. (1967). Observations on natural clones in *Holcus mollis. New Phytol.,* 66, 401-408.

Harder, W., and L. Dijkhuizen. (1982). Strategies of mixed substrate utilization in microorganisms. *Phil. Trans. Roy. Soc. Lond. B,* 297, 459-480.

Harder, W., L. Dijkhuizen, and H. Veldkamp. (1984). Environmental regulation of microbial metabolism. D.P. Kelly, and N.G. Carr (eds.), *The Microbe 1984. Part II. Prokaryotes and Eukaryotes (Proc. 36th Sympos. Soc. Gen. Microbiol.).* (pp. 51-95). Cambridge, UK: Cambridge Univ. Press.

Hardwick, R. C. (1987). The nitrogen content of plants and the self-thinning rule of plant ecology: A test of the core-skin hypothesis. *Ann. Bot.,* 60, 439-446.

Harley, C. B., A.B. Futcher, and C.W. Greider. (1990). Teleomeres shorten during ageing of human fibroblasts. *Nature (London),* 345, 458-460.

Harley, J. L. (1971). Fungi in ecosystems. *J. Ecol.,* 59, 653-668.

Harper, J. L. (1964). The individual in the population. *J. Ecol., 52 (Suppl.),* 149-158.

Harper, J. L. (1968). The regulation of numbers and mass in plant populations. R.C. Lewontin (ed.), *Population Biology and Evolution.* (pp. 139-158). Syracuse, NY: Syracuse Univ. Press.

Harper, J. L. (1977). *Population Biology of Plants.* NY: Academic Press.

Harper, J. L. (1981a). The concept of population in modular organisms. R.M. May (ed.), *Theoretical Ecology: Principles and Applications, 2nd ed.* (pp. 53-77). Oxford, UK: Blackwell.

Harper, J. L. (1981b). The meanings of rarity. H. Synge (ed.), *The Biological Aspects of Rare Plant Conservation.* (pp. 189-203). NY: Wiley.

Harper, J. L. (1982). After description. E.I. Newman (ed.), *The Plant Community as a Working Mechanism.* (pp. 11-25). Oxford, UK: Blackwell.

Harper, J. L. (1984). Comments in the forward to the book. R. Dirzo, and J. Sarukhan (eds.), *Perspectives on Plant Population Ecology.* (pp. xv-xviii). Sunderland, Mass.: Sinauer Assoc.

Harper, J. L. (1985). Modules, branches, and the capture of resources. J.B.C. Jackson, L.W. Buss, and R.E. Cook (eds.), *Population Biology and Evolution of Clonal Organisms.* (pp. 1-33). New Haven, Conn.: Yale Univ. Press.

Harper, J. L., and A.D. Bell. (1979). The population dynamics of growth form in organisms with modular construction. R.M. Anderson, B.D. Turner, and L.R. Taylor (eds.), *Population Dynamics.* (pp. 29-52). Oxford, UK: Blackwell.

Harper, J. L., B.R. Rosen, and J. White (eds.). (1986). The growth and form of modular organisms. *Proc. Roy. Soc. Lond. B,* 313, 1-250.

Hartl, D. L. (1985). Engineered organisms in the environment: Inferences from population genetics. H.O. Halvorson, D. Pramer, and M. Rogul (eds.), *Engineered Organisms in the Environment: Scientific Issues.* (pp. 83-98). Washington, DC: Amer. Soc. Microbiol.

Hartl, D. L. (1988). *A Primer of Population Genetics, 2nd ed.* Sunderland, Mass.: Sinauer Assoc.

Hartl, D. L., and D.E. Dykhuizen. (1984). The population genetics of *Escherichia coli. Annu. Rev. Genet.,* 18, 31-68.

Hartley, C. L., H.M. Clements, and K.B. Linton. (1977). *Escherichia coli* in the faecal flora of man. *J. Appl. Bacteriol., 43,* 261-269.

Haselkorn, R., J.W. Golden, P.J. Lammers, and M.E. Mulligan. (1987). Rearrangement of *nif* genes during cyanobacterial heterocyst differentiation. *Phil. Trans. Roy. Soc. Lond. B, 317,* 173-181.

Haxo, F. T., and L.R. Blinks. (1950). Photosynthetic action spectra of marine algae. *J. Gen. Physiol., 33,* 389-422.

Hayes, W. E. (1976). *The Genetics of Bacteria and Their Viruses: Studies in Basic Genetics and Molecular Biology.* Oxford, UK: Blackwell.

Hayflick, L. (1965). The limited in vitro lifetime of human diploid cell strains. *Exp. Cell Res., 37,* 614-636.

Hayflick, L. (1977). The cellular basis for biological aging. C.E. Finch, and L. Hayflick (eds.), *Handbook of the Biology of Aging.* (pp. 159-186). NY: Van Nostrand.

Heal, O. W., and S.F. Maclean, Jr. (1975). Comparative productivity in ecosystems—secondary productivity. W.H. van Dobben, and R.H. Lowe-McConnell (eds.), *Unifying Concepts in Ecology.* (pp. 89-108). The Hague: W. Junk.

Hedrick, P. W. (1986). Genetic polymorphism in heterogeneous environments: A decade later. *Annu. Rev. Ecol. System., 17,* 535-566.

Hedrick, P. W., M.E. Ginevan, and E.P. Ewing. (1976). Genetic polymorphism in heterogeneous environments. *Annu. Rev. Ecol. Syst., 7,* 1-32.

Heinrich, B. (1976). The foraging specializations of individual bumblebees. *Ecol. Monogr., 46,* 105-128.

Helling, R. B., and M.I. Lomax. (1978). The molecular cloning of genes. General procedures. A.M. Chakrabarty (ed.), *Genetic Engineering.* (pp. 1-30). Boca Raton, Florida: CRC Press.

Helling, R. B., T. Kinney, and J. Adams. (1981). The maintenance of plasmid-containing organisms in populations of *E. coli. J. Gen. Microbiol., 123,* 129-141.

Hickey, D. A. (1982). Selfish DNA: A sexually-transmitted nuclear parasite. *Genetics, 101,* 519-531.

Hickey, D. A., and M.R.Rose. (1988). The role of gene transfer in the evolution of eukaryotic sex. R.E. Michod, and B.R. Levin (eds.), *The Evolution of Sex: An Examination of Current Ideas.* (pp. 161-175). Sunderland, Mass.: Sinauer Assoc.

Hill, A. V. (1950). The dimensions of animals and their muscular dynamics. *Sci. Progr. (London), 38,* 209-230.

Hillis, D. M. (1987). Molecular versus morphological approaches to systematics. *Annu. Rev. Ecol. Syst., 18,* 23-42.

Hirano, S. S., and C.D. Upper. (1989). Diel variation in population size and ice nucleation activity of *Pseudomonas syringae* on snap bean leaflets. *Appl. Environ. Microbiol., 55,* 623-630.

Hoffman, H. (1964). Morphogenesis of bacterial aggregations. *Annu. Rev. Microbiol., 18,* 111-130.

Holliday, R. (1969). Errors in protein synthesis and clonal senescence in fungi. *Nature (London), 221,* 1224-1228.

Holliday, R. (1988). A possible role for meiotic recombination in germ line reprogramming and maintenance. R.E. Michod, and B.R. Levin (eds.), *The Ev-*

olution of Sex: An Examination of Current Ideas. (pp. 45-55). Sunderland, Mass.: Sinauer Assoc.

Hollings, M. (1978). Mycoviruses: Viruses that infect fungi. *Adv. Vir. Res., 22*, 1-53.

Holmberg, S. D., M.T. Osterholm, K.A. Senger, and M.L. Cohen. (1984). Drug-resistant *Salmonella* from animals fed antibiotics. *New Engl. J. Med., 311*, 617-622.

Horn, H. S. (1971). *The Adaptive Geometry of Trees.* Princeton, NJ: Princeton Univ. Press.

Horn, H. S. (1978). Optimal tactics of reproduction and life-history. J.R. Krebs, and N.B. Davies (eds.), *Behavioural Ecology: An Evolutionary Approach.* (pp. 411-429). Oxford, UK: Oxford Univ. Press.

Horn, H. S., and R.M. May. (1977). Limits to similarity among coexisting competitors. *Nature (London), 270*, 660-661.

Horsfield, K. (1980). Are diameter, length and branching ratios meaningful in the lung? *J. Theor. Biol., 87*, 773-784.

Horton, R. E. (1945). Erosional development of streams and their drainage basins: hydrophysical approach to quantitative morphology. *Bull. Geol. Soc. Amer., 56*, 275-370.

Huffaker, C. B. (1964). Fundamentals of biological weed control. P. DeBach (ed.), *Biological Control of Insect Pests and Weeds.* (pp. 631-649). NY: Reinhold.

Hughes, R. N. (1987). The functional ecology of clonal animals. *Funct. Ecol., 1*, 63-69.

Hughes, R. N., and J.M. Cancino. (1985). An ecological overview of cloning in metazoa. J.B.C. Jackson, L.W. Buss, and R.E.Cook (eds.), *Population Biology and Ecology of Clonal Organisms.* (pp. 153-186). New Haven, Conn.: Yale Univ. Press.

Hungate, R. E. (1975). The rumen microbial ecosystem. *Annu. Rev. Ecol. System., 6*, 39-66.

Hutchinson, G. E. (1959). Homage to Santa Rosalia, or why are there so many kinds of animals? *Amer. Nat., 93*, 145-159.

Hutchinson, G. E. (1975). *A Treatise on Limnology, vol. III. Limnological Botany.* NY: Wiley.

Hutchison, C. A. III, S.C. Hardies, D.D. Loeb, W.R. Shehee, and M.H. Edgell. (1989). LINEs and related transposons: Long interspersed repeated sequences in the eukaryotic genome. D.E. Berg, and M.M. Howe (eds.), *Mobile DNA.* (pp. 593-617). Washington, DC: Amer. Soc. Microbiol.

Huxley, J. S. (1958). Evolutionary processes and taxonomy with special reference to grades. *Uppsala Univ. Arsskr.*, no. 6, 21-38.

Huxley, T. H. (1852). Upon animal individuality. *Proc. Roy. Instit., 1*, 184-189.

Ingold, C. T. (1946). Size and form in agarics. *Trans. Brit. Mycol. Soc., 29*, 108-113.

Ingold, C. T. (1965). *Spore Liberation.* London: Clarendon Press of Oxford Univ. Press.

Ingold, C. T. (1971). *Fungal Spores: Their Liberation and Dispersal.* London: Clarendon Press of Oxford Univ. Press.

Istock, C. A. (1967). The evolution of complex life cycle phenomena: an ecological perspective. *Evolution, 21,* 592-605.

Istock, C. A. (1970). Natural selection in ecologically and genetically defined populations. *Behav. Sci., 15,* 101-115.

Istock, C. A. (1984). Boundaries to life history variation and evolution. P.W. Price, C.N. Slobodchikoff, and W.S. Gaud (eds.), *A New Ecology. Novel Approaches to Interactive Systems.* (pp. 143-168). NY: Wiley.

Jackson, H. S. (1931). Present evolutionary tendencies and the origin of life cycles in the Uredinales. *Mem. Bull. Torrey Bot. Club, 18,* 5-108.

Jackson, J. B. C. (1979). Morphological strategies of sessile animals. G. Larwood, and B.R. Rosen (eds.), *Biology and Systematics of Colonial Organisms.* (pp. 499-555). NY: Academic Press.

Jackson, J. B. C. (1985). Distribution and ecology of clonal and aclonal benthic invertebrates. J.B.C. Jackson, L.W. Buss, and R.E. Cook (eds.), *Population Biology and Evolution of Clonal Organisms.* (pp. 297-355). New Haven, Conn.: Yale Univ. Press.

Jackson, J. B. C., and A.G. Coates. (1986). Life cycles and evolution of clonal (modular) animals. *Phil. Trans. Roy. Soc. B, 313,* 7-22.

Jackson, J. B. C., L.W. Buss, and R.E. Cook (eds.). (1985). *Population Biology and Evolution of Clonal Organisms.* New Haven, Conn.: Yale Univ. Press.

Jacob, F. (1982). *The Possible and the Actual.* NY: Pantheon.

Jaenisch, R. (1988). Transgenic animals. *Science (USA), 240,* 1468-1474.

Jannasch, H. W. (1984). Microbes in the oceanic environment. D.P. Kelly, and N.G. Carr (eds.), *The Microbe 1984. Part II. Prokaryotes and Eukaryotes (Proc. 36th Sympos. Soc. Gen. Microbiol.).* (pp. 97-122). Cambridge, UK: Cambridge Univ. Press.

Jannasch, H. W. (1989). Chemosynthetically sustained ecosystems in the deep sea. H.G. Schlegel, and B. Bowien (eds.), *Autotrophic Bacteria.* (pp. 147-166). New York: Springer-Verlag.

Jannasch, H. W., and C.O. Wirsen. (1979). Chemosynthetic primary production at East Pacific sea floor spreading centers. *BioScience, 29,* 592-598.

Janzen, D. H. (1977). Why fruits rot, seeds mold, and meat spoils. *Amer. Nat., 111,* 691-713.

Jennings, D. H. (1986). Morphological plasticity in fungi. D.H. Jennings, and A.J. Trewavas (eds.), *Plasticity in Plants (Proc. 40th Sympos. Soc. Exptl. Biol.).* (pp. 329-346). Cambridge, UK: Company of Biologists, Cambridge Univ.

Jennings, D. H. (1982). The movement of *Serpula lacrimans* from substrate to substrate over nutritionally inert surfaces. J. Frankland, J. Hedger, and M. Swift (eds.), *Decomposer Basidiomycetes.* (pp. 91-108). Cambridge, UK: Cambridge Univ. Press.

Jennings, D. H., and A.D.M. Rayner (eds.). (1984). *The Ecology and Physiology of the Fungal Mycelium (Proc. Sympos. Brit. Mycol. Soc.).* Cambridge, UK: Cambridge Univ. Press.

Jennings, D. H., and A.J. Trewavas (eds.). (1986). *Plasticity in Plants (Proc. 40th Sympos. Soc. Exptl. Biol.).* Cambridge, UK: Company of Biologists, Cambridge Univ.

Jinks, J. L. (1952). Heterokaryosis: a system of adaptation in wild fungi. *Proc. Roy. Soc. Lond. B, 140,* 83-99.

Jinks, J. L. (1959). Lethal suppressive cytoplasms in aged clones of *Aspergillus glaucus. J. Gen. Microbiol., 21,* 397-409.

Johnson, T. E. (1990). Increased life-span of *age-1* mutants in *Caenorhabditis elegans* and lower Gompertz rate of aging. *Science (USA), 249,* 908-912.

Jones, A. P. (1985). Altering gene expression with 5-azacytidine. *Cell, 40,* 485-486.

Jones, E. W., and G.R. Fink. (1983). Regulation of amino acid and nucleotide biosynthesis in yeast. J.N. Strathern, E.W. Jones, and J.R. Broach (eds.), *The Molecular Biology of the Yeast Saccharomyces: Metabolism and Gene Expression.* (pp. 181-299). Cold Spring Harbor, NY: Cold Spring Harbor Laboratory.

Jorgensen, B. B. (1977). Distribution of colorless sulfur bacteria (*Beggiatoa* spp) in a coastal marine sediment. *Mar. Biol., 41,* 19-28.

Judson, H. F. (1979). *The Eighth Day of Creation.* NY: Simon and Schuster.

Jukes, T. H. (1987). Transitions, transversions, and the molecular evolutionary clock. *J. Mol. Evol., 26,* 87-98.

Kahl, M. P., Jr. (1964). Food ecology of the wood stork (*Mycteria americana*) in Florida. *Ecol. Monogr., 34,* 97-117.

Kays, S., and J.L. Harper. (1974). The regulation of plant and tiller density in a grass sward. *J. Ecol., 62,* 97-105.

Kemperman, J. A., and B.V. Barnes. (1976). Clone size in American aspens. *Can. J. Bot., 54,* 2603-2607.

Kerr, A., and K. Htay. (1974). Biological control of crown gall through bacteriocin production. *Physiol. Plant Pathol., 4,* 37-44.

Kimura, M. (1987). Molecular evolutionary clock and the neutral theory. *J. Molec. Evol., 26,* 24-33.

Kingsman, A. J., K.F. Chater, and S.M. Kingsman (eds.). (1988). *Transposition.* (*Proc. 43rd Sympos. Soc. Gen. Microbiol.*). Cambridge, UK: Cambridge Univ. Press.

Kinkel, L. L., J.H. Andrews, and E.V. Nordheim. (1989). Fungal immigration dynamics and community development on apple leaves. *Microb. Ecol., 18,* 45-58.

Kinkel, L. L., J.H. Andrews, F.M. Berbee, and E.V. Nordheim. (1987). Leaves as islands for microbes. *Oecologia, 71,* 405-408.

Kirk, J. T. O. (1975a). A theoretical analysis of the contribution of algal cells to the attenuation of light within natural waters. I. General treatment. *New Phytol., 75,* 11-20.

Kirk, J. T. O. (1975b). A theoretical analysis of the contribution of algal cells to the attenuation of light in natural waters. II. Spherical cells. *New Phytol., 75,* 21-36.

Kirk, J. T. O. (1976). A theoretical analysis of the contribution of algal cells to the attenuation of light within natural waters. III. Cylindrical and spheroidal cells. *New Phytol., 77,* 341-358.

Kirk, T. K., and P. Fenn. (1982). Formation and action of the lignolytic system in basidiomycetes. J.C. Frankland, J.N. Hedger, and M.J. Swift (eds.), *Decomposer Basidiomycetes: Their Biology and Ecology.* (pp. 67-90). Cambridge, UK: Cambridge Univ. Press.

Kirkwood, T. B. L. (1981). Repair and its evolution: Survival versus reproduction. C.R. Townsend, and P. Calow (eds.), *Physiological Ecology: An Evolutionary Approach to Resource Use.* (pp. 165-189). Sunderland, Mass.: Sinauer Assoc.

Kirkwood, T. B. L. (1984). In vitro ageing of animal cells. I. Davies, and D.C. Sigee (eds.), *Cell Ageing and Cell Death.* (pp. 55-72). Cambridge, UK: Cambridge Univ. Press.

Kirkwood, T. B. L., and R. Holliday. (1986). Ageing as a consequence of natural selection. A.H. Bittles, and K.J. Collins (eds.), *The Biology of Human Ageing.* (pp. 1-16). Cambridge, U.K.: Cambridge Univ. Press.

Kirkwood, T. B. L., and R. Holliday. (1979). The evolution of ageing and longevity. *Proc. Roy. Soc. Lond. B, 205,* 531-546.

Klekowski, E. J. Jr. (1988). *Mutation, Developmental Selection, and Plant Evolution.* NY: Columbia Univ. Press.

Klekowski, E. J. Jr, and P.J. Godfrey. (1989). Ageing and mutation in plants. *Nature (London), 340,* 389-391.

Kluyver, A. J. (1931). *The Chemical Activities of Microorganisms.* London: Univ. London Press.

Koch, A. L. (1971). The adaptive responses of *Escherichia coli* to a feast and famine existence. *Adv. Microb. Physiol., 6,* 147-217.

Koch, A. L. (1976). How bacteria face depression, recession and derepression. *Perspect. Biol. Med., 20,* 44-63.

Koch, A. L. (1981). Evolution of antibiotic resistance gene function. *Microbiol. Rev., 45,* 355-378.

Koch, A. L. (1986). The variability and individuality of the bacterium. F.C. Neidhardt (ed.), Escherichia coli and Salmonella typhimurium. *Cellular and Molecular Biology, vol. 2.* (pp. 1606-1614). Washington, DC: Amer. Soc. Microbiol.

Koch, A. L. (1987). Why *Escherichia coli* should be renamed *Escherichia ilei.* A. Torriani-Gorini, F. Rothman, S. Silver, A. Wright, and E. Yagil (eds.), *Phosphate Metabolism and Cellular Regulation in Microorganisms.* (pp. 300-305). Washington, DC: Amer. Soc. Microbiol.

Koch, A. L., and M. Schaechter. (1984). The world and ways of E. coli. A.L. Demain (ed.), *Biology of Industrial Microorganisms. vol. 1.* (pp. 1-25). Reading, Mass.: Addison-Wesley.

Koehl, M. A. R. (1986). Seaweeds in moving water: Form and mechanical function. T. Givnish (ed.), *On the Economy of Plant Form and Function.* (pp. 603-634). Cambridge, UK: Cambridge Univ. Press.

Koshland, D. E., Jr. (1979). Bacterial chemotaxis. J.R. Sokatch, and L.N. Ornston (eds.), *The Bacteria, vol. 7. Mechanisms of Adaptation.* (pp. 111-166). NY: Academic Press.

Krebs, C. J. (1985). *Ecology: The Analysis of Distribution and Abundance, 3rd ed.* NY: Harper and Row.

Krebs, H. (1981). The evolution of metabolic pathways. M.J. Carlile, J.F. Collins, and B.E.B. Moseley (eds.), *Molecular and Cellular Aspects of Microbial Evolution (Proc. 32nd Sympos. Soc. Gen. Microbiol.).* (pp. 215-228). Cambridge, UK: Cambridge Univ. Press.

Krebs, J. R., and N.B. Davies. (1984). *Evolutionary Ecology, 2nd ed.* Sunderland, Mass.: Sinauer Assoc.

Kubiena, W. L. (1938). *Micropedology*. Ames, Iowa: Collegiate Press.

Kubiena, W. (1932). Uber Fruchtkorperbildung und engere Standortwahl von Pilzen in Bodenhohlraumen. *Arch. f. Mikrobiol.*, *3(4)*, 507-542.

Kurten, B. (1959). Rates of evolution in fossil mammals. *Cold Spring Harbor Sympos. Quant. Biol.*, *24*, 205-215.

Kushner, P. J., L.C. Blair, and I. Herskowitz. (1979). Control of yeast cell types by mobile genes: a test. *Proc. Natl. Acad. Sci. (USA)*, *76*, 5264-5268.

Laanbroek, H. J., A.J. Smit, G. Klein Nulend, and H. Veldkamp. (1979). Competition for L-glutamate between specialized and versatile *Clostridium species*. *Arch. Microbiol.*, *120*, 61-66.

Lack, D. (1971). *Ecological Isolation in Birds*. Oxford, UK: Blackwell.

Lambert, M. E., J.F. McDonald, and I.B. Weinstein (eds.). (1988). *Eukaryotic Transposable Elements as Mutagenic Agents. (Banbury Report no. 30)*. Cold Spring Harbor, NY: Cold Spring Harbor Laboratory.

Larwood, G., and B.R. Rosen (eds.). (1979). *Biology and Systematics of Colonial Organisms*. NY: Academic Press.

Lauer, A. R. (1952). Age and sex in relation to accidents. R.H. Baldock (Chairman, Highway Res. Bd.), *Road-User Characteristics. Highway Res. Bd. Bull. no. 60.* (pp. 25-35). Washington, DC: Nat. Acad. Sci.

Lederberg, J. (1989). Replica plating and indirect selection of bacterial mutants: Isolation of preadaptive mutants in bacteria by sib selection. *Genetics, 121*, 395-399.

Lehmann, H., and R.G. Huntsman. (1961). Why red blood cells are the shape they are. R.G. MacFarlane, and A.H.T. Robb (eds.), *Functions of the Blood.* (pp. 73-148). NY: Academic Press.

Leopold, A. C. (1980). Aging and senescence in plants. K.V. Thimann (ed.), *Senescence in Plants.* (pp. 1-12). Boca Raton, Florida: CRC Press.

Leopold, L. B. (1971). Trees and streams: the efficiency of branching patterns. *J. Theor. Biol.*, *31*, 339-354.

Lerner, I. M. (1968). *Heredity, Evolution, and Society*. San Francisco: Freeman.

Leversee, G. (1976). Flow and feeding in fan-shaped colonies of the gorgonian coral, *Leptogorgia*. *Biol. Bull.*, *151*, 344-356.

Levin, B. R. (1981). Periodic selection, infectious gene exchange and the genetic structure of *Escherichia coli* populations. *Genetics, 99*, 1-23.

Levin, B. R. (1988). The evolution of sex in bacteria. R.E. Michod, and B.R. Levin (eds.), *The Evolution of Sex: An Examination of Current Ideas.* (pp. 194-211). Sunderland, Mass.: Sinauer Assoc.

Levins, R. (1968). *Evolution in Changing Environments: Some Theoretical Explorations*. Princeton, NJ: Princeton Univ. Press.

Levins, R. (1969). Dormancy as an adaptive strategy. H.W. Woolhouse (ed.), *Dormancy and Survival (Proc. 23rd Sympos. Soc. Exptl. Biol.).* (pp. 1-10). Cambridge, UK: Cambridge Univ. Press.

Levy, S. B., and R.V. Miller (eds.) (1989). *Gene Transfer in the Environment*. NY: McGraw-Hill.

Lewis, C. M., and R. Holliday. (1970). Mistranslation and ageing in *Neurospora*. *Nature (London)*, *228*, 877-880.

Lewontin, R. C. (1978). Adaptation. *Sci. Amer., 239(3)*, 212-230.

Lewontin, R. C. (1965). Selection for colonizing ability. H.G. Baker, and G.L. Stebbins (eds.), *The Genetics of Colonizing Species.* (pp. 77-91). NY: Academic Press.

Lewontin, R. C. (1983). Gene, organism and environment. D.S. Bendall (ed.), *Evolution from Molecules to Men.* (pp. 273-285). Cambridge, UK: Cambridge Univ. Press.

Lewontin, R. C., and J.L. Hubby. (1966). A molecular approach to the study of genic heterozygosity in natural populations. II. Amount of variation and degree of heterozygosity in natural populations of *Drosophila pseudoobscura. Genetics, 54*, 595-609.

Lindstedt, S. L., and S.D. Swain. (1988). Body size as a constraint of design and function. M.S. Boyce (ed.), *Evolution of Life Histories of Mammals.* (pp. 93-105). New Haven, Conn.: Yale Univ. Press.

Lindstedt, S. L., and W.A. Calder III. (1981). Body size, physiological time, and longevity of homeothermic animals. *Quart. Rev. Biol., 56*, 1-16.

Linton, A. H., B. Handley, and A.D. Osborne. (1978). Fluctuations in *Escherichia coli* O-serotypes in pigs throughout life in the presence and absence of antibiotic treatment. *J. Appl. Bacteriol., 44*, 285-298.

Lofts, B. (1970). *Animal Photoperiodism. (Studies in Biology no. 25)*. London: Arnold.

Lonsdale, W. M. (1990). The self-thinning rule: dead or alive? *Ecology, 71*, 1373-1388.

Loomis, W. F. (1988). *Four Billion Years: An Essay on the Evolution of Genes and Organisms.* Sunderland, Mass.: Sinauer Assoc.

Lovett Doust, L. (1981). Population dynamics and local specialization in a clonal perennial (*Ranunculus repens*). 1. The dynamics of ramets in contrasting habitats. *J. Ecol., 69*, 743-755.

Lubchenco, J., and J. Cubit. (1980). Heteromorphic life histories of certain marine algae as adaptations to variations in herbivory. *Ecology, 61*, 676-687.

MacArthur, R. H. (1972). *Geographical Ecology: Patterns in the Distribution of Species.* Princeton, NJ: Princeton Univ. Press.

MacArthur, R. H., and E.R. Pianka. (1966). On optimal use of a patchy environment. *Amer. Nat., 100*, 603-609.

MacArthur, R. H., and E.O. Wilson (1967). *The Theory of Island Biogeography.* Princeton, NJ: Princeton Univ. Press.

MacArthur, R. H., and J.H. Connell (1966). *The Biology of Populations.* NY: Wiley.

MacArthur, R. H., and R. Levins. (1964). Competition, habitat selection, and character displacement in a patchy environment. *Proc. Natl. Acad. Sci. (USA), 51*, 1207-1210.

MacArthur, R. H., and R. Levins. (1967). The limiting similarity, convergence, and divergence of coexisting species. *Amer. Nat., 101*, 377-385.

MacDonald, N. (1983). *Trees and Networks in Biological Models.* NY: Wiley.

Mace, M. E., A.A. Bell, and C.H. Beckman (eds.). (1981). *Fungal Wilt Diseases of Plants.* NY: Academic Press.

Marcou, D. (1961). Notion de longevite et nature cytoplasmique du determinant de la senescence chez quelques champignons. *Annales des Sciences Naturelles (Botanique), 12(2)*, 653-764.

Margulis, L., and D. Sagan (1986). *Origins of Sex: Three Billion Years of Recombination* New Haven, Conn.: Yale Univ. Press.

Margulis, L., D. Sagan, and L. Olendzenski. (1985). What is sex? H.O. Halvorson, and A. Monroy (eds.), *The Origin and Evolution of Sex*. (pp. 69-85). NY: Liss.

Margulis, L., J.O. Corliss, M. Melkonian, and D.J. Chapman (1990). *Handbook of Protoctista*. Boston: Jones and Bartlett.

Margulis, L., and K.V. Schwartz (1988). *Five Kingdoms, 2nd ed.* NY: Freeman.

May, R. M. (1978). The dynamics and diversity of insect faunas. L.A. Mound, and N. Waloff (eds.), *Diversity of Insect Faunas (Proc. 9th Sympos. Roy. Entomol. Soc.)*. (pp. 188-204). Oxford, UK: Blackwell.

May, R. M. (1986). The search for patterns in the balance of nature: advances and retreats. *Ecology, 67*, 1115-1126.

May, R. M. (1988). How many species are there on earth? Science (USA), 241, 1441-1449.

Maynard Smith, J. (1976). Group selection. *Quart. Rev. Biol., 51*, 277-283.

Maynard Smith, J. (1978). *The Evolution of Sex*. Cambridge, UK: Cambridge Univ. Press.

Mayr, E. (1970). *Populations, Species, and Evolution*. Cambridge, Mass.: Harvard Univ. Press.

Mayr, E. (1982). *The Growth of Biological Thought. Diversity, Evolution and Inheritance*. Cambridge, Mass.: Belknap Press of Harvard Univ. Press.

McClintock, B. (1956). Controlling elements and the gene. *Cold Spring Harbor Sympos. Quant. Biol., 21*, 197-216.

McDonald, J. F. (1990). Macroevolution and retroviral elements. *BioScience, 40*, 183-191.

McDonald, J. F., D.J. Strand, M.R. Brown, S.M. Paskewitz, A.M. Csink, and S.H. Voss. (1988). Evidence of host-mediated regulation of retroviral element expression at the post-transcriptional level. M.E. Lambert, J.F. McDonald, and I.B. Weinstein (eds.), *Eukaryotic Transposable Elements as Mutagenic Agents. (Banbury Report no. 30)*. (pp. 219-234). Cold Spring Harbor, NY: Cold Spring Harbor Laboratory.

McFadden, C. S. (1986). Colony fission increases particle capture rates of a soft coral: Advantages of being a small colony. *J. Exptl. Mar. Biol. Ecol., 103*, 1-20.

McIntosh, R. P. (1985). *The Background of Ecology: Concept and Theory*. Cambridge, UK: Cambridge Univ. Press.

McKinney, F. K., and J.B.C. Jackson (1989). *Bryozoan Evolution*. Boston: Unwin Hyman.

McMahon, T. (1973). Size and shape in biology. *Science (USA), 179*, 1201-1204.

McMahon, T. A., and J.T. Bonner (1983). *On Size and Life*. NY: Scientific American Books.

McMahon, T. A., and R.E. Kronauer. (1976). Tree structures: deducing the principle of mechanical design. *J. Theor. Biol., 59*, 443-466.

Medawar, P. B. (1946). Old age and natural death. *Modern Quart., 1*, 1-30.

Medawar, P. B. (1952). *An Unsolved Problem in Biology*. London: Lewis.

Meselson, M. S., and C.M. Radding. (1975). A general model for genetic recombination. *Proc. Natl. Acad. Sci. (USA), 72*, 358-361.

Meyer, A. (1989). Cost of morphological specialization: feeding performance of the two morphs in the trophically polymorphic cichlid fish, *Cichlasoma citrinellum. Oecologia, 80*, 431-436.

Michod, R. E., and B.R. Levin (eds.). (1987). *The Evolution of Sex: An Examination of the Current Ideas.* Sunderland, Mass.: Sinauer Assoc.

Mitsuhashi, S. (ed.). (1971). *Transferable Drug Resistance Factor R.* Baltimore, Maryland: Univ. Park Press.

Mooney, H. A., and J.A. Drake (eds.). (1986). *Ecology of Biological Invasions of North America and Hawaii.* NY: Springer-Verlag.

Moore, D., L.A. Casselton, D.A. Wood, and J.C. Frankland (eds.). (1985). *Developmental Biology of Higher Fungi.* Cambridge, UK: Cambridge Univ. Press.

Mortimer, R. K., and J.R. Johnston. (1959). Life span of individual yeast cells. *Nature (London), 183*, 1751-1752.

Moss, D. N. (1964). Optimum lighting of leaves. *Crop Sci., 4*, 131-136.

Muller, C. H. (1966). The role of chemical inhibition (allelopathy) in vegetational composition. *Bull. Torrey Bot. Club, 93*(332-351).

Munkres, K. D. (1976). Ageing of *Neurospora crassa*. III. Induction of cellular death and clonal senescence of an inositolless mutant by inositol-starvation and the protective effect of dietary antioxidants. *Mechan. Ageing Dev., 5*, 163-169.

Murashige, T. (1974). Plant propagation through tissue cultures. *Annu. Rev. Plant Physiol., 25*, 135-166.

Murdoch, W. W. (1975). Diversity, complexity, stability and pest control. *J. Appl. Ecol., 12*, 795-807.

Negus, N. C., and P.J. Berger. (1988). Cohort analysis: Environmental cues and diapause in microtine rodents. M.S. Boyce (ed.), *Evolution of Life Histories of Mammals.* (pp. 65-74). New Haven, Conn.: Yale Univ. Press.

Neidhardt, F. E. (ed.-in-chief). (1987). Escherichia coli and Salmonella typhimurium: *Cellular and Molecular Biology, vols. 1 and 2.* Washington, DC: Amer. Soc. Microbiol.

Nester, E. W., C.E.Roberts, M.E. Lidstrom, N.N. Pearsall, and M.T. Nester (1983). *Microbiology, 3rd ed.* NY: Saunders.

Neushul, M. (1972). Functional interpretation of benthic marine algal morphology. I.A. Abbott, and M. Kurogi (eds.), *Contributions to the Systematics of Benthic Marine Algae of the North Pacific.* (pp. 47-70). Kobe, Japan: Japanese Soc. Phycol.

Newell, N. D. (1949). Phyletic size increase, an important trend illustrated by fossil invertebrates. *Evolution, 3*, 103-124.

Nye, P. H., and P.B. Tinker (1977). *Solute Movement in the Soil-Root System.* Oxford, UK: Blackwell.

O'Brien, W. J., H.I. Browman, and B.I. Evans. (1990). Search strategies of foraging animals. *Amer. Sci., 78*, 152-160.

Ochman, H., and A.C. Wilson. (1987). Evolution in bacteria: Evidence for a universal substitution rate in cellular genomes. *J. Mol. Evol., 26*, 74-86.

Oinonen, E. (1967). Sporal regeneration of bracken in Finland in light of the dimensions and age of its clones. *Acta For. Fenn., 83*, 3-96.

Oliver, S. G., and A.P.J. Trinci. (1985). Modes of growth of bacteria and fungi. A.T. Bull, and H. Dalton (eds.), *Comprehensive Biotechnology, vol. 1. The Principles of Biotechnology: Scientific Fundamentals.* (pp. 159-187). Oxford, UK: Pergamon.

Orgel, L. E. (1963). The maintenance of the accuracy of protein synthesis and its relevance to ageing. *Proc. Natl. Acad. Sci. (USA), 49*, 517-521.

Orgel, L. E., and F.H.C. Crick. (1980). Selfish DNA: The ultimate parasite. *Nature (London), 284*, 604-607.

Orshan, G. (1963). Seasonal dimorphism of desert and Mediterranean chamaephytes and its significance as a factor in their water economy. A.J. Rutter, and F.W. Whitehead (eds.), *The Water Relations of Plants.* (pp. 207-222). Oxford, UK: Blackwell.

Orton, T. J. (1984). Genetic variation in somatic tissues: Method or madness? D.S. Ingram, and P.H. Williams (eds.), *Advances in Plant Pathology, vol. 2.* (pp. 153-189). NY: Academic Press.

Pace, N. R., G.J. Olsen, and C.R. Woese. (1986). Ribosomal RNA phylogeny and the primary lines of evolutionary descent. *Cell, 45*, 325-326.

Palmer, J. D. (1985). Evolution of chloroplast and mitochondrial DNA in plants and algae. R.J. MacIntyre (ed.), *Molecular Evolutionary Genetics.* (pp. 131-240). NY: Plenum.

Palumbi, S. R., and J.B.C. Jackson. (1983). Aging in modular organisms: Ecology of zooid senescence in *Steginoporella* sp. (Bryozoa: Cheilostomata). *Biol. Bull., 164*, 267-278.

Pardee, A. B. (1961). Response of enzyme synthesis and activity to environment. G.C. Meynell, and H. Gooder (eds.), *Microbial Reaction to Environment (Proc. 11th Sympos. Soc. Gen. Microbiol.).* (pp. 19-40). Cambridge, UK: Cambridge Univ. Press.

Park, D. (1976). Carbon and nitrogen levels as factors influencing fungal decomposers. J.M. Anderson, and A. Macfadyen (eds.), *The Role of Terrestrial and Aquatic Organisms in Decomposition Processes.* (pp. 41-59). Oxford, UK: Blackwell.

Park, D. (1985a). Does Horton's law of branch length apply to open branching systems? *J. Theor. Biol., 112*, 299-313.

Park, D. (1985b). Application of Horton's first and second laws of branching to fungi. *Trans. Brit. Mycol. Soc., 84*, 577-584.

Park, D., and P.M. Robinson. (1967). A fungal hormone controlling internal water distribution normally associated with ageing in fungi. H.W. Woolhouse (ed.), *Aspects of the Biology of Ageing (Proc. 21st Sympos. Soc. Exptl. Biol.).* (pp. 323-335). Cambridge, UK: Cambridge Univ. Press.

Parker, G. A., R.R. Baker, and V.G.F. Smith. (1972). The origin and evolution of gamete dimorphism and the male-female phenomenon. *J. Theor. Biol., 36*, 529-553.

Pearl, R. (1940). *Introduction to Medical Biometry and Statistics, 3rd ed.* Philadelphia: Saunders.

Pegg, G. F. (1985). Life in a black hole—the micro-environment of the vascular pathogen. *Trans. Brit. Mycol. Soc., 85*, 1-20.

Penry, D. L., and P.A. Jumars. (1987). Modeling animal guts as chemical reactors. *Amer. Nat., 129*, 69-96.

Pentecost, A. (1980). Aspects of competition in saxicolous lichen communities. *The Lichenologist, 12*, 135-144.

Pereira-Smith, O. M., and J.R. Smith. (1988). Genetic analysis of indefinite division in human cells: Identification of four complementation groups. *Proc. Natl. Acad. Sci. (USA), 85*, 6042-6046.

Perkins, D. D., and B.C. Turner. (1988). *Neurospora* from natural populations: Toward the population biology of a haploid eukaryote. *Exp. Mycol., 12*, 91-131.

Perkins, D. D., B.C. Turner, and E.G. Barry. (1976). Strains of *Neurospora* collected from nature. *Evolution, 30*, 281-313.

Perkins, J. M., E.G. Eglington, and J.L. Jinks. (1971). The nature of permanently induced changes in *Nicotiana rustica. Heredity, 27*, 441-457.

Person, C. (1966). Genetic polymorphism in parasitic systems. *Nature (London), 212*, 266-267.

Peters, R. H. (1983). *The Ecological Implications of Body Size.* Cambridge, UK: Cambridge Univ. Press.

Pfennig, N. (1984). Microbial behaviour in natural environments. D.P. Kelly, and N.G. Carr (eds.), *The Microbe 1984. Part II. Prokaryotes and Eukaryotes (Proc. 36th Sympos. Soc. Gen. Microbiol.).* (pp. 23-50). Cambridge, UK: Cambridge Univ. Press.

Pfennig, N. (1989). Ecology of photosynthetic purple and green sulfur bacteria. H.G. Schlegel, and B. Bowien (eds.), *Autotrophic Bacteria.* (pp. 97-116). NY: Springer-Verlag.

Pianka, E. R. (1970). On r- and K-selection. *Amer. Nat., 104*, 592-597.

Pianka, E. R. (1976). Natural selection of optimal reproductive tactics. *Amer. Zool., 16*, 775-784.

Pianka, E. R. (1988). *Evolutionary Ecology, 4th ed.* NY: Harper and Row.

Pielou, E. C. (1969). *An Introduction to Mathematical Ecology.* NY: Wiley.

Pimm, S. (1982). *Food Webs.* London: Chapman and Hall.

Pimm, S. L., H.L. Jones, and J. Diamond. (1988). On the risk of extinction. *Amer. Nat., 132*, 757-785.

Pirie, N. W. (1973). "On being the right size". *Annu. Rev. Microbiol., 27*, 119-132.

Pirt, S. J. (1975). *Principles of Microbe and Cell Cultivation.* London: Blackwell.

Plomin, R. (1990). The role of inheritance in behavior. *Science (USA), 248*, 183-188.

Plomley, N. J. B. (1959). Formation of the colony in the fungus *Chaetomium. Aust. J. Biol. Sci., 12*, 53-64.

Poindexter, J. S. (1981a). The caulobacters: ubiquitous unusual bacteria. *Microbiol. Rev., 45*, 123-179.

Poindexter, J. S. (1981b). Oligotrophy. Fast and famine existence. *Adv. Microb. Ecol., 5*, 63-89.

Pontecorvo, G. (1956). The parasexual cycle in fungi. *Annu. Rev. Microbiol., 10*, 393-400.

Postgate, J. R. (1976). Death in macrobes and microbes. T.R.G. Gray, and J.R. Postgate (eds.), *The Survival of Vegetative Microbes (Proc. 26th Sympos. Soc. Gen. Microbiol.).* (pp. 1-18). Cambridge, UK: Cambridge Univ. Press.

Postgate, J. R., and P.H. Calcott. (1985). Ageing and death in microbes. A.T. Bull, and H. Dalton (eds.), *Comprehensive Biotechnology. The Principles, Applications and Regulations of Biotechnology in Industry, Agriculture and Medicine.* (pp. 239-249). Oxford, UK: Pergamon.

Price, P. W. (1977). General concepts on the evolutionary biology of parasites. *Evolution, 31,* 405-420.

Price, P. W. (1980). *Evolutionary Biology of Parasites.* Princeton, NJ: Princeton Univ. Press.

Price, P. W. (1984). Communities of specialists: vacant niches in ecological and evolutionary time. D.R. Strong, D. Simberloff, L.G. Abele, and A.B. Thistle (eds.), *Ecological Communities: Conceptual Issues and the Evidence.* (pp. 510-523). Princeton, NJ: Princeton Univ. Press.

Price, P.W. (1990). Host populations as resources defining parasite community organization. G.W. Esch, A.O. Bush, and J.M. Aho (eds.), *Parasite Communities. Patterns and Processes.* (pp. 21-40). N.Y: Chapman and Hall.

Prosser, J. I. (1983). Hyphal growth patterns. J.E. Smith (ed.), *Fungal Differentiation. A Contemporary Synthesis.* (pp. 357-396). NY: Dekker.

Ptashne, M. (1989). How gene activators work. *Sci. Amer., 260* (1), 41-47.

Puhalla, J. E., and A.E. Bell. (1981). Genetics and biochemistry of wilt pathogens. M.E. Mace, A.A. Bell, and C.H. Beckman (eds.), *Fungal Wilt Diseases of Plants.* (pp. 145-192). NY: Academic Press.

Puhalla, J. E., and J.E. Mayfield. (1974). The mechanism of heterokaryotic growth in *Verticillium dahliae. Genetics, 76,* 411-422.

Pulliam, H. R. (1988). Sources, sinks, and population regulation. *Amer. Nat., 132,* 652-661.

Pyke, G. H. (1984). Optimal foraging theory: a critical review. *Annu. Rev. Ecol. Syst., 15,* 523-575.

Pyke, G. H., H.R. Pulliam, and E.L. Charnov. (1977). Optimal foraging: A selective review of theory and facts. *Quart. Rev. Biol., 52,* 137-154.

Raff, R. A., and T.C. Kaufman (1983). *Embryos, Genes, and Evolution.* NY: Macmillan.

Raper, J. R. (1954). Life cycles, sexuality, and sexual mechanisms in the fungi. D.H. Wenrich, I.F. Lewis, and J.R. Raper (eds.), *Sex in Microorganisms.* (pp. 42-81). Washington, D.C.: Amer. Assoc. Advancemt. Sci.

Raper, J. R., and A.S. Flexer. (1970). The road to diploidy with emphasis on a detour. H.P. Charles, and B.C.J.G. Knight (eds.), *Organization and Control in Prokaryotic and Eukaryotic Cells (Proc. 20th Sympos. Soc. Gen. Microbiol.).* (pp. 401-432). Cambridge, UK: Cambridge Univ. Press.

Rathcke, B., and E.P. Lacey. (1985). Phenological patterns of terrestrial plants. *Annu. Rev. Ecol. Syst., 16,* 179-214.

Raven, J. A. (1986a). Evolution of plant life forms. T. Givnish (ed.), *On the Economy of Plant Form and Function.* (pp. 421-492). Cambridge, UK: Cambridge Univ. Press.

Raven, J. A. (1986b). Physiological consequences of extremely small size for autotrophic organisms in the sea. T. Platt, and W.K.W. Li (eds.), *Photosynthetic Picoplankton. (Proc. NATO Adv. Study Inst.).* (pp. 1-70). Ottawa, Ontario: Can. Dept. Fisheries and Oceans (Can. Bull. Fish. Aquat. Sci., vol. 214).

Raven, P. H., and G.B. Johnson (1986). *Biology*. St. Louis, Missouri: Times Mirror/ Mosby College Publishing.

Rayner, A. D. M., D. Coates, A.M. Ainsworth, T.J.H. Adams, E.N.D. Williams, and N.K. Todd. (1984). The biological consequences of the individualistic mycelium. D.H. Jennings, and A.D.M. Rayner (eds.), *The Ecology and Physiology of the Fungal Mycelium. (Proc. Sympos. Brit. Mycol. Soc.)*. (pp. 509-540). Cambridge, UK: Cambridge Univ. Press.

Rayner, A. D. M., and D. Coates. (1987). Regulation of mycelial organization and responses. A.D.M. Rayner, C.M. Brasier, and D. Moore (eds.), *Evolutionary Biology of the Fungi. (Proc. Sympos. Brit. Mycol. Soc.)*. (pp. 115-136). Cambridge, UK: Cambridge Univ. Press.

Rayner, A. D. M., R. Watling, and J.C. Frankland. (1985a). Resource relations— an overview. D. Moore, L.A. Casselton, D.A. Wood, and J.C. Frankland (eds.), *Developmental Biology of Higher Fungi*. (pp. 1-40). Cambridge, UK: Cambridge Univ. Press.

Rayner, A. D. M., K.A. Powell, W. Thompson, and D.H. Jennings. (1985b). Morphogenesis of vegetative organs. D. Moore, L.A. Casselton, D.A. Wood, and J.C. Frankland (eds.), *Developmental Biology of Higher Fungi*. (pp. 249-279). Cambridge, UK: Cambridge Univ. Press.

Rayner, A. D. M., and N.R. Franks. (1987). Evolutionary and ecological parallels between ants and fungi. *Trends Ecol. Evol. (TREE)*, 2 127-132.

Razin, A., and A.D. Riggs. (1980). DNA methylation and gene function. *Science (USA)*, 210, 604-610.

Reanney, D. (1976). Extrachromosomal elements as possible agents of adaptation and development. *Bact. Rev.*, 40, 552-590.

Reanney, D. C., P.C. Gowland, and J.H. Slater. (1983). Genetic interaction among microbial communities. J.H. Slater, R. Whittenbury, and J.W.T. Wimpenny (eds.), *Microbes in Their Natural Environments. (Proc. 34th Sympos. Soc. Gen. Microbiol.)*. (pp. 379-421). Cambridge, UK: Cambridge Univ. Press.

Rensch, B. (1960). *Evolution Above the Species Level*. NY: Columbia Univ. Press.

Rettenmeyer, C. W. (1963). Behavioral studies of army ants. *Univ. Kansas Sci. Bull.*, 44, 281-465.

Reuter, G., M. Giarre, J. Farah, J. Gausz, A. Spierer, and P. Spierer. (1990). Dependence of position-effect variegation in *Drosophila* on dose of a gene encoding an unusual zinc-finger protein. *Nature (London)*, 344, 219-223.

Ricklefs, R. E. (1979). *Ecology, 2nd ed*. NY: Chiron Press.

Ricklefs, R. E. (1990). *Ecology, 3rd ed*. NY: Freeman.

Riedl, R. (1978). *Order in Living Systems: A Systems Analysis of Evolution (trans. by R.P.S. Jefferies)*. NY: Wiley.

Robertson, E., A. Bradley, M. Kuehn, and M. Evans. (1986). Germ-line transmission of genes introduced into cultured pluripotential cells by retroviral vector. *Nature (London)*, 323, 445-448.

Roelfs, A. P. (1985). Wheat and rye stem rust. A.P. Roelfs, and W.R. Bushnell (eds.), *The Cereal Rusts, vol. II. Diseases, Distribution, Epidemiology, and Control*. (pp. 3-37). NY: Academic Press.

Roelfs, A. P., and W.R. Bushnell (eds.). (1985). *The Cereal Rusts, vol. II. Diseases, Distribution, Epidemiology, and Control*. NY: Academic Press.

Rogers, J. H. (1985). The origin and evolution of retroposons. D.C. Reanney, and P. Chambon (eds.), *Genome Evolution in Prokaryotes and Eukaryotes. Internat. Rev. Cytol., vol. 93.* (pp. 187-279). NY: Academic Press.

Rogers, J. (1986). The origin of retroposons. *Nature (London), 319,* 725.

Romano, A. H. (1966). Dimorphism. G.C. Ainsworth, and A.S. Sussman (eds.), *The Fungi: An Advanced Treatise.* (pp. 181-209). NY: Academic Press.

Roper, J. A. (1966). Mechanisms of inheritance. 3. The parasexual cycle. G.C. Ainsworth, and A.S. Sussman (eds.), *The Fungi: An Advanced Treatise.* (pp. 589-617). NY: Academic Press.

Rose, M., and B. Charlesworth. (1980). A test of evolutionary theories of senescence. *Nature (London), 287,* 141-142.

Rose, M. R. (1985). The evolution of senescence. P.G. Greenwood, P.H. Harvey, and M. Slatkin (eds.), *Evolution: Essays in Honour of John Maynard Smith.* (pp. 117-128). Cambridge, UK: Cambridge Univ. Press.

Rose, M. R., and B. Charlesworth. (1981). Genetics of life history in *Drosophila melanogaster.* II. Exploratory selection experiments. *Genetics, 97,* 187-196.

Rosen, B. R. (1986). Modular growth and form of corals: a matter of metamers? *Phil. Trans. Roy. Soc. Lond. B, 313,* 115-142.

Rosen, R. (1967). *Optimality Principles in Biology.* NY: Plenum Press.

Rosengarten, R., and K.S. Wise. (1990). Phenotypic switching in mycoplasmas: phase variation of diverse surface lipoproteins. *Science (USA), 247,* 315-318.

Ross, C. (1988). The intrinsic rate of natural increase and reproductive effort in primates. *J. Zool. Lond., 214,* 199-219.

Roszak, D. B., D.J. Grimes, and R.R. Colwell. (1984). Viable but non-recoverable stages of *Salmonella enteritidis* in aquatic systems. *Can. J. Microbiol., 30,* 334-338.

Roth, V. L. (1981). Constancy in the size ratios of sympatric species. *Amer. Nat., 118,* 394-404.

Rothrock, C. S., and D. Gottlieb. (1984). Role of antibiosis in antagonism of *Streptomyces hygroscopicus var. geldanus to Rhizoctonia solani* in soil. *Can. J. Microbiol., 30,* 1440-1447.

Roughgarden, J. (1971). Density-dependent natural selection. *Ecology, 51,* 453-468.

Roughgarden, J., S. Gaines, and H. Possingham. (1988). Recruitment dynamics in complex life cycles. *Science (USA), 241,* 1460-1466.

Roughgarden, J., Y. Iwasa, and C. Baxter. (1985). Demographic theory for an open marine population with space-limited recruitment. *Ecology, 66,* 54-67.

Ryan, F. J., G.W. Beadle, and E.L. Tatum. (1943). The tube method of measuring the growth rate of *Neurospora. Amer. J. Bot., 30,* 784-799.

Sackville Hamilton, N. R., B. Schmid, and J.L. Harper. (1987). Life history concepts and the population biology of clonal organisms. *Proc. Roy. Soc. Lond. B, 232,* 35-57.

Sackville Hamilton, N. R., and J.L. Harper. (1989). The dynamics of *Trifolium repens* in a permanent pasture. I. The population dynamics of leaves and nodes per shoot axis. *Proc. Roy. Soc. Lond. B, 237,* 133-173.

Sager, R. (1985). Microbial sexual eukaryotes. H.O. Halvorson, and A. Monroy (eds.), *The Origin and Evolution of Sex.* (pp. 93-96). NY: Liss.

Sager, R. (1989). Tumor suppressor genes: The puzzle and the promise. *Science (USA), 246,* 1406-1412.

Sakaguchi, K. (1990). Invertrons, a class of structurally and functionally related genetic elements that includes linear DNA plasmids, transposable elements, and genomes of adeno-type viruses. *Microbiol. Rev., 54,* 66-74.

Salisbury, E. J. (1942). *The Reproductive Capacity of Plants. Studies in Quantitative Biology.* London: Bell and Sons.

Sanders, F. E. (1971). *Effect of root and soil properties on the uptake of nutrients by competing roots.* Unpublished doctoral dissertation, Oxford University, Oxford, U.K.

Savage, D. C. (1977). Microbial ecology of the gastro-intestinal tract. *Annu. Rev. Microbiol., 31,* 107-133.

Savageau, M. (1983). *Escherichia coli* habitats, cell types, and molecular mechanisms of gene control. *Amer. Nat., 122,* 732-744.

Savile, D. B. O. (1953). Short season adaptations in the rust fungi. *Mycologia, 45,* 75-87.

Savile, D. B. O. (1955). A phylogeny of the basidiomycetes. *Can. J. Bot., 33,* 60-104.

Savile, D. B. O. (1971a). Coevolution of the rust fungi and their hosts. *Quart. Rev. Biol., 46,* 211-218.

Savile, D. B. O. (1971b). Co-ordinated studies of parasitic fungi and flowering plants. *Nat. Can., 98,* 535-552.

Savile, D. B. O. (1976). Evolution of the rust fungi (Uredinales) as reflected by their ecological problems. *Evol. Biol., 9,* 137-207.

Schaffer, W. M. (1974). Optimal reproductive effort in fluctuating environments. *Amer. Nat., 108,* 783-798.

Schlegel, H. G., and B. Bowien (eds.). (1989). *Autotrophic Bacteria.* NY: Springer-Verlag.

Schmid, B. (1986). Spatial dynamics and integration within clones of grassland perennials with different growth forms. *Proc. Roy. Soc. Lond. B, 228,* 173-186.

Schmid, B., G.M. Puttick, K.H. Burgess, and F.A. Bazzaz. (1988). Correlations between genet architecture and some life history features in three species of *Solidago. Oecologia, 75,* 459-464.

Schmidt-Nielsen, K. (1984). *Scaling.* Cambridge, UK: Cambridge Univ. Press.

Schneider, E. L., and Y. Mitsui. (1976). The relationship between *in vitro* cellular aging and *in vivo* human age. *Proc. Natl. Acad. Sci. (USA), 73,* 3584-3588.

Schoener, T. W. (1969). Models of optimal size for solitary predators. *Amer. Nat., 103,* 277-313.

Schopf, J. W. (1978). The evolution of the earliest cells. *Sci. Amer., 239 (3)* 85-103.

Schopf, J. W. (ed.) (1983). *Earth's Earliest Biosphere: Its Origin and Evolution.* Princeton, NJ: Princeton Univ. Press.

Schopf, T. J., D.M. Raup, S.J. Gould, and D.S. Simberloff. (1975). Genomic versus morphologic rates of evolution: Influence of morphologic complexity. *Paleobiology, 1,* 63-70.

Sebens, K. P. (1979). The energetics of asexual reproduction and colony formation in benthic marine invertebrates. *Amer. Zool., 19,* 683-697.

Sebens, K. P. (1980). The control of asexual reproduction and indeterminate body size in the sea anemone, *Anthopleura elegantissima* (Brandt). *Biol. Bull., 158,* 370-382.

Sebens, K. P. (1982). The limits to indeterminate growth: an optimal size model applied to passive suspension feeders. *Ecology, 63,* 209-222.

Sebens, K. P. (1983). Asexual reproduction in *Anthopleura elegantissima* (Brandt) (Anthozoa: Actiniaria): Seasonality and spatial extent of clones. *Ecology, 63,* 434-444.

Sebens, K. P. (1987). The ecology of indeterminate growth in animals. *Annu. Rev. Ecol. Syst., 18,* 371-407.

Seilacher, A. (1967). Fossil behavior. *Sci. Amer., 217*(8), 72-80.

Selander, R. K., D.A. Caugant, and T.S. Whittam. (1987). Genetic structure and variation in natural populations of *Escherichia coli.* F.C. Neidhardt (ed.-in-chief.), *Escherichia coli* and *Salmonella typhimurium, vol.2. Cellular and Molecular Biology.* (pp. 1625-1648). Washington, DC: Amer. Soc. Microbiol.

Seshadri, T., and J. Campisi. (1990). Repression of c-*fos* transcription and an altered genetic program in senescent human fibroblasts. *Science (USA), 247,* 205-209.

Sewell, G. W. F., and J.F. Wilson. (1964). Occurrence and dispersal of *Verticillium* conidia in xylem sap of the hop (*Humulus lupulus L.*). *Nature (London), 204,* 901.

Sexton, R., and H.W. Woolhouse. (1984). Senescence and abscission. M.B. Wilkins (ed.), *Advanced Plant Physiology.* (pp. 469-497). London: Pitman.

Shamel, A. D., and C.S. Pomeroy. (1936). Bud mutations in horticultural crops. *J. Hered., 27,* 486-494.

Shapiro, J. A. (ed.). (1983). *Mobile Genetic Elements.* NY: Academic Press.

Shapiro, J. A. (1985). Intercellular communication and genetic change in bacterial populations. H.O. Halvorson, D. Pramer, and M. Rogul (eds.), *Engineered Organisms in the Environment: Scientfic Issues.* (pp. 63-69). Washington, DC: Amer. Soc. Microbiol.

Shapiro, J. A. (1988). Bacteria as multicellular organisms. *Sci. Amer., 258* (6), 82-89.

Sharp, P. M., D.C. Shields, K.H. Wolfe, and W.-H. Li. (1989). Chromosomal location and evolutionary rate variation in eubacterial genes. *Science (USA), 246,* 808-810.

Shepherd, M. G. (1988). Morphogenetic transformation of fungi. M.R. McGinnis (ed.), *Current Topics in Medical Mycology.* (pp. 278-304). NY: Springer-Verlag.

Shields, W. G., and J.G. Bockheim. (1981). Deterioration of trembling aspen clones in the Great Lakes region. *Can. J. For. Res., 11,* 530-537.

Sibly, R. M., and P. Calow (1986). *Physiological Ecology of Animals: An Evolutionary Approach.* Oxford, UK: Blackwell.

Silander, J. A., Jr. (1985). Microevolution in clonal plants. J.B.C. Jackson, L.W. Buss, and R.E. Cook (eds.), *Population Biology and Evolution of Clonal Organisms.* (pp. 107-152). New Haven, Conn.: Yale Univ. Press.

Simon, M., J. Zieg, M. Silverman, G. Mandel, and R. Doolittle. (1980). Phase variation: evolution of a controlling element. *Science (USA), 209,* 1370-1374.

Singer, S. (1988). Clonal populations with special reference to *Bacillus sphaericus*. *Adv. Appl. Microbiol., 33*, 47-74.

Sistrom, W. R. (1969). *Microbial Life, 2nd ed.* NY: Holt, Rinehart and Winston.

Slade, N. A., and R.J. Wassersug. (1975). On the evolution of complex life cycles. *Evolution, 29*, 568-571.

Slater, J. H. (1985). Gene transfer in microbial communities. H.O. Halvorson, D. Pramer, and M. Rogul (eds.), *Engineered Organisms in the Environment: Scientific Issues* (pp. 89-98). Washington, DC: Amer. Soc. Microbiol.

Slatkin, M. (1985). Somatic mutations as an evolutionary force. P.J. Greenwood, P.H. Harvey, and M. Slatkin (eds.), *Evolution: Essays in Honour of John Maynard Smith.* (pp. 19-30). Cambridge, UK: Cambridge Univ. Press.

Slobodkin, L. B. (1968). Toward a predictive theory of evolution. R.C. Lewontin (ed.), *Population Biology and Evolution* (pp. 187-205). Syracuse, NY: Syracuse Univ. Press.

Slobodkin, L. B. (1988). Intellectual problems of applied ecology. *BioScience, 38*, 337-342.

Slutsky, B., J. Buffo, and D.R. Soll. (1985). High frequency switching of colony morphology in *Candida albicans. Science (USA), 230*, 666-669.

Slutsky, B., M. Staebell, J. Anderson, L. Risen, M. Pfaller, and D.R. Soll. (1987). "White-opaque transition": a second high-frequency switching system in *Candida albicans. J. Bact., 169*, 189-197.

Smith, A. J., and D.S. Hoare. (1977). Specialist phototrophs, lithotrophs, and methyltrophs: a unity among a diversity of procaryotes. *Bact. Rev., 41*, 419-448.

Smith, A. P., and J.O. Palmer. (1976). Vegetative reproduction and close packing in a successional plant species. *Nature (London), 261*, 232-233.

Smith, F. E. (1954). Quantitative aspects of population growth. E.J. Boell (ed.), *Dynamics of Growth Processes. (Proc. 11th Growth Sympos.; Soc. for the Study of Development and Growth).* (pp. 277-294). Princeton, NJ: Princeton Univ. Press.

Smith, J. N. M. (1974). The food searching behaviour of two European thrushes. II. The adaptiveness of the search patterns. *Behaviour, 49*, 1-61.

Smith, J. E. (1983). *Fungal Differentiation. A Contemporary Synthesis.* NY: Dekker.

Smith, J. E., and D.R. Berry (eds.). (1978). *The Filamentous Fungus. Vol. 3. Developmental Mycology.* London: Arnold.

Smith, J. C., T. Platt, W.K.W. Li, E.P.W. Horne, W.G. Harrison, D.V. Subba Rao, and B.D. Irwin. (1985). Arctic marine photoautotrophic picoplankton. *Mar. Ecol. Progr. Ser., 20*, 207-220.

Smith, J. R. (1978). Genetics of aging in lower organisms. E.L. Schneider (ed.), *The Genetics of Aging.* (pp. 137-149). NY: Plenum Press.

Smith-Sonneborn, J. (1981). Genetics and ageing in protozoa. *Int. Rev. Cytol., 73*, 319-354.

Sober, E. (1984). *The Nature of Selection. Evolutionary Theory in Philosophical Focus.* Cambridge, Mass.: MIT Press.

Sokal, R. R. (1970). Senescence and genetic load: Evidence from *Tribolium. Science (USA), 167*, 1733-1734.

Soll, D. R., and B. Kraft. (1988). A comparison of high frequency switching in the yeast *Candida albicans* and the slime mold *Dictyostelium discoideum. Develop. Genet., 9*, 615-628.

Sonea, S., and M. Panisset. (1983). *The New Bacteriology.* Boston: Jones and Bartlett.

Sonneborn, T. M. (1954). The relation of autogamy to senescence and rejuvenescence in *Paramecium aurelia. J. Protozool., 1*, 38-53.

Soriano, P., and R. Jaenisch. (1986). Retroviruses as probes for mammalian development: Allocation of cells to the somatic and germ lineages. *Cell, 46*, 19-29.

Sournia, A. (1982). Form and function in marine phytoplankton. *Biol. Rev., 57*, 347-394.

Southwood, T. R. E. (1977). Habitat, the templet for ecological strategies? *J. Anim. Ecol., 46*, 337-365.

Southwood, T. R. E. (1988). Tactics, strategies and templets. *Oikos, 52*, 3-18.

Stanier, R. Y. (1953). Adaptation, evolutionary and physiological: or Darwinism among the micro-organisms. R. Davies, and E.F. Gale (eds.), *Adaptation in Micro-organisms (Proc. 3rd Sympos. Soc. Gen. Microbiol.).* (pp. 1-14). Cambridge, UK: Cambridge Univ. Press.

Stanier, R. Y., and G. Cohen-Bazire. (1957). The role of light in the microbial world: Some facts and speculations. R.E.O. Williams, and C.C. Spicer (eds.), *Microbial Ecology (Proc. 7th Sympos. Soc. Gen. Microbiol.).* (pp. 56-89). Cambridge, UK: Cambridge Univ. Press.

Stanley, S. M. (1973). An explanation for Cope's rule. *Evolution, 27*, 1-26.

Stanley, S. M. (1979). *Macroevolution, Pattern and Process.* San Francisco: Freeman.

Stearns, S. (1982). The role of development in the evolution of life histories. J.T. Bonner (ed.), *Evolution and Development.* (pp. 237-258). NY: Springer-Verlag.

Stearns, S. C. (1976). Life-history tactics: a review of these ideas. *Quart. Rev. Biol., 51*, 3-47.

Stearns, S. C. (1977). The evolution of life history traits: A critique of the theory and a review of the data. *Annu. Rev. Ecol. Syst., 8*, 145-171.

Stearns, S. C. (1983). The influence of size and phylogeny on patterns of covariation among life-history traits in the mammals. *Oikos, 41*, 173-187.

Stearns, S. C. (1989a). Trade-offs in life history evolution. *Funct. Ecol., 3*, 259-268.

Stearns, S. C. (1989b). The evolutionary significance of phenotypic plasticity. *BioScience, 39*, 436-445.

Stebbins, G. L. (1968). Integration of development and evolutionary progress. R.C. Lewontin (ed.), *Population Biology and Evolution.* (pp. 17-36). Syracuse, NY: Syracuse Univ. Press.

Steele, E. J. (1981). *Somatic Selection and Adaptive Evolution, 2nd ed.* Chicago: Univ. Chicago Press.

Stephens, D. W., and J.R. Krebs. (1986). *Foraging Theory.* Princeton, NJ: Princeton Univ. Press.

Stevens, G. C. (1989). The latitudinal gradient in geographical range: How so many species coexist in the tropics. *Amer. Nat., 133*, 240-256.

Stevens, P. S. (1974). *Patterns in Nature.* Boston: Little, Brown and Co.

Stout, J. E., V.L. Yu, and M.G. Best. (1985). Ecology of *Legionella pneumophila* within water distribution systems. *Appl. Environ. Microbiol., 49,* 221-228.

Strahler, A. N. (1952). Hypsomatric (area-altitude) analysis of erosional topography. *Bull. Geol. Soc. Amer., 63,* 1117-1141.

Strathmann, R. (1974). The spread of sibling larvae of sedentary marine invertebrates. *Amer. Nat., 108,* 29-44.

Strobeck, C. (1975). Selection in a fine-grained environment. *Amer. Nat., 109,* 419-425.

Strong, D. R., Jr., and T.S. Ray, Jr. (1975). Host tree location behavior of a tropical vine (*Monstera gigantea*) by skototropism. *Science (USA), 190,* 804-806.

Sugawara, O., M. Oshimura, M. Koi, L.A. Annab, and J.C. Barrett. (1990). Induction of cellular senescence in immortalized cells by human chromosome 1. *Science (USA), 247,* 707-710.

Sugihara, G., and R.M. May. (1990). Applications of fractals in ecology. *Trends Ecol. Evol. (TREE), 5,* 79-86.

Swedmark, B. (1964). The interstitial fauna of marine sand. *Biol. Rev., 39,* 1-42.

Swift, M. J. (1976). Species diversity and the structure of microbial communities in terrestrial habitats. J.M. Anderson, and A. Macfadyen (eds.), *The Role of Terrestrial and Aquatic Organisms in Decomposition Processes.* (pp. 185-222). Oxford, UK: Blackwell.

Swift, M. J. (1982). Microbial succession during the decomposition of organic matter. R.G. Burns, and J.H. Slater (eds.), *Experimental Microbial Ecology.* (pp. 164-177). Oxford, UK: Blackwell.

Sytsma, K. J. (1990). DNA and morphology: Inference of plant phylogeny. *Trends Ecol. Evol. (TREE), 5,* 104-110.

Szaniszlo, P. J. (ed.). (1985). *Fungal Dimorphism.* NY: Plenum.

Takagi, Y., T. Nobuoka, and M. Doi. (1987). Clonal lifespan of *Paramecium tetraurelia*: effect of selection on its extension and use of fissions for its determination. *J. Cell Sci., 88,* 129-138.

Takahashi, M., and P.K. Bienfang. (1983). Size structure of phytoplankton biomass and photosynthesis in subtropical Hawaiian waters. *Mar. Biol., 76,* 203-211.

Templeton, A. R., and E.D. Rothman. (1974). Evolution in heterogeneous environments. *Amer. Nat., 108,* 409-428.

Thompson, D. W. (1961). *On Growth and Form.* (Abridged edition, edited by J.T. Bonner). Cambridge, UK: Cambridge Univ. Press.

Thompson, K., and D. Rabinowitz. (1989). Do big plants have big seeds? *Amer. Nat., 133,* 722-728.

Thompson, W. (1984). Distribution, development and functioning of mycelial cord systems of decomposer basidiomycetes of the deciduous woodland floor. D.H. Jennings, and A.D.M. Rayner (eds.), *The Ecology and Physiology of the Fungal Mycelium.* (pp. 185-214). Cambridge, UK: Cambridge Univ. Press.

Thompson, W., and A.D.M. Rayner. (1982). Structure and development of mycelial cord systems of *Phanerochaete laevis* in soil. *Trans. Brit. Mycol. Soc., 78,* 193-200.

Thornsberry, C., A. Balows, J.C. Feeley, and W.Jakubowski (eds.). (1984). *Legionella.* (*Proc. 2nd Internat. Sympos.*). Washington, DC: Amer. Soc. Microbiol.

Thorson, G. (1950). Reproductive and larval ecology of marine bottom invertebrates. *Biol. Rev., 25,* 1-45.

Tilman, D. (1982). *Resource Competition and Community Structure.* Princeton, NJ: Princeton Univ. Press.

Tison, D. L., D.H. Pope, W.B. Cherry, and C.B. Fliermans. (1980). Growth of *Legionella pneumophila* in association with blue-green algae (cyanobacteria). *Appl. Environ. Microbiol., 39,* 456-459.

Tomlinson, P. B. (1987). Architecture of tropical plants. *Annu. Rev. Ecol. Syst., 18,* 1-21.

Townsend, C. R., and P. Calow (eds.) (1981). *Physiological Ecology: An Evolutionary Approach to Resource Use.* Sunderland, Mass.: Sinauer Assoc.

Trager, W. (1986). *Living Together. The Biology of Animal Parasitism.* NY: Plenum Press.

Traub, R., and C.L. Wisseman, J. (1974). The ecology of chigger-borne rickettsiosis (scrub typhus). *J. Med. Entomol., 11,* 237-303.

Trevors, J. T., and J.D. van Elsas. (1989). A review of selected methods in environmental microbial genetics. *Can. J. Microbiol., 35,* 895-902.

Trevors, J. T., T. Barkay, and A.W. Bourquin. (1987). Gene transfer among bacteria in soil and aquatic environments: a review. *Can. J. Microbiol., 33,* 191-198.

Trinci, A. P. J. (1984). Regulation of hyphal branching and hyphal orientation. D.H. Jennings, and A.D.M. Rayner (eds.), *The Ecology and Physiology of the Fungal Mycelium.* (pp. 23-52). Cambridge, UK: Cambridge Univ. Press.

Trinci, A. P. J., and C.F. Thurston. (1976). Transition to the non-growing state in eukaryotic micro-organisms. T.R.G. Gray, and J.R. Postgate (eds.), *The Survival of Vegetative Microbes. (Proc. 24th Sympos. Soc. Gen. Microbiol.).* (pp. 55-79). Cambridge, UK: Cambridge Univ. Press.

Trinci, A. P. J., and E.G. Cutter. (1986). Growth and form in lower plants and the occurrence of meristems. *Phil. Trans. Roy. Soc. B, 313,* 95-113.

Tudzynski, P. (1982). DNA plasmids in eukaryotes with emphasis on mitochondria. *Progr. Bot., 44,* 297-307.

Tuomi, J., and T. Vuorisalo. (1989a). Hierarchical selection in modular organisms. *Trends. Ecol. Evol. (TREE), 4,* 209-213.

Tuomi, J., and T. Vuorisalo. (1989b). What are the units of selection in modular organisms? *Oikos, 54,* 227-233.

Twain, M. (1874). *Life on The Mississippi.* NY: Harper and Row.

Valentine, J. W. (1978). The evolution of multicellular plants and animals. *Sci. Amer., 239(3),* 104-117.

van Gemerden, H. (1974). Coexistence of organisms competing for the same substrate: An example among the purple sulfur bacteria. *Microb. Ecol., 1,* 104-119.

van Valen, L. (1973). A new evolutionary law. *Evol. Ecol., 1,* 1-30.

van Valen, L. (1978). Arborescent animals and other colonids. *Nature (London), 276,* 318.

Vanderplank, J. E. (1986). Specific susceptibility and specific feeding in gene-for-gene systems. D.S. Ingram, and P.H. Williams (eds.), *Advances in Plant Pathology, vol. 5.* (pp. 199-223). NY: Academic Press.

Varmus, H. E. (1983). Retroviruses. D.A. Shapiro (ed.), *Mobile Genetic Elements.* (pp. 411-503). NY: Academic Press.

Varmus, H. (1988). Retroviruses. *Science (USA), 240,* 1427-1435.

Varmus, H., and P. Brown. (1989). Retroviruses. D.E. Berg, and M.M. Howe (eds.), *Mobile DNA.* (pp. 53-108). Washington, DC: Amer. Soc. Microbiol.

Vasek, F. C. (1980). Creosote bush: Long-lived clones in the Mohave Desert. *Amer. J. Bot., 67,* 246-255.

Verhulst, P. F. (1838). Notice sur la loi que la population suit dans son accroissement. *Corresp. Math. Phys., 10,* 113-121.

Via, S., and R. Lande. (1985). Genotype—environment interaction and the evolution of phenotypic plasticity. *Evolution, 39,* 505-522.

Vidal, G. (1984). The oldest eukaryotic cells. *Sci. Amer., 250(2),* 48-57.

Waddington, C. H. (1942). Canalization of development and the inheritance of acquired characters. *Nature (London), 150,* 563-565.

Wainwright, S. A., and J.R. Dillon. (1969). On the orientation of sea fans (Genus *Gorgonia*). *Biol. Bull., 136,* 130-139.

Wainwright, S. A., W.D. Biggs, J.D. Currey, and J.M. Gosline (1982). *Mechanical Design in Organisms.* Princeton, NJ: Princeton Univ. Press.

Walbot, V., and C.A. Cullis. (1983). The plasticity of the plant genome—is it a requirement for success? *Plant Mol. Biol. Report., 1(4),* 3-11.

Walbot, V., and C.A. Cullis. (1985). Rapid genomic change in higher plants. *Annu. Rev. Plant Physiol., 36,* 367-396.

Walford, R. L. (1983). *Maximum Life Span.* NY: Norton.

Waller, D. M., and D.A. Steingraeber. (1985). Branching and modular growth: Theoretical models and empirical patterns. J.B.C. Jackson, L.W. Buss, and R.E. Cook (eds.), *Population Biology and Evolution of Clonal Organisms.* (pp. 225-257). New Haven, Conn: Yale Univ. Press.

Ward, D. M., R. Weller, and M.M. Bateson. (1990). 16S rRNA sequences reveal numerous uncultured microorganisms in a natural community. *Nature (London), 345,* 63-65.

Wassersug, R. J. (1974). Evolution of anuran life cycles. *Science (USA), 185,* 377-378.

Watkinson, A. R., and J. White. (1985). Some life history consequences of modular construction in plants. *Phil. Trans. Roy. Soc. Lond. B, 313,* 31-51.

Watson, J. D., N.H. Hopkins, J.W. Roberts, J.A. Steitz, and A.M. Weiner (1987). *Molecular Biology of the Gene, 4th ed.* Menlo Park, Calif.: Benjamin Cummings.

Wattiaux, J. M. (1968). Cumulative parental age effects in *Drosophila subobscura. Evolution, 22,* 406-421.

Weismann, A. (1885). *Die Kontinuitat des Keimplasmas als Grundlage einer Theorie der Vererbung.* Jena, Germany: Gustav Fischer.

Werner, E. E. (1977). Species packing and niche complementarity in three sunfishes. *Amer. Nat., 111,* 553-578.

Werner, E. E., and D.J. Hall. (1977). Competition and habitat shift in two sunfishes (Centrarchidae). *Ecology, 58,* 869-876.

Werner, E. E., and J.F. Gilliam. (1984). The ontogenetic niche and species interactions in size-structured populations. *Annu. Rev. Ecol. Syst., 15,* 393-425.

Wherry, E. T. (1972). Box-huckleberry as the oldest living protoplasm. *Castanea*, 37, 94-95.

White, J. (ed.) (1985). *The Population Structure of Vegetation*. Dordrecht, The Netherlands: Junk.

Whitham, T. G., and C.N. Slobodchikoff. (1981). Evolution by individuals, plant-herbivore interactions, and mosaics of genetic variability: The adaptive significance of somatic mutations in plants. *Oecologia*, 49, 287-292.

Whittenbury, R., and D.P. Kelly. (1977). Autotrophy: A conceptual phoenix. B.A. Haddock, and W.A. Hamilton (eds.), *Microbial Energetics. (Proc. 27th Sympos. Soc. Gen. Microbiol.).* (pp. 121-159). Cambridge, UK: Cambridge Univ. Press.

Whittingham, W. F., and K.B. Raper. (1960). Nonviability of stalk cells in *Dictyostelium. Proc. Natl. Acad. Sci. (USA)*, 46, 642-649.

Wickner, R. B., A. Hinnebusch, A.M. Lambowitz, I.C. Gunsalus, and A. Hollaender. (1986). *Extrachromosomal Elements in Lower Eukaryotes*. NY: Plenum Press.

Wiklund, C., B. Karlsson, and J. Forsberg. (1987). Adaptive versus constraint explanations for egg-to-body size relationships in two butterfly families. *Amer. Nat.*, 130, 828-838.

Wilbur, H. M. (1980). Complex life cycles. *Annu. Rev. Ecol. Syst.*, 11, 67-93.

Wilbur, H. M. (1984). Complex life cycles and community organization in amphibians. P.W. Price, C.N. Slobodchikoff, and W.S. Gaud (eds.), *A New Ecology: Novel Approaches to Interactive Systems*. (pp. 195-224). NY: Wiley.

Wilbur, H. M., and J.P. Collins. (1973). Ecological aspects of amphibian metamorphosis. *Science (USA)*, 182, 1305-1314.

Wilbur, H. M., D.W. Tinkle, and J.P. Collins. (1974). Environmental uncertainty, trophic level and resource availability in life history evolution. *Amer. Nat.*, 108, 805-817.

Wilkinson, H. T., R.D. Miller, and R.L. Miller. (1981). Infiltration of fungal and bacterial propagules into soil. *Soil Sci. Soc. Amer. J.*, 45, 1034-1039.

Williams, F. M. (1980). On understanding predator–prey interactions. D.C. Ellwood, J.N. Hedger, M.J. Latham, J.M. Lynch, and J.H. Slater (eds.), *Contemporary Microbial Ecology*. (pp. 349-375). NY: Academic Press.

Williams, G. C. (1957). Pleiotropy, natural selection, and the evolution of senescence. *Evolution*, 11, 398-411.

Williams, G. C. (1966). *Adaptation and Natural Selection*. Princeton, NJ: Princeton Univ. Press.

Williams, G. C. (1975). *Sex and Evolution*. Princeton, NJ: Princeton Univ. Press.

Wilson, A. C., S.S. Carlson, and T.J. White. (1977). Biochemical evolution. *Annu. Rev. Biochem.*, 46, 573-639.

Wilson, E. O. (1975). *Sociobiology*. Cambridge, Mass.: Belknap Press of Harvard Univ. Press.

Wilson, E. O. (1980). Caste and division of labor in leaf-cutter ants (Hymenoptera: Formicidae: Atta). II. The ergonomic optimization of leaf cutting. *Behav. Ecol. Sociobiol.*, 7, 157-165.

Wilson, E. O. (1983). Caste and division of labor in leaf-cutter ants (Hymenoptera: Formicidae: Atta). III. Ergonomic resiliency in foraging by *A. cephalotes. Behav. Ecol. Sociobiol.*, 14, 47-54.

Wilson, E. O., and W.H. Bossert (1971). *A Primer of Population Biology*. Sunderland, Mass.: Sinauer Assoc.

Wilson, E. O., T. Eisner, W.R. Briggs, R.E. Dickerson, R.L. Metzenberg, R.D. O'Brien, M. Sussman, and W.E. Boggs (1978). *Life on Earth, 2nd ed.* Sunderland, Mass.: Sinauer Assoc.

Winogradsky, S. (1949). *Microbiologie du Sol; Problemes et Methodes*. Paris: Mason.

Woese, C. R. (1987). Bacterial evolution. *Microb. Rev., 51*, 221-271.

Woese, C. R., O. Kandler, and M.L. Wheelis. (1990). Towards a natural system of organisms: Proposal for the domains Archaea, Bacteria, and Eucarya. *Proc. Natl. Acad. Sci. (USA), 87*, 4576-4579.

Wolda, H. (1989). Seasonal cues in tropical organisms. Rainfall? Not necessarily! *Oecologia, 80*, 437-442.

Wolfe, M. S., and C.E. Caten (eds.). (1987). *Populations of Plant Pathogens: Their Dynamics and Genetics*. Oxford, UK: Blackwell.

Woolhouse, H. W. (1967). The nature of senescence in plants. H.W. Woolhouse (ed.), *Aspects of the Biology of Ageing. (Proc. 21st Sympos. Soc. Exptl. Biol.)*. (pp. 179-213). Cambridge, UK: Cambridge Univ. Press.

Woolhouse, H. W. (1984). Senescence in plant cells. I. Davis, and D.C. Sigee (eds.), *Cell Ageing and Cell Death*. (pp. 123-153). Cambridge, UK: Cambridge Univ. Press.

Wright, F. L. (1953). *The Future of Architecture*. NY: Horizon Press.

Wright, R. M., and D.J. Cummings. (1983). Integration of mitochondrial gene sequences within the nuclear genome during senescence in a fungus. *Nature (London), 302*, 86-88.

Wright, S. (1956). Modes of selection. *Amer. Nat., 90*, 5-24.

Wright, S. (1978). *Evolution and the Genetics of Populations. vol. 4. Variability Within and Among Natural Populations*. Chicago: Univ. Chicago Press.

Yanagita, T. (1977). Cellular age in micro-organisms. T. Ishikawa, T.Maruyama, and H. Matsumiya (eds.), *Growth and Differentiation in Micro-organisms*. (pp. 1-36). Baltimore, Maryland: Univ. Park Press.

Yang, D., Y. Oyaizu, H. Oyaizu, G.J. Olsen, and C.R. Woese. (1985). Mitochondrial origins. *Proc. Natl. Acad. Sci. (USA), 82*, 4443-4447.

Yarwood, C. E. (1956). Generation time and the biological nature of viruses. *Amer. Nat., 40*, 97-102.

Yoda, K., T. Kira, H. Ogawa, and K. Hozumi. (1963). Self-thinning in overcrowded pure stands under cultivated and natural conditions. *J. Biol. Osaka City Univ., 14*, 107-129.

Zambryski, P. (1989). Agrobacterium—plant cell DNA transfer. D.E. Berg, and M.M. Howe (eds.), *Mobile DNA*. (pp. 309-33). Washington, D.C.: Amer. Soc. Microbiol.

Zamenhof, S., and H.H. Eichhorn. (1967). Study of microbial evolution through loss of biosynthetic functions: Establishment of "defective" mutants. *Nature (London), 216*, 456-458.

Zuckerkandl, E., and L. Pauling. (1965). Evolutionary divergence and convergence in proteins. V. Bryson, and H.J. Kogel (eds.), *Evolving Genes and Proteins*. (pp. 97-166). NY: Academic Press.

Index

289